TV TUNER

SCHEMATIC/SERVICING MANUAL

By Robert Goodman

TAB BOOKS
Blue Ridge Summit, Pa. 17214

FIRST EDITION

FIRST PRINTING—OCTOBER 1974

Copyright ©1974 by TAB BOOKS

Printed in the United States of America

Reproduction or publication of the content in any manner, without express permission of the publisher, is prohibited. No liability is assumed with respect to the use of the information herein.

Hardbound Edition: International Standard Book No. 0-8306-4696-5

Paperbound Edition: International Standard Book No. 0-8306-3696-X

Library of Congress Card Number: 74-83670

Preface

There are two schools of thought about TV tuner repair. One says: Don't do it; send the tuner in to a rebuilder. Another says: Why share your profit with a rebuilder? Fix the tuner yourself.

Probably the greatest wisdom lies between these positions. If you suspect tuner troubles, clean the contacts; then treat them with a good-quality tuner lubricant. Seven times out of ten, this will solve the problem. In many other cases a few minutes of troubleshooting will disclose the problem and pay important dividends in time saved—and money. The information in this book will enable you to perform the troubleshooting with the greatest success and efficiency possible.

This book includes an introductory chapter of principles common to all VHF and UHF tuners. Another chapter discusses, in detail, the principles of varactor tuners. Other chapters are devoted to the tuners of specific manufacturers; Tuners of all major domestic manufacturers are covered, plus those of Sony.

The specific information includes theory of operation, alignment procedures, trap coil adjustments, tuner diagrams, and schematics. I sincerely hope that all this information will enable you to repair TV tuners rapidly—and profitably.

In the appendix, you'll find a compendium of manufacturers' service hints, special notes, and other helpful information pertaining to the tuner portions of those sets included in this volume. Before making a repair on the tuner of any manufacturer included, be sure to check theat manufacturer's notes in the appendix to see if there are specific trouble areas you should be on the lookout for.

Robert Goodman

Contents

Chapter 1

VHF and UHF Tuner Characteristics

The VHF tuner (also referred to as the **front end**) of a TV receiver operates as a station channel selector, rf amplifier, mixer, and converter; it accepts an rf signal from the antenna and produces an output at the receiver's intermediate frequency (i-f). The schematic in Fig. 1-1 shows the usual stages found in the VHF tuner system.

The first stage is the rf amplifier. Tuned to the desired TV station frequency, this stage amplifies the incoming signal.

The second stage, the mixer-converter section, receives the amplified rf TV channel signal and beats it with a fixed signal from the local oscillator (third stage). The difference between the TV signal and the local oscillator signal is the i-f, and it is this i-f signal that appears at the tuner output.

The oscillator generates a different frequency for each channel. The new signal frequencies obtained are a result of a heterodyne process that includes the harmonic combinations of the local oscillator and the incoming TV channel signal. The plate coil circuit of the mixer is tuned to the difference, or intermediate frequency, and feeds this signal to the i-f amplifiers. VHF tuners are typically designed to receive 12 channels; one other position is used to couple in the UHF tuner.

The two most-used VHF tuners are the turret and bandswitch types. A few of the newer color TV chassis are using varactors (electrically tuned capacitors) in fully solid-state tuners.

An exploded view of a conventional tube-type turret tuner is shown in Fig. 1-2; a photo of a bandswitching type is illustrated in Fig. 1-3.

VHF TUNERS (Tube-Type)

As shown in the diagram of Fig. 1-4, the bandswitch tuner has three stages. The rf amplifier uses a tiny 2HA5 while the mixer and oscillator share the 5KZ8 tube. The pentode section of a 5KZ8 serves as a mixer and the triode section functions as an oscillator.

The rf amplifier is a neutralized tetrode which has the input grid and plate circuits tuned incrementally by a trapped inductor mounted on a wafer switch. The input impedance is a conventional balanced 300 ohms. A pi-section trap circuit to reject FM-band signals and interference near the i-f range is incorporated between the antenna coil and rf tuning assembly. The negative agc voltage is applied to the rf amplifier grid to prevent strong signal overload.

The plate circuit inductance is basically the primary of a coupling transformer. The transformer secondary is the mixer grid inductance. Close physical mounting on adjacent wafers completes the transformer action.

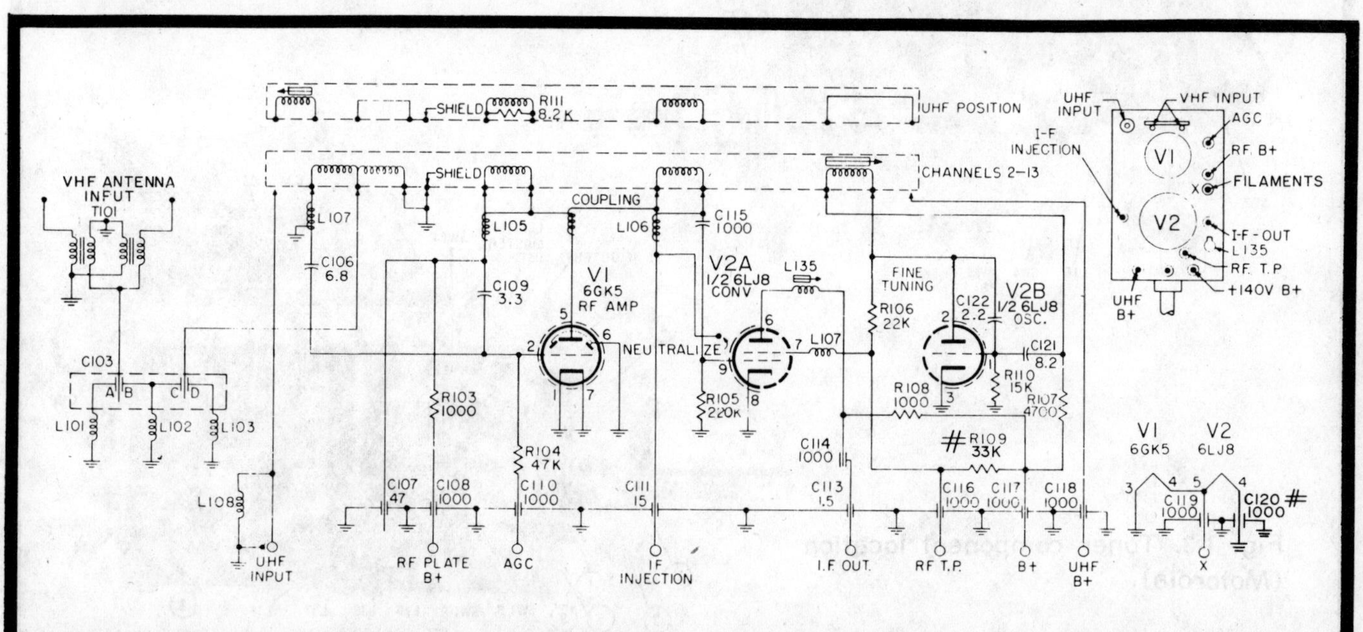

Fig. 1-1. VHF tuner schematic (GE).

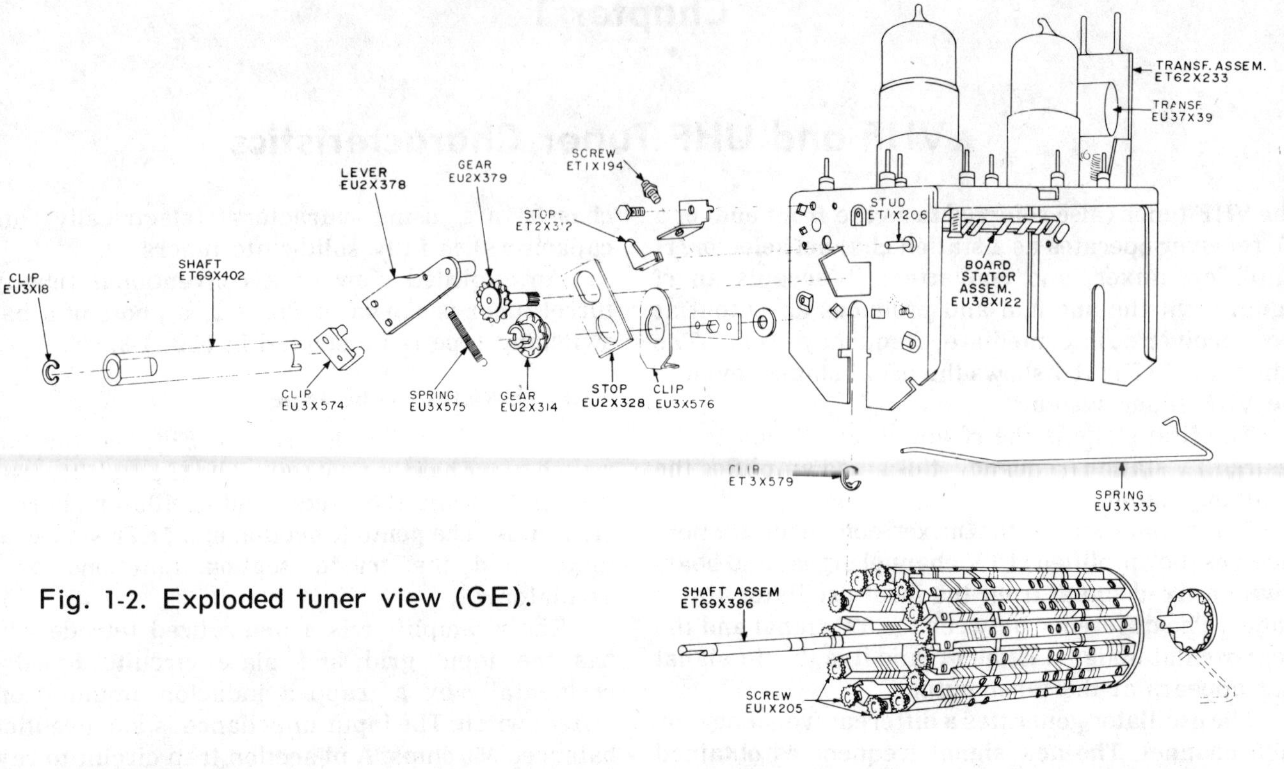

Fig. 1-2. Exploded tuner view (GE).

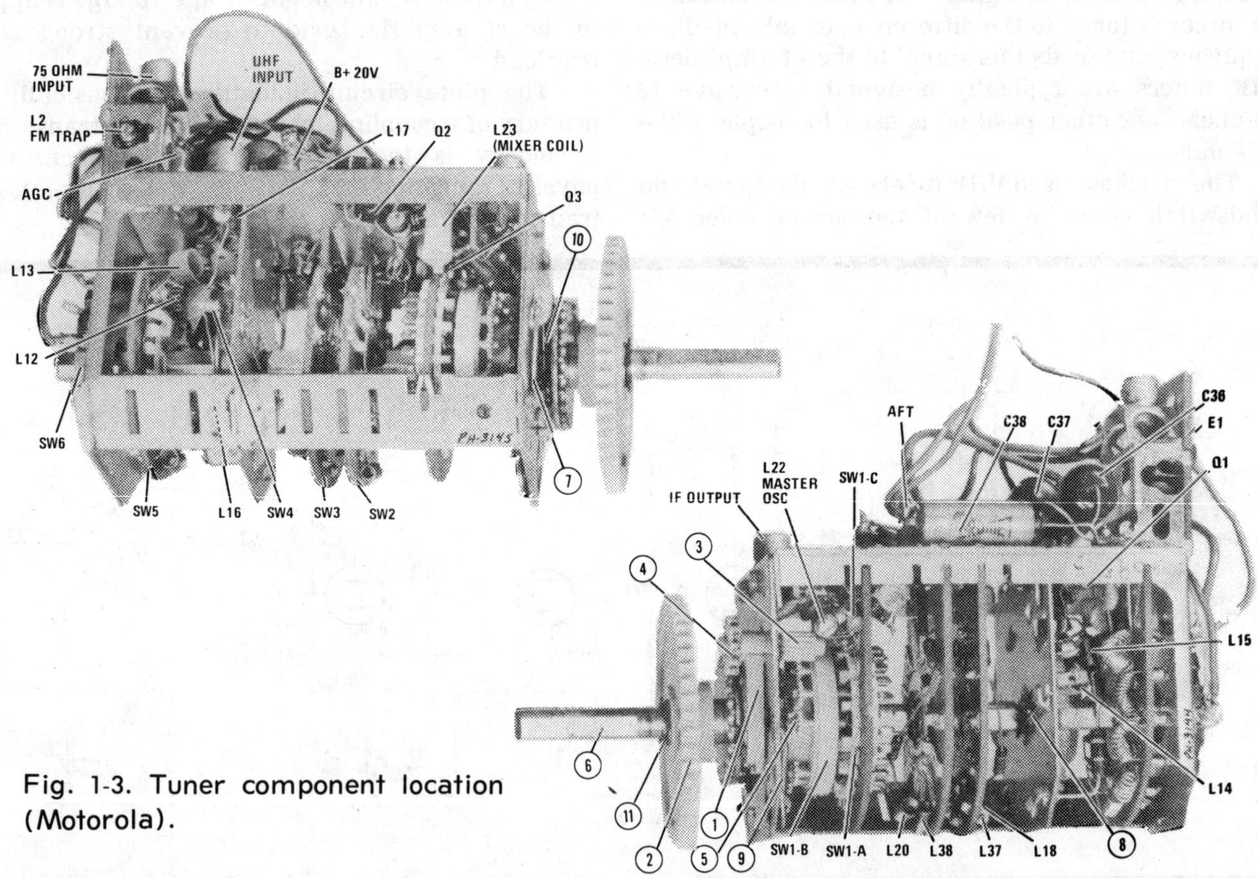

Fig. 1-3. Tuner component location (Motorola).

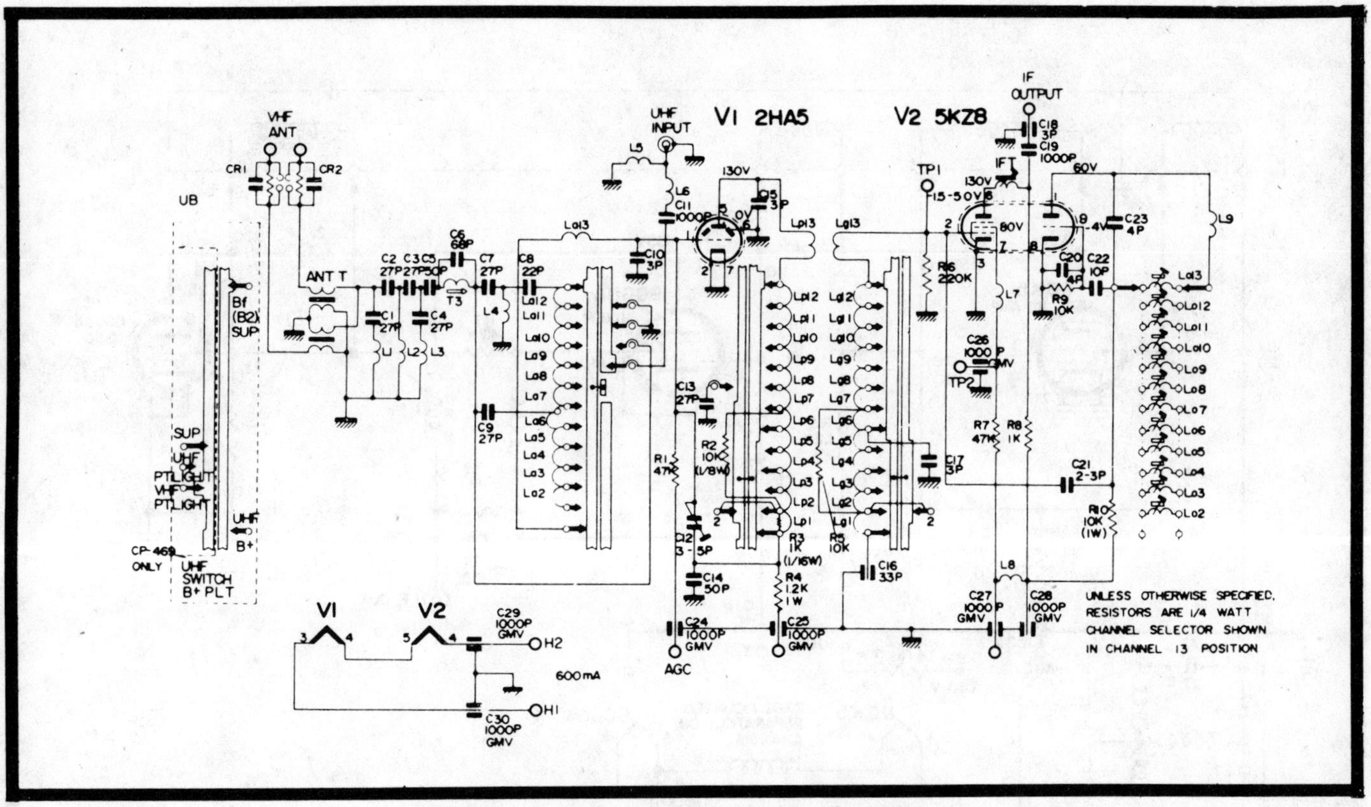

Fig. 1-4. Motorola VHF tuner schematic.

When the wafer is switched to the UHF position, the rf amplifier becomes an amplifier at 44 MHz, the UHF tuner's output frequency.

The rf signal is accepted at the mixer grid (pentode section of 5KZ8), where the video and sound modulation are detected. A continuous-wave carrier generated by a local oscillator is inserted at this point (via a capacitor) as a replacement for the original rf carrier. This superheterodyne principle creates the 45.75 MHz picture i-f carrier along with the 41.25 MHz sound carrier which are both selected by the video i-f for amplification. The mixer output is low-side link-coupled to the transformer primary in the plate circuit.

The mixer becomes a 44 MHz amplifier when the wafer switch is positioned for UHF reception. At the same time, the local oscillator is disabled to prevent further heterodyne action which would produce other sum and difference frequencies.

The local oscillator (triode section of 5KZ8) is tuned by switching separate oscillator coils into the circuit as various channels are selected. In addition, these individual coils are manually adjusted from the front of the receiver for fine tuning. By switching in separate coils for every channel, the oscillator is tuned to a frequency which is always higher than the received rf carrier by 45.75 MHz for the picture and 41.25 MHz for the sound carrier. There is no oscillator coil mounted in the UHF (channel 1) position. Therefore, the oscillator is disabled in this switch setting.

Signal Path

In the strip-type tuner circuit of Fig. 1-5, when a VHF antenna is connected to the 300-ohm tuner input terminals, the incoming signal goes through Z1, C1, and a channel strip coil to the grid of the rf amplifier. Z1 is an impedance-matching transformer and a 44 MHz trap. The amplified rf signal is developed across the plate coil (channel strip contacts 3 and 4) and mutually coupled to the mixer grid coil (channel strip contacts 5 and 6). Since interelectrode capacitance is quite high in triodes, two neutralizing capacitors in series are placed in the plate circuit. Capacitor C6 has a fixed value while C4 is variable. Another trimmer capacitor (C5) in the plate circuit is adjustable on channel 10 for an optimum response curve.

An agc voltage is fed to the grid of the rf amplifier through R1. The agc connection point on top of most tuners is an important test point. The voltage here is usually between —1 and —4V (under certain conditions it may be zero), depending on the strength of the received signal.

The rf signal on the mixer grid coil is coupled through C8 to the grid of the mixer. Trimmer capacitor C7 may be adjusted for an optimum response curve.

The oscillator circuit operates at a different frequency for each channel. Every time the channel is switched, a different tunable coil is connected between contacts 7 and 8. Each channel is adjusted with a slug-tuned coil. Another coil in parallel with

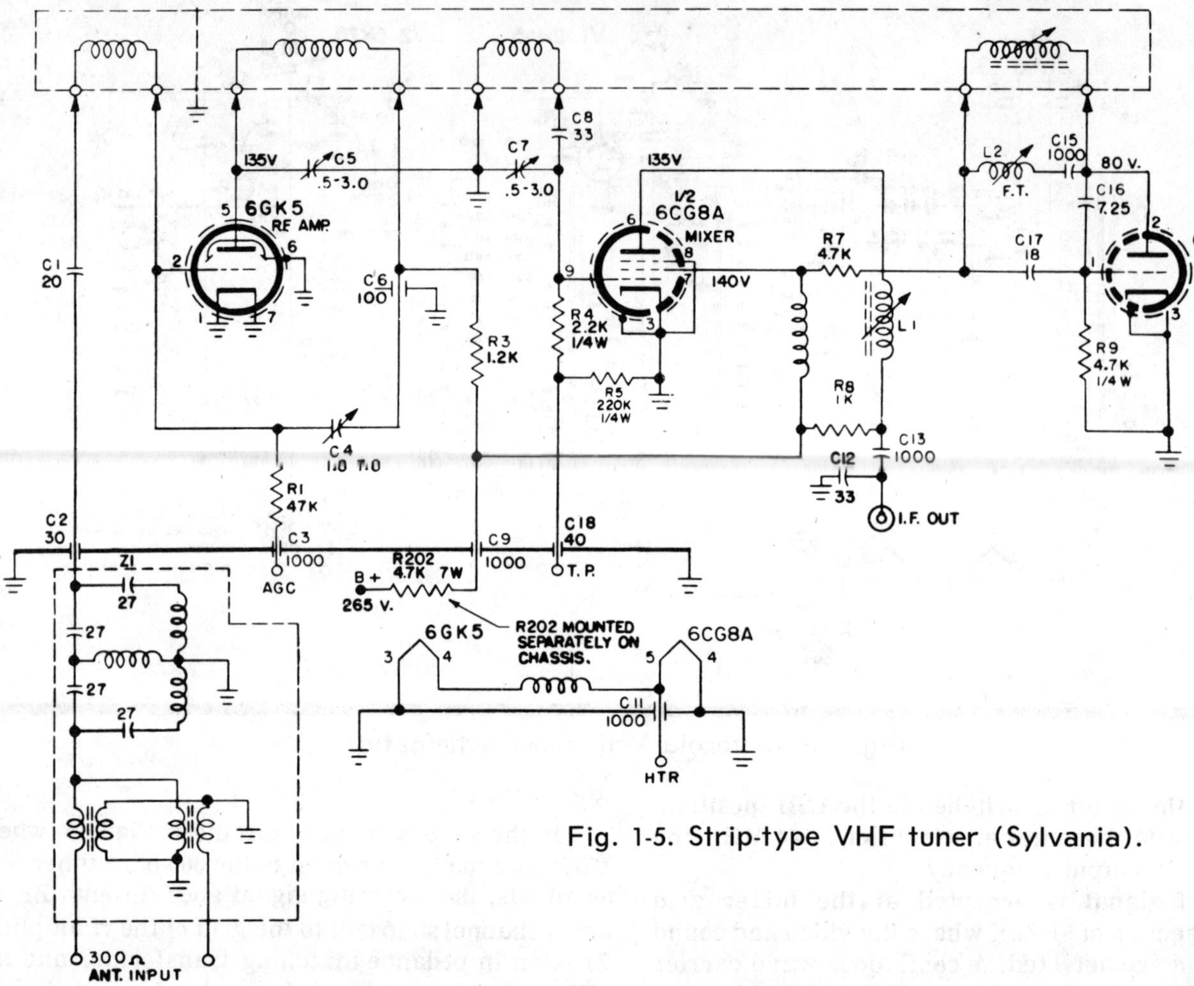

Fig. 1-5. Strip-type VHF tuner (Sylvania).

the oscillator coil, is included for fine-tuning of each channel. The signal from the oscillator, 44 MHz higher than the incoming rf signal, is coupled by the oscillator coil to the mixer grid coil. The mixing process produces a 44 MHz i-f output at the plate of the mixer. L1, the mixer plate coil, is adjusted for maximum amplitude at the i-f center frequency.

Automatic Fine Tuning Control

The automatic fine tuning circuit is the same as an automatic frequency control (AFC). The AFC circuitry in the tuner shown in Fig. 1-6 has as its main element a varactor which controls the AFC action. (On the schematic, the varactor, numbered BA142, is shown at the extreme right.) When reverse biased, the diode acts like a capacitance which is inversely proportional to the applied bias. If the voltage is high, the capacitance is low and vice versa.

The diode is connected to the plate of oscillator V2B through a 10 pF capacitor so that the maximum capacitance effect of the diode is less than 10 pF. An AFC control voltage is applied to the anode of the varactor. Any change in the control voltage will change the diode's capacitance value and therefore the frequency of the oscillator.

A fixed back-bias level is applied to the cathode of the varactor. The bias voltage is derived from bleeder network R118-R119, connected between +140V and ground. Resistor R116 provides isolation against oscillator radiation.

The AFC control voltage is also applied through a varactor to the UHF tuner. The two diodes are then fed in parallel. When the VHF tuner is switched into the back-bias circuit in parallel with R119, it increases the back bias on BA142 so the varactor is ineffective during UHF operation.

The block diagram in Fig. 1-7 shows the complete operation of the AFC system. An incoming picture signal mixes with the local oscillator in the tuner to produce an i-f video signal. This signal is then amplified by the video i-f amplifier. This signal is taken from the plate of the third i-f amplifier and fed to the input of the AFC circuit. The output of the AFC circuit is a dc voltage used to correct and trim the frequency of a tuned oscillator. This correction

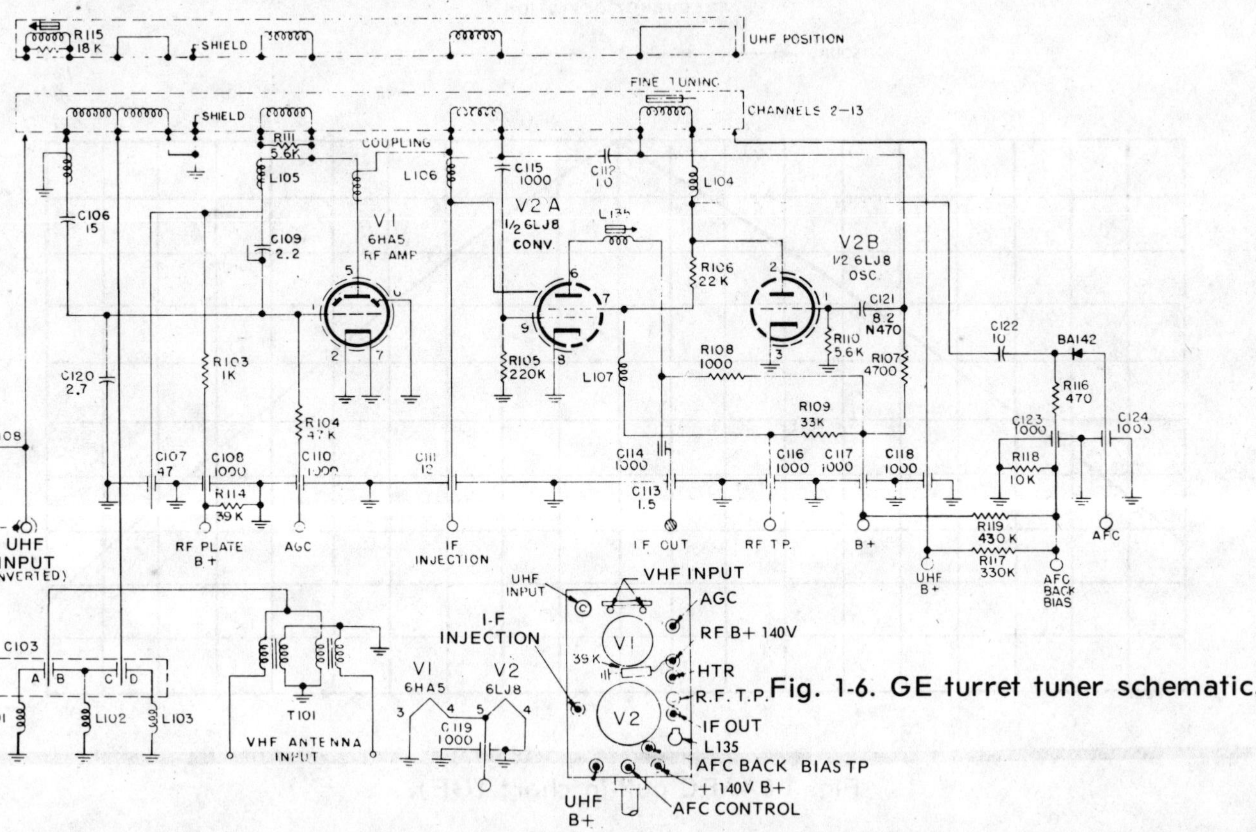

Fig. 1-6. GE turret tuner schematic.

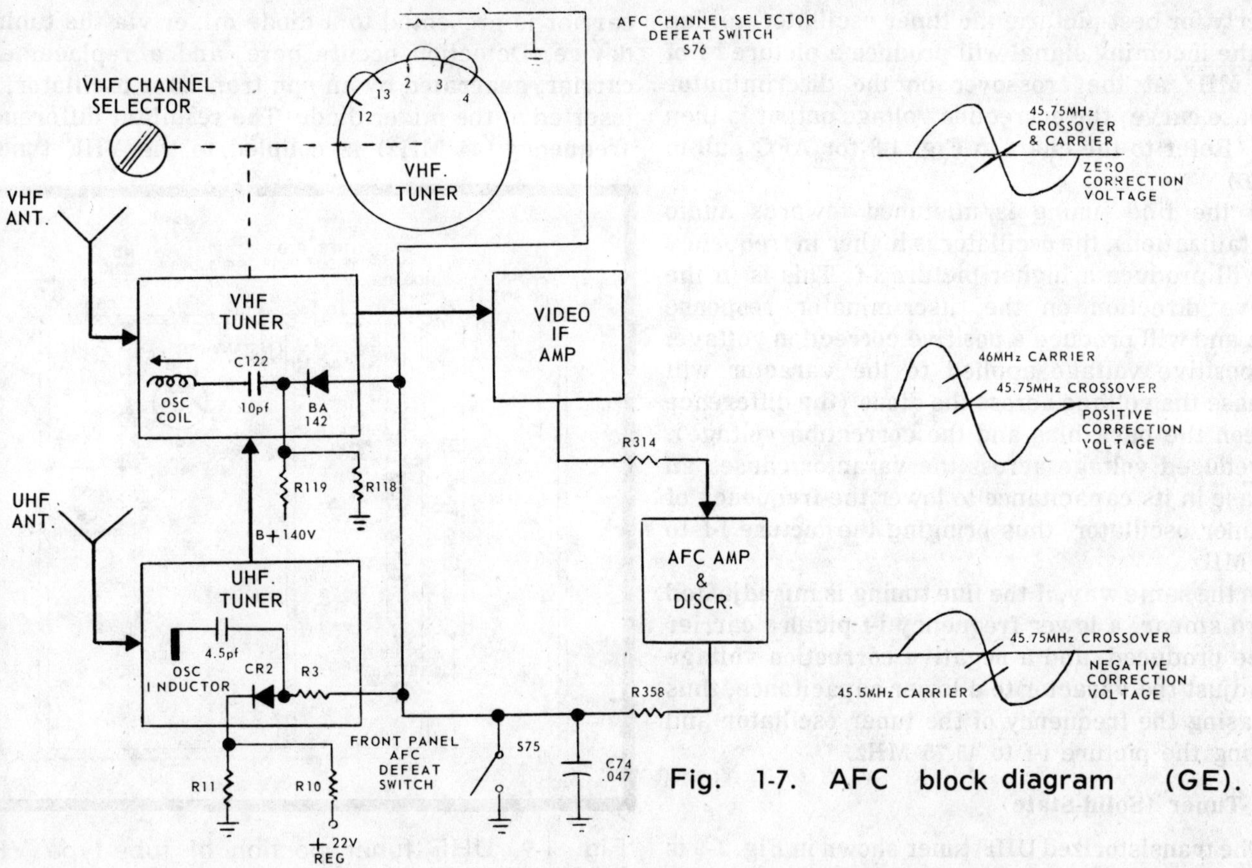

Fig. 1-7. AFC block diagram (GE).

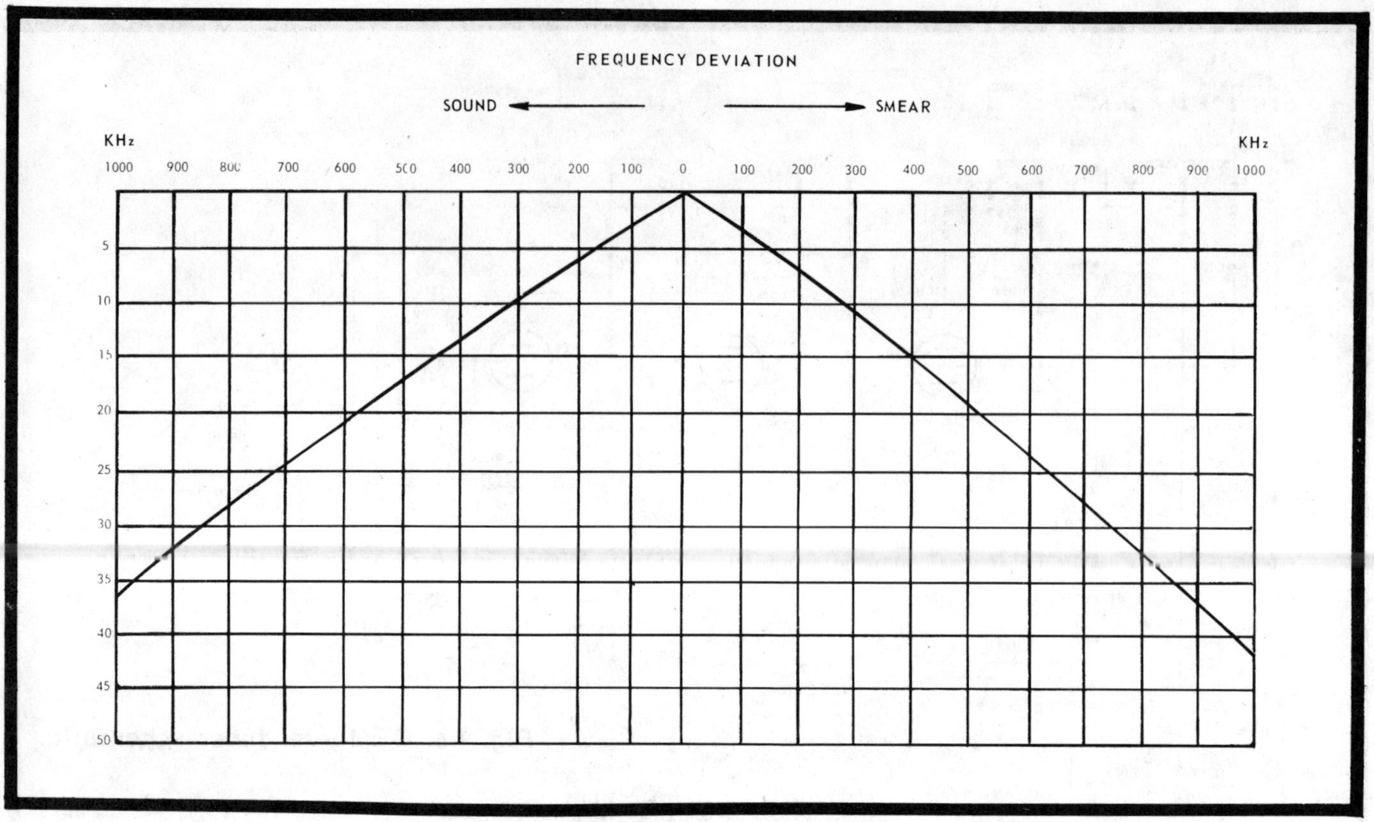

Fig. 1-8. AFC pull-in chart (GE).

voltage is fed to the back-biased varactors in the VHF and UHF tuners.

If the fine tuning control has been adjusted properly for best picture, the tuner oscillator mixing with the incoming signal will produce a picture i-f of 45.75 MHz at the crossover on the discriminator response curve; the correction voltage output is then zero. (Refer to the chart in Fig. 1-8 for AFC pull-in range.)

If the fine tuning is mistuned towards audio (crystallization), the oscillator is higher in frequency and will produce a higher picture i-f. This is in the positive direction on the discriminator response curve and will produce a positive correction voltage. The positive voltage applied to the varactor will decrease the voltage across the diode (the difference between the back bias and the correction voltage). The reduced voltage across the varactor causes an increase in its capacitance to lower the frequency of the tuner oscillator, thus bringing the picture i-f to 45.75 MHz.

In the same way, if the fine tuning is misadjusted toward smear, a lower frequency i-f picture carrier will be produced, and a negative correction voltage will adjust the varactor to a lower capacitance, thus increasing the frequency of the tuner oscillator and bringing the picture i-f to 45.75 MHz.

UHF Tuner (Solid-State)

The transistorized UHF tuner shown in Fig. 1-9 is a companion to the tube-type VHF tuner of Fig. 1-4. It has two stages of rf preselection which are tuned lines to select the desired rf carrier. There is no rf amplification at these ultrahigh frequencies, so the rf carrier is presented to a diode mixer via the tuning device. Detection occurs here, and a replacement carrier, generated by an npn transistor oscillator, is inserted at the mixer diode. The resultant difference frequency (44 MHz) is coupled to the VHF tuner,

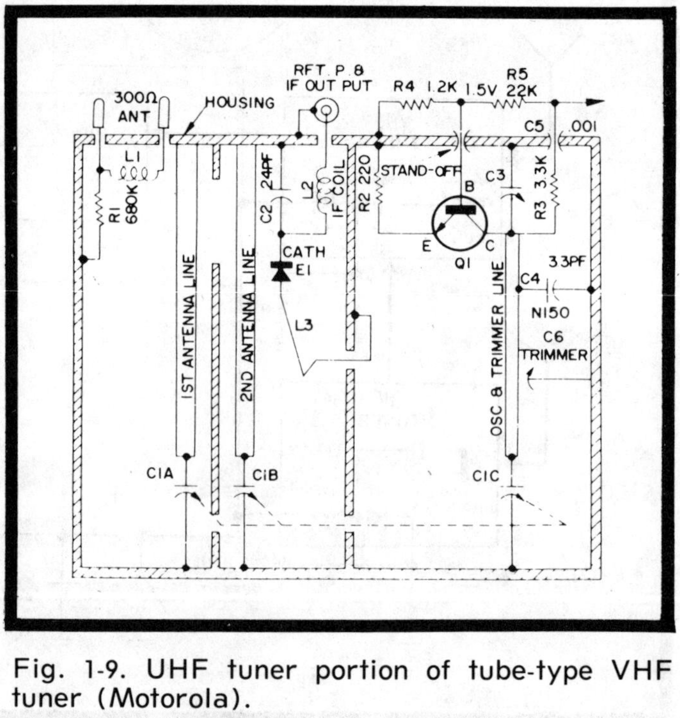

Fig. 1-9. UHF tuner portion of tube-type VHF tuner (Motorola).

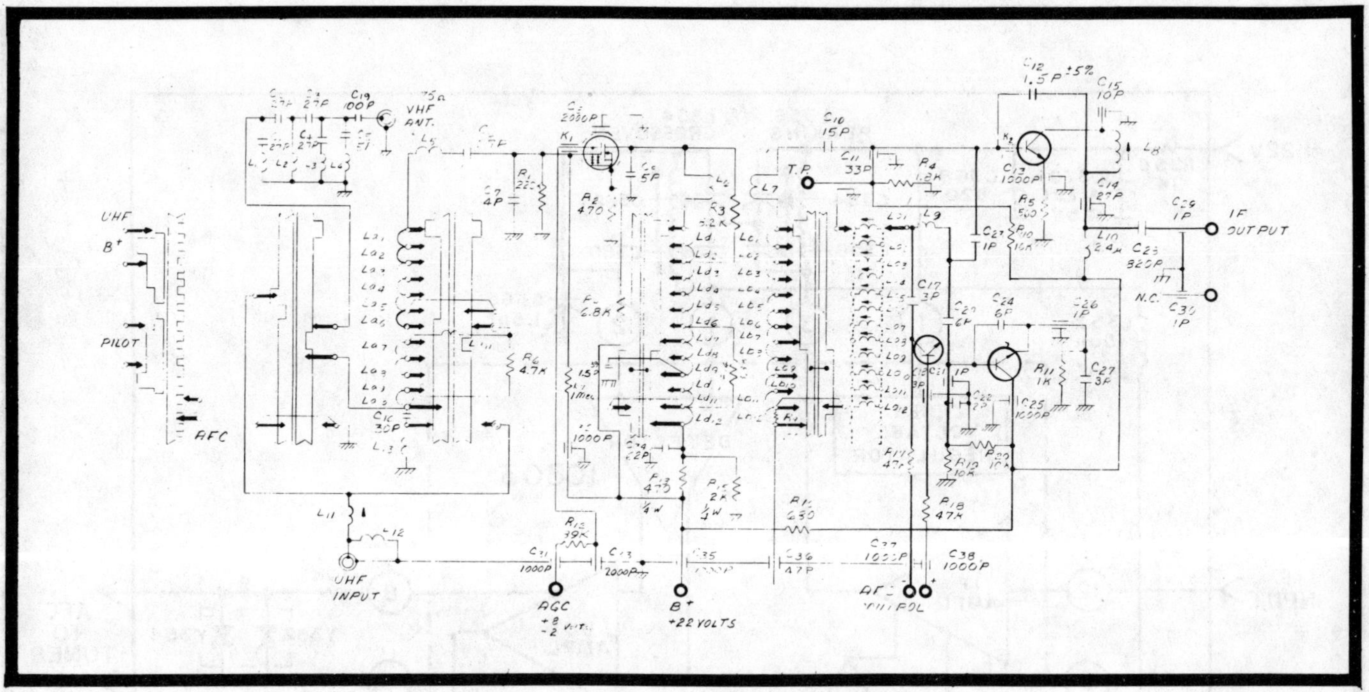

Fig. 1-10. GE solid-state tuner.

which becomes an amplifier at this frequency, compensating for the lack of rf amplification at UHF. The tuning devices in this UHF tuner are generally of the continuous type.

SOLID-STATE VHF TUNER AND AFC SYSTEM

The VHF tuner shown in Fig. 1-10 is a wafer-switch type using three transistors. A MOSFET is used for the rf amplifier, and npn bipolar transistors are used in the oscillator and mixer stages. The impedance of the VHF input terminal is 75 ohms. The input terminal is connected by means of a shielded cable to a 75-to-300-ohm balun. The use of shielded cable between the antenna terminals and the VHF input minimizes the possibility of cochannel or harmonic interference, particularly in CATV systems. This also provides a means of using shielded antenna cable from the antenna to the tuner.

When the tuner is in the UHF position, the mixer and rf amplifier stages function as 45 MHz amplifiers.

The tuner agc action is controlled in the MOSFET. The agc voltage is normally positive with respect to chassis ground, but the gain does not decrease with a more positive voltage as it does in some transistorized tuners. On weak signals, the agc voltage is about 8V and the tuner runs at maximum gain. On strong signals, the agc voltage will decrease to about 4V, and the gain will be reduced. A very strong signal will further reduce the agc voltage toward zero, and tuner gain will be reduced accordingly.

The AFC system (Fig. 1-11) is on a module which contains integrated circuit IC303, tuning coils L354 and L356, and a handful of other components. IC303 contains an i-f amplifier, agc amplifier, detector, dc amplifier, and a voltage regulator. A 45.75 MHz i-f signal fed to terminal 12 of the IC is amplified by the i-f amplifier section and appears at terminal 3. The signal is transferred to the detector stage by means of "tapped" capacitance in tuned circuits L354-C60-C362, and L356-C358-C364. The detector circuit operates similar to the common AFC-AFT discriminator circuits used in many color TV receivers. L356 is the tuning adjustment and L354 is the crossover adjustment. The detector outputs are dc voltages whose amplitudes are proportional to the amount by which the input signal has shifted from 45.75 MHz. These dc outputs, amplified by the dc amplifier, appear at terminals 5 and 8 of the IC. The voltages at terminals 5 and 8 center at around 6V; that is, if the incoming signal is exactly 45.75 MHz, both voltages will be about 6V. If the frequency of the incoming signal shifts from 45.75 MHz, one of the voltages will become more positive than 6V and the other voltage will become less positive than 6V.

These voltages, coupled to the anode and cathode of a varactor in the tuner, cause the diode to change capacitance in a direction which will correct the frequency shift. Because one of the voltages is coupled to the cathode of the varactor and the other is coupled to the anode, the varactor "sees" twice the voltage change that it would if the correction voltage were single-ended. Hence, the tuning accuracy and response is greatly improved.

RCA KRK 142 SOLID-STATE VHF TUNER

The KRK 142 is a fully solid-state VHF tuner incorporating a MOSFET as an rf amplifier and a

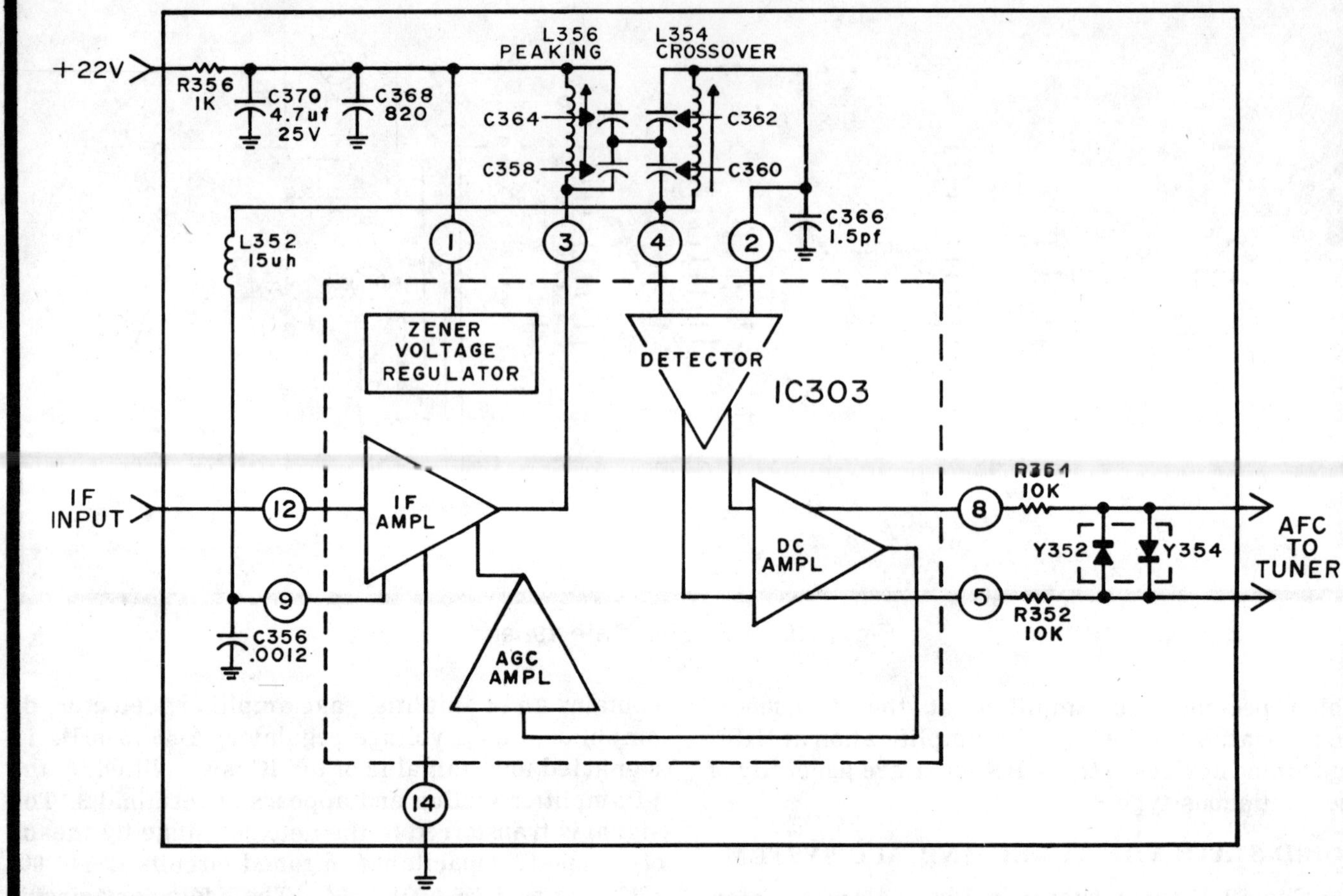

Fig. 1-11. GE AFC module.

"cascode" mixer stage. The separate oscillator employs a Colpitts circuit which is manually preset and thereafter automatically fine-tuned. The relationship of the various stages used in this tuner is illustrated in the block diagram of Fig. 1-12.

A dual-gate MOSFET is used as an rf amplifier in this tuner. The dual-gate device may be considered to be two separate transistors within a common envelope, which in circuit applications are arranged in a cascode configuration, as illustrated in Fig. 1-13. Transistor Q1 is a common-source amplifier to which the signal input is applied (at its gate). This amplifier directly drives amplifier Q2, arranged in the common-gate configuration. The signal output is taken from the drain circuit.

This circuit arrangement exhibits many of the characteristics and therefore has many of the advantages of two vacuum tubes arranged in the well known **cascode** configuration. Internal feedback from the Miller effect, which otherwise would tend to cause amplifier instability or oscillation, is loaded by the relatively low impedance of the driven stage (Q2), whose grounded gate serves as effective insurance against gate-current flow. As a result, it is not necessary to neutralize this amplifier in the VHF range.

The rf gain can be controlled and obtained by reverse biasing gate G2. This reverse gain control voltage will reduce the amplifier gain by reducing the drain current in both transistors. In combination with the chassis circuitry, this voltage control forms the rf automatic gain control.

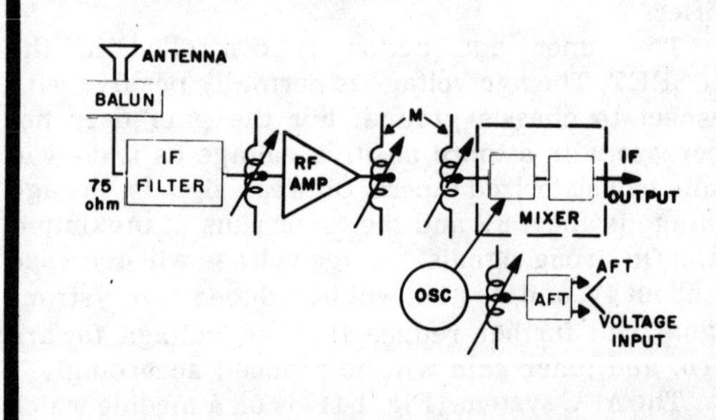

Fig. 1-12. Block diagram of VHF tuner (RCA).

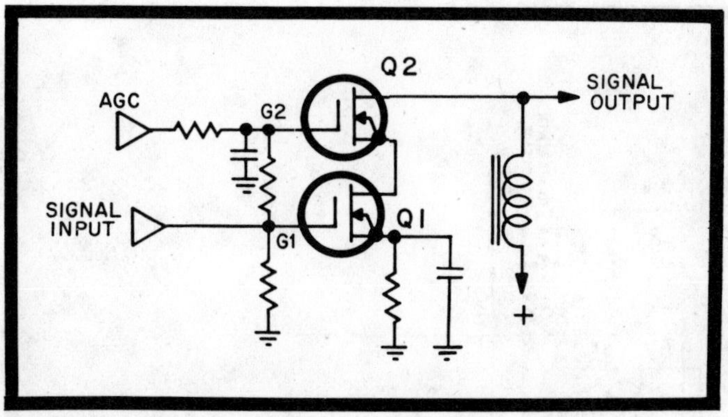

Fig. 1-13. Basic dual-gate MOSFETs in a cascode tuner circuit (RCA).

The rf amplifier circuitry is shown in the complete tuner schematic of Fig. 1-14. The signal from a 300-ohm antenna input is impedance-matched to a 75-ohm balun transformer mounted on the back of the receiver. The signal is then fed to gate 1 of the FET via the high-pass filter network and the rf tuned circuitry. Gate 2 receives the rf agc voltage from the chassis agc circuitry. This voltage can range from —5V (strong signal) to +6.7V (weak signal). The bias for G1 is derived from a portion of the agc voltage via R16 plus the voltage developed across source resistor R11.

Using this circuit, the source and gate 1 voltages will track with applied agc, thereby helping to reduce input capacitance variations. Printed-wiring techniques are used to form the inductances which couple the rf amplifier output signal to the mixer input circuitry. The use of printed inductances optimizes reliability of mechanical assembly and electrical alignment.

The mixer circuitry illustrated in Fig. 1-15 also uses a cascode circuit arrangement. However, two separate transistors are used. Transistor Q1 operates in a common-emitter configuration, the output of which drives a common-base amplifier. As with the rf amplifier, the principal advantage of this cascode arrangement is its inherent stability offered by the enhanced input-output isolation of the grounded-base transistor.

The base biasing network for Q1 consists of resistors R5-R8. This bias is set at a point which will allow the mixing action to be accomplished most efficiently. The base of transistor Q2 is biased via the divider network of R6, R7, and R8.

The resulting i-f signal is coupled from the mixer by a circuit arrangement commonly called "low side C" coupling. The i-f signal coupled to the i-f amplifier input is developed across capacitor C30, connected from inductor L30 of the tuned output circuit (L30 and C4) to ground (hence the name "low side C"). The advantage of this coupling arrangement is to minimize the amount of oscillator energy that can be coupled into the i-f section. At oscillator frequencies, inductor L30 acts as an rf choke, thus greatly reducing the amount of oscillator energy that can be developed across C30.

The local oscillator used in this tuner employs a Colpitts (split-capacitor) circuit basically arranged as illustrated in Fig. 1-16.

Energy at the desired frequency of oscillation is developed across the tank circuit. The inductance of this tank circuit consists of L3 and the oscillator coil, which is changed for each channel. The capacitance is supplied primarily by capacitor C9 and the AFT transistor. Oscillation is sustained by a capacitive voltage divider feedback network comprised of the transistor internal capacitance existing between collector and emitter and between emitter and base. Capacitor C8 couples the oscillator transistor to the tank circuit. This capacitive coupling minimizes the effects of the transistor capacitance on the tuning of the tank circuit.

A more complete schematic of the oscillator circuit is shown in Fig. 1-17. Bias voltages are applied through R4 and L4. Transistor base bias is determined by the voltage-divider network consisting of R1 and R3. Inductors L4 and L1 prevent rf energy from being coupled to ground.

The oscillator frequency changing that is needed for automatic fine tuning (AFT) is provided by transistor Q4. The internal capacitance of this transistor varies in proportion to the AFT voltage applied to its collector and base terminals. In this manner, the AFT system controls a portion of the oscillator tuning capacity and thus the oscillator frequency.

TURRET TUNER SERVICING

When the one- or two-piece tuner cover is removed most of the components can be checked. One feature of the turret tuner is that by removing a clip on each end of the tuner shaft, the entire shaft-and-drum assembly can be pulled out. This exposes all of the circuit components, which makes repair much easier.

The most common problem with tuners is dirty contacts. The channel strip buttons and the wiper contacts in the turret tuner should be cleaned with a good tuner cleaner spray. Occasionally, a tube socket pin may become dirty, but this may also be flushed out with the same tuner cleaner.

Isolating the trouble in a TV set to one stage is the first step in troubleshooting. Since the tuner, i-f section, video detector, and agc circuits are tied together, the circuit has to be broken somewhere. The agc connection point is a good place to begin.

Trouble Tips

Weak, snowy, or no picture? Any of these three troubles may be caused by the tuner. If the tuner tubes have been checked and the picture is still snowy, short the agc terminal on the tuner to ground. If the picture begins to return to normal or becomes

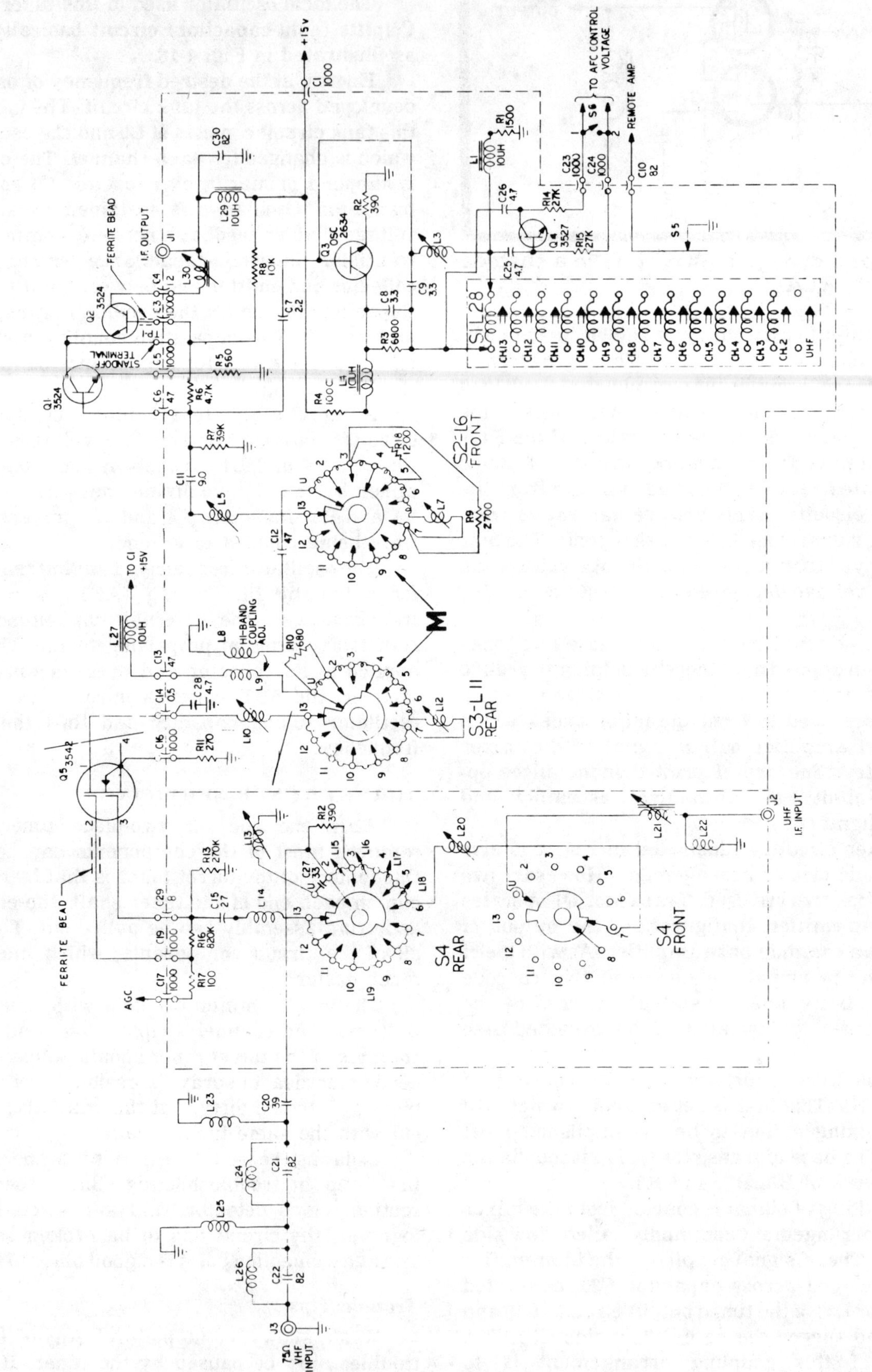

Fig. 1-14. RCA KRK-142 VHF tuner.

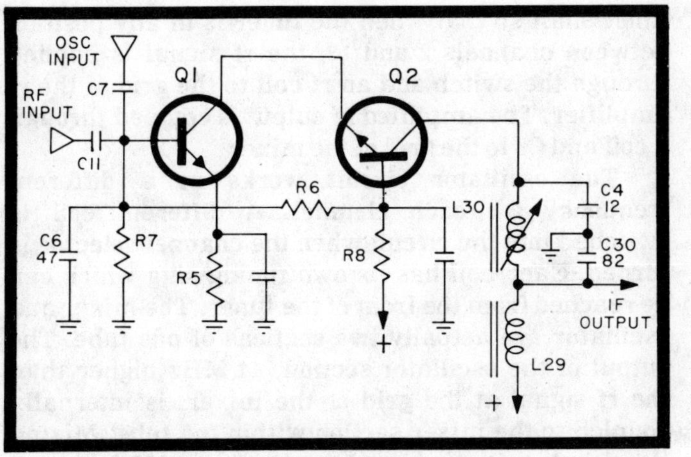

Fig. 1-15. Cascode transistors offer good input-output isolation in mixer service (RCA).

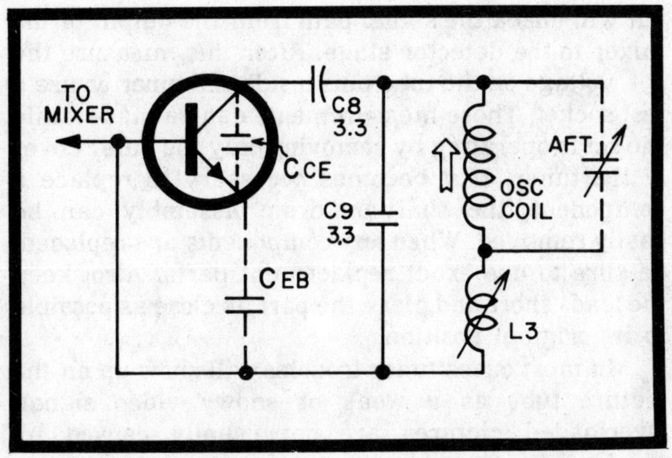

Fig. 1-16. Basic tuner oscillator (RCA).

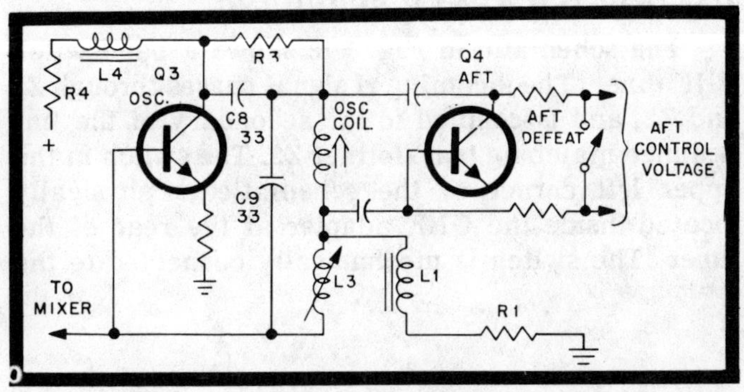

Fig. 1-17. Oscillator circuit (RCA).

overloaded, the trouble is probably in the agc circuit. If the picture does not begin to return to normal, the trouble is probably in the tuner or i-f stages. A good way to isolate the faulty section: Use a sweep generator tuned to about 44 MHz and connect a scope at the video detector. This method saves a lot of guessing and gives a good indication of the condition of the i-f system.

Feed a 44 MHz sweep signal to the mixer tube by loosely coupling it to a shield around the tube. This will not serve as a check of the entire mixer circuit,

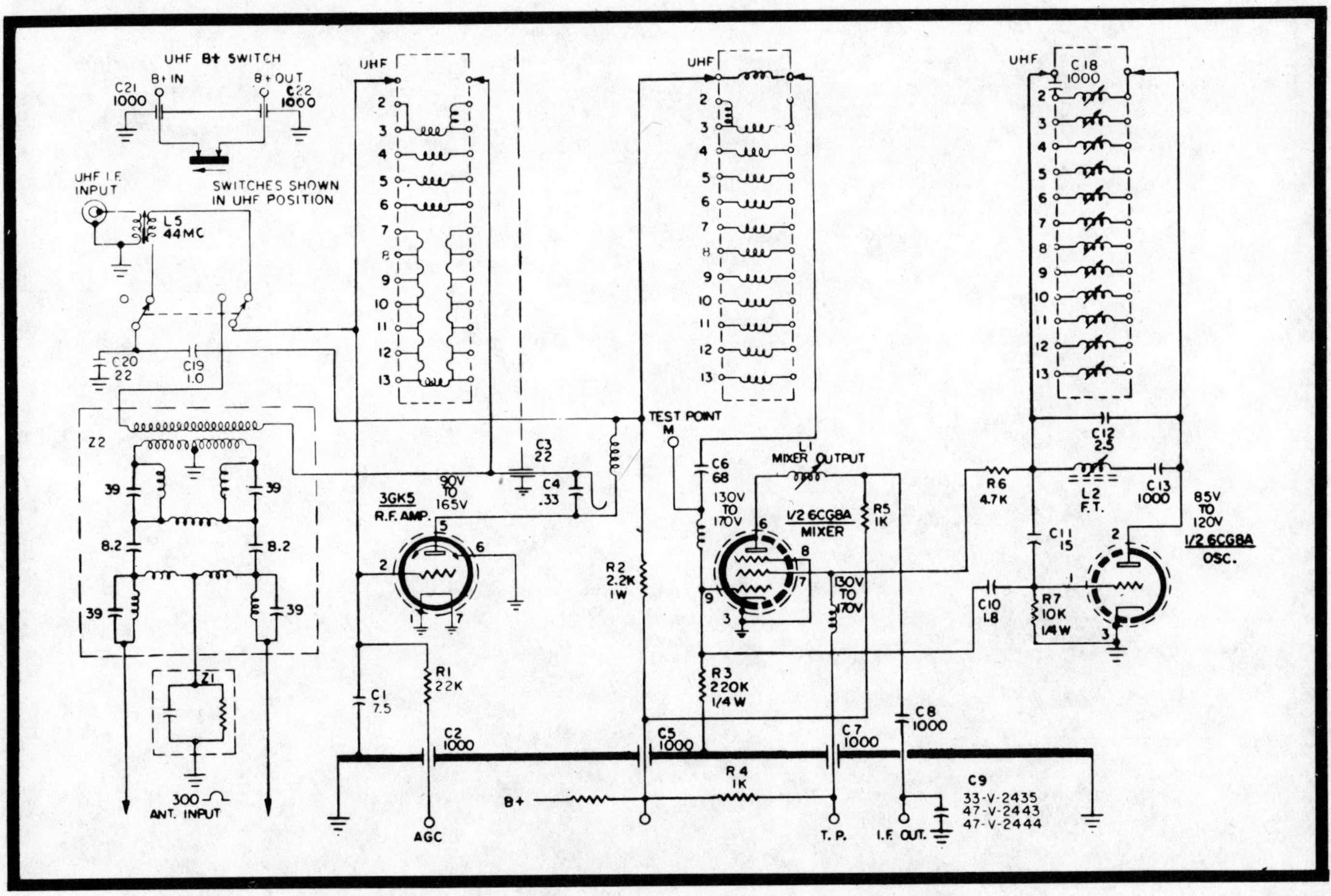

Fig. 1-18. Bandswitch-type tuner (Sylvania).

but will check the signal path from the output of the mixer to the detector stage. After this, measure the B+ voltage on the tube pins inside the tuner or use a test socket. These measurements can be made while the set is operating by removing only the outer cover of the tuner. If it becomes necessary to replace a component, the shaft-and-drum assembly can be easily removed. When any components are replaced, be sure to use exact replacement parts. Also, keep the leads short and place the part as close as possible to its original position.

In most cases tuner troubles will show up on the picture tube as a weak or snowy video signal. Overloaded pictures are not usually caused by troublesome tuners.

BANDSWITCH TUNER SERVICING

The schematic in Fig. 1-18 shows a bandswitch VHF tuner. The incoming rf signal passes through Z1 and Z2, and is coupled to the secondary of the impedance-matching transformer Z2. The switch in the upper left corner of the schematic is physically located inside the UHF adapter at the rear of the tuner. The switch is mechanically connected to the tuner shaft so that when the tuner is in any position between channels 2 and 13, the rf signal is applied through the switch and an rf coil to the grid of the rf amplifier. The amplified rf output is coupled through a coil and C6 to the grid of the mixer.

The oscillator circuit works at a different frequency for each channel. A different coil is switched into the circuit when the channel selector is turned. Each coil has its own tuning slug which can be reached from the front of the tuner. The mixer and oscillator are actually two sections of one tube. The output of the oscillator section, 44 MHz higher than the rf signal at the grid of the mixer, is internally coupled to the mixer section within the tube. Mixing the rf and oscillator signals produces a 44 MHz output which is coupled by the tunable mixer output coil to the i-f output terminal on the tuner top.

An agc voltage is fed to the grid of the rf amplifier through R1. When the tuner is in the channel 1 position, the switch in the UHF adapter connects B+ to the UHF tuner. It also applies the 44 MHz output of the UHF tuner to the grid of the rf amplifier. During UHF reception, the VHF tuner's rf amplifier and mixer tubes act as 44 MHz i-f amplifiers.

Chapter 2

RCA's Electronic VHF Tuner

The VHF tuner for the CTC 47 (KRK 158) is so radically different from tuners used in earlier models that a separate chapter describing it is warranted. Several advantages with a tuner of this type are immediately apparent; there are no switch contacts or gear trains—indeed, there are no moving parts at all! No torque is required, channel changing is almost instantaneous, and channel selection is completely silent. As a consequence, tuners of this type will likely never need servicing. In the interest of completeness, though, a theory discussion is presented here.

Tuning is accomplished by applying a dc voltage to one of 13 points, corresponding to 12 VHF channels plus channel 1 (for UHF). This use of digital rather than analog control simplifies the design and precludes several difficulties (such as long-term drift) normally encountered in analog systems.

In signal processing, the KRK 158 is similar to the KRK 142, which is used in the CTC 40 chassis. In signal selection, the KRK 158 has no counterpart, since other tuners are tuned mechanically. Stated another way, the signal processing portion of the KRK 158 performs the basic function of any other VHF tuner: rf amplification, mixing, conversion, and detection.

SIGNAL PROCESSING

When operating on channels 2 through 13, the VHF antenna input is fed to J2, the 75-ohm unbalanced input (Fig. 2-1). As with the CTC 40 and CTC 38, the balun is mounted on the rear of the cabinet rather than on the tuner. This allows the use of 75-ohm coaxial cable between the balun and the tuner to minimize pickup of unwanted off-frequency signals. The components connected between J2 and C2304 comprise an i-f rejection filter which prevents any possible radiation from the i-f amplifiers from feeding back into the input.

The rf amplifier stage of the KRK 158 uses a dual-gate MOSFET. Unlike a conventional bipolar transistor, which has low to moderate input impedance, the input impedance of the MOSFET is extremely high. Because of this characteristic, its operation is more closely analogous to that of a tube than to a bipolar transistor. There is one difference between the characteristics of the MOSFET gate and a vacuum-tube grid. The grid of a vacuum tube will draw considerable current if it is driven positive with respect to the cathode, whereas the gate current of a MOSFET remains extremely small regardless of its potential relative to the source (within the design limits of the device). Obviously, if the gate-to-source potential becomes great enough, substantial gate current will flow, but this is destructive to the device, since it constitutes a breakdown of the insulation between the gate and the channel. (The various elements of a MOSFET are identified in Fig. 2-2.) Once this breakdown occurs, usually the result of excessive voltage being applied to the gate, the device is irreparably damaged and must be replaced.

At one time, this type of failure was common in MOSFETs because the insulation-breakdown voltage between gate and channel is quite small. However, the device used in the KRK 158 has zener-protected gates, making it much less subject to damage from mishandling. The integral zener diodes limit gate potential to about 10V. This allows the MOSFET to be handled, using only the basic precautions recommended for any transistor.

CASCODE AMPLIFIERS

As with any other active device, tube or transistor, a MOSFET may be connected into three basic

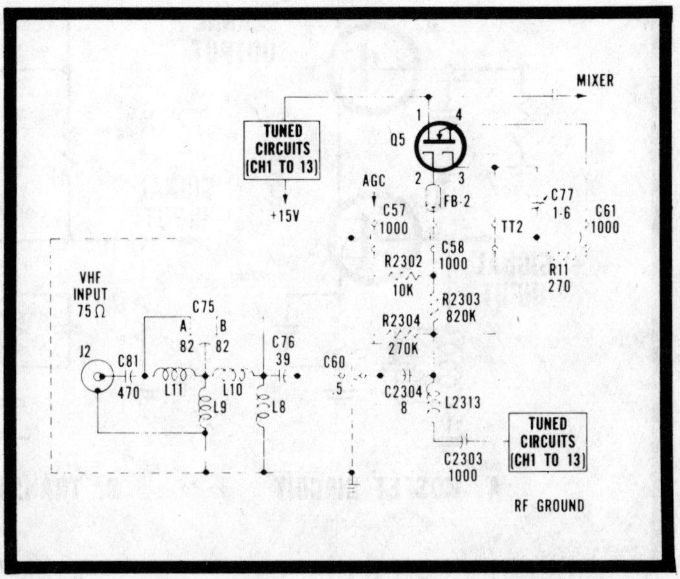

Fig. 2-1. Simplified rf amplifier circuit (RCA).

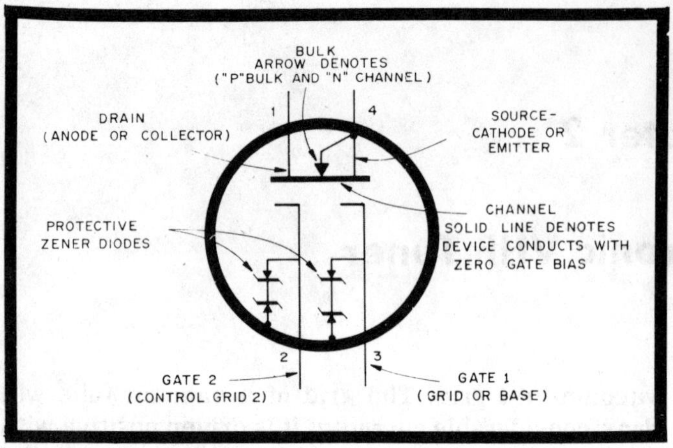

Fig. 2-2. N-channel, depletion-mode MOSFET with protected gates (RCA).

circuit configurations. These are common source, common-gate, or common-drain (analogous to common-emitter or common-cathode, common-base or common-grid, and emitter-follower or cathode-follower). When two MOSFET devices are connected with one connected common-source and the other connected common-gate, the circuit arrangement exhibits many of the characteristics, and therefore many of the advantages, of two vacuum tubes arranged in the cascode configuration. Simplified cascode circuits using tubes, transistors, and MOSFETs are shown in Fig. 2-3.

In the MOSFET circuit (Fig. 2-3A), Q1 is operated as a common-source amplifier; signal is applied to the gate and recovered from the drain. Since the device has high transconductance, a small change in gate voltage produces a relatively large change in drain current, resulting in a signal voltage appearing at the drain.

The transistorized cascode amplifier (Fig. 2-3B) and the vacuum-tube cascode amplifier (Fig. 2-3C) function in much the same way as the MOSFET cascode amplifier. In each case the signal is fed into the usual element, base or grid, and output is derived from the collector or plate. The output of the first stage of each is heavily loaded by the input to the second stage, which is operated common-base (grounded-grid). The use of vacuum-tube cascode amplifiers has been fairly widespread. Tuners which used dual triodes (6BQ7 and equals) were of the cascode type. Transistorized cascode circuits also are common.

The cascode circuit is used so often for two reasons: It has a good noise figure with almost optimum gain, and it does not require neutralization. Neutralization in conventional high-gain single-stage amplifiers is necessary because of the capacitance which always exists between input and output circuit elements. To prevent feedback and signal instability because of the **Miller effect** and the undesirably high **interelectrode capacitance**, a portion of the output may be deliberately fed back to the input at a phase angle which is exactly opposite to the phase of the signal fed back via the internal capacitance of the device. The two signals effectively cancel each other,

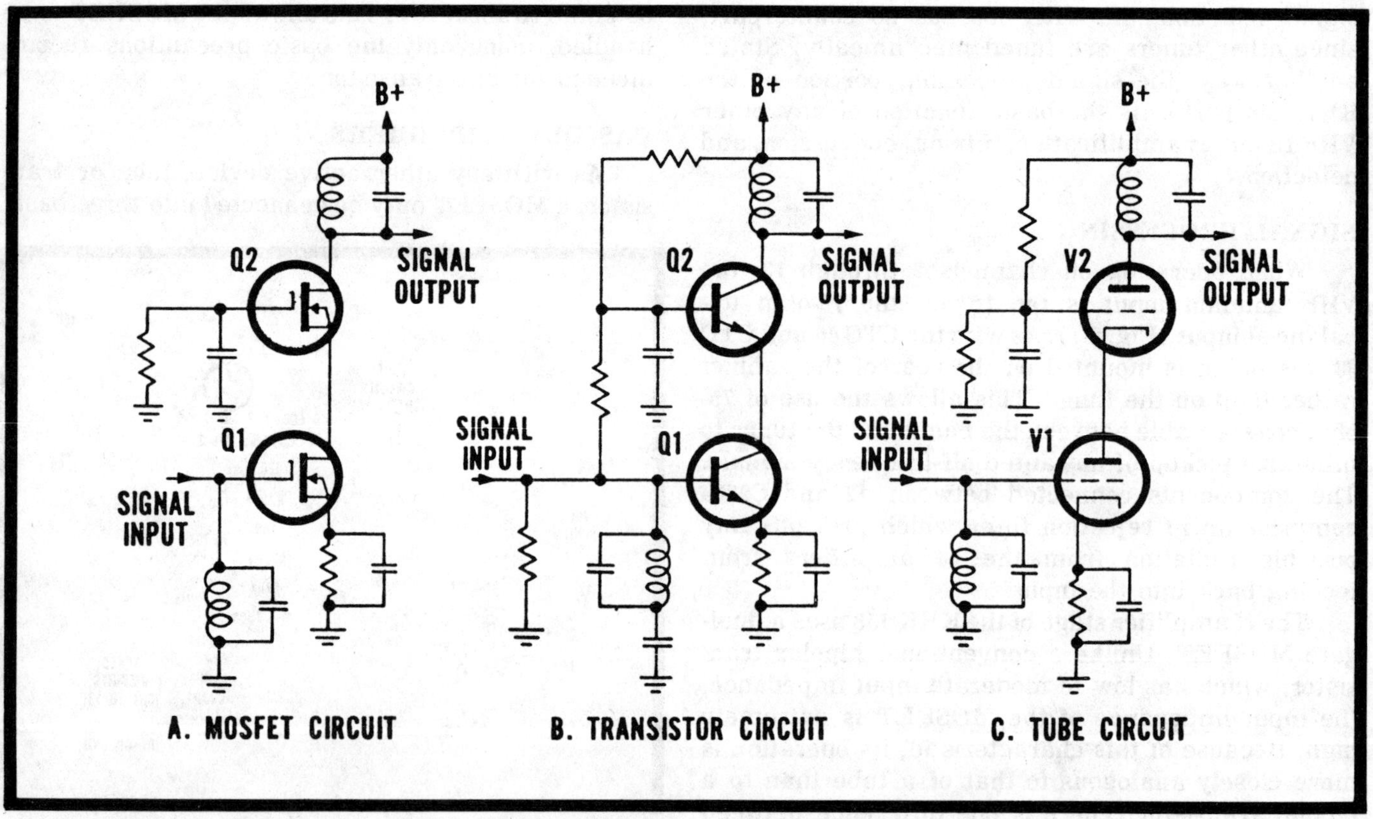

Fig. 2-3. Simplified cascode configurations (RCA)

and the circuit is said to be "neutralized." However, where broad bandwidth is required, neutralization is not necessarily the best approach, because the neutralizing capacitance value will vary, depending on frequency. Here is where cascode operation comes in.

In the cascode circuit, the voltage gain of the first stage is quite high, but the output loading is relatively heavy. The second stage operates with a grounded gate, which acts as a barrier to possible buildup of interelectrode capacitances and effectively shields the input circuit from the output circuit.

The cascode circuit of Fig. 2-3A shows two separate MOSFETs for ease of explanation. In practice, both devices may be constructed as a single unit. Since the drain of Q1 is directly connected to the source of Q2, it is feasible to extend the length of the channel material and place two gate structures over it.

Returning to Fig. 2-1, the dual-gate MOSFET used in the KRK 158 tuner has signal fed to the first gate and the second is held at rf ground by C58. Although no signal is fed to the second gate, agc bias voltage is connected to it via R2302; and, by the divider network consisting of R2303 and R2304, a portion of the agc bias also is applied to the first gate. Variation in agc voltage moves the operating point of the MOSFET amplifier in a manner which optimizes its performance.

CROSS MODULATION

Cross modulation in the rf amplifier is undesirable because it produces interference patterns or "beats" in the picture. Although it may occur at other points, the tuner is often the point of origin of cross modulation. An amplifier which is perfectly linear does not cross modulate, because all signals fed into it pass through unchanged except for amplitude. If the amplifier is not linear, each input signal will distort every other output signal.

Suppose that a 2V p-p sine wave at a frequency of 83.25 MHz (channel 6 pix carrier) is fed into an amplifier. Let the amplifier be biased at zero volts with a gain figure at this bias of 10. If the amplifier is to be gain-controlled by changing the bias, then its amplification must be greater than 10 with +1V of bias and less than 10 if the bias is —1V. For the sake of simplicity, assume that gain is 11 with +1V of bias, 10 at zero bias, and 9 with —1V of bias.

The instantaneous gain of this amplifier is determined by the instantaneous sum of the bias and the signal. Therefore, if a bias level of zero volts is maintained, the instantaneous voltage at the input will be +1V at the positive crest of the signal and —1V at the negative crest. At the output, the negative crest will be 11 times the positive input crest (11V) while the positive crest will be only 9V. The peak-to-peak amplitude of the output is 10 times the input, but the negative half-cycles are larger than the positive half-cycles; the signal has been distorted.

An analysis of the distortion which may occur in the amplifier is beyond the scope of this discussion; however, if it occurs the result is that harmonics of the channel 6 input are generated. This is known as harmonic distortion, a familiar term in audio work. Now assume that, in addition to the channel 6 signal, a second signal having the same amplitude but a frequency of 61.25 MHz (channel 3) also is present at the amplifier input. If the amplifier is not linear, harmonics of this signal also will be generated.

The harmonics of channel 6 and channel 3 are of minor consequence in themselves, unlike their action on each other. Since the gain of the amplifier constantly varies, depending on the instantaneous total of all voltages present at the input, the channel 6 output will be amplitude-modulated by the channel 3 signal. As a result, some of the channel 3 signal will appear in the channel 6 video to be viewed on the CRT. This is a very simple example of how cross modulation is produced. Various combinations of the harmonics also can produce spurious signals within the bandpass of the receiver. For instance, consider the second harmonics of channel 3 and channel 6. These are 122.5 MHz and 166.5 MHz, respectively. Neither of these is likely to pass through the i-f amplifier, but the beat between them is 44 MHz, which is within the i-f response curve, and this signal could appear in the picture.

The explanation given above actually exaggerates the problem for the sake of illustration. In practice, the level of the on-channel signal probably will not exceed 0.1V under most conditions; therefore, the amount of harmonic energy generated would be correspondingly less. Furthermore, the unwanted-channel energy is attenuated by the tuned circuit in the input of the amplifier, making it and its harmonics much weaker than the desired signal. Finally, the circuit parameters of the amplifiers are chosen so that the dynamic range of the agc bias is large in comparison with the signal level, and the device itself is designed to have a constant rate of gain change throughout the range of the agc bias voltage.

AGC DELAY

Agc bias voltage is fed to the second gate of Q5 to control the gain of the stage. Under no-signal conditions this is about +6.7V, which provides maximum gain. At input-signal levels of about 100 mV, the agc bias is driven to its maximum negative value of —5V. Figure 2-4 shows the agc voltage supplied to the tuner and the i-f strip as a function of input signal.

The rf agc bias is delayed in the agc system of the CTC 47 chassis so that the tuner develops maximum gain up to an input level of about 300 uV. The signal level at which the agc begins to decrease the gain of

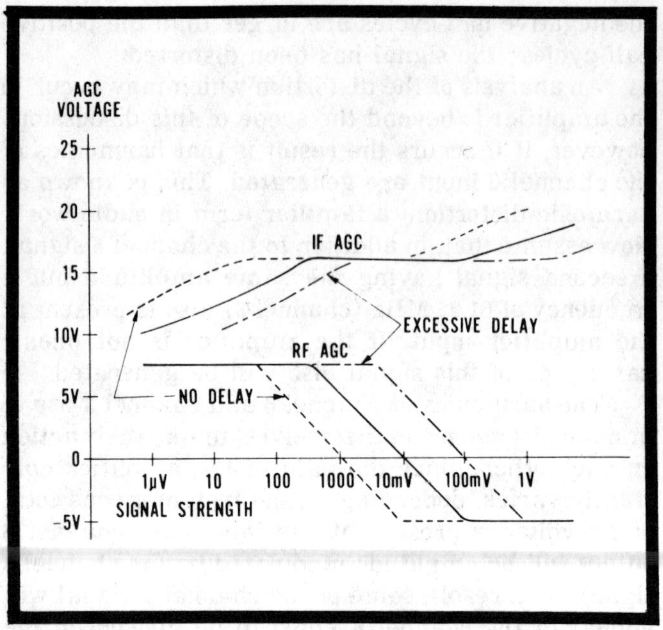

Fig. 2-4. Rf and i-f agc voltage vs signal strength (RCA).

the rf amplifier is adjustable by means of a "noise control" in the agc system. At one extreme of the control, there is very little agc delay and the tuner bias begins to drop from +6.7V even on very weak signals. At the other extreme of the control, the tuner is operated at maximum gain until the signal level is considerably greater than 300 mV.

Since cross modulation in the mixer stage is aggravated by very strong signals, it is desirable to operate the tuner with very little delay. On the other hand, receiver noise may be reduced (sensitivity improved) by using near-maximum delay. By use of the noise control, the agc delay may be adjusted for best possible operation under various local conditions of signal strengths.

NOISE AND AGC DELAY

To understand how tuner agc affects receiver noise, it is necessary to remember that any high-quality television receiver has sufficient gain to produce nearly full output from the video detector, even with no input signal. This noise, of course, is

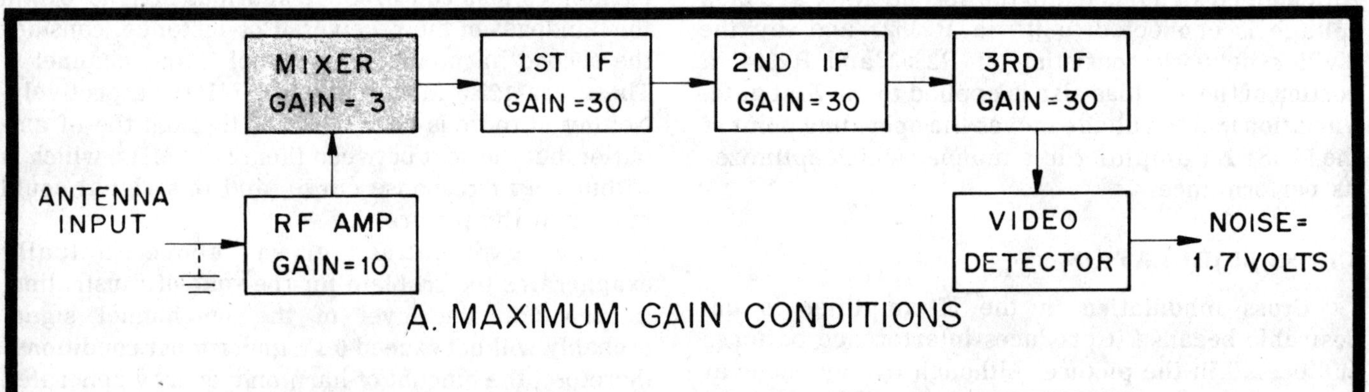

Fig. 2-5. The effects of agc delay on receiver noise (RCA).

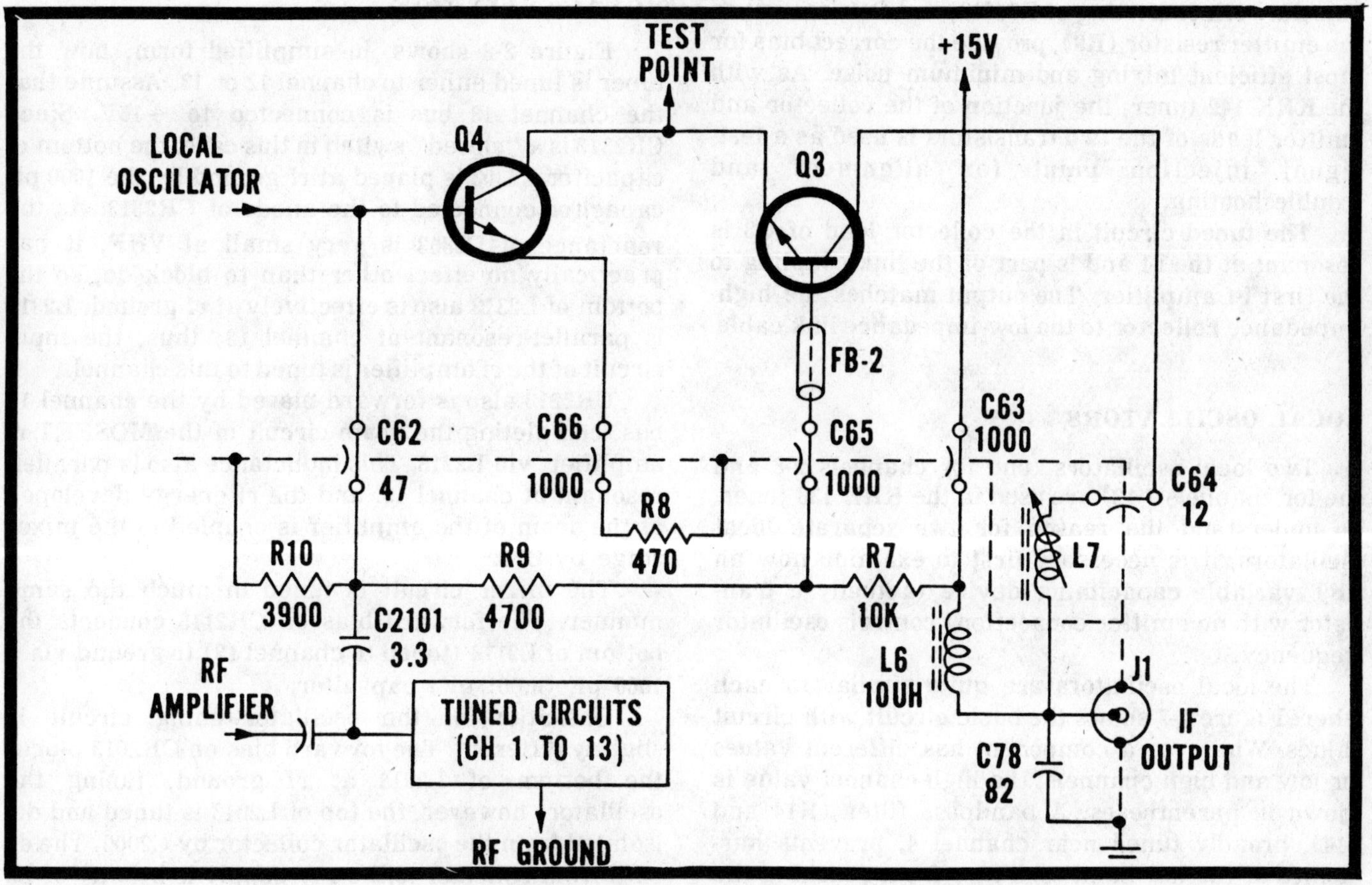

Fig. 2-6. Simplified mixer circuit (RCA).

generated within the receiver itself. When a signal is present at the antenna terminals, the agc reduces the receiver gain to maintain the same level of output from the video detector; but as the receiver gain is reduced, the amount of noise at the video detector also is reduced. Therefore, at some level of input signal, the ratio between signal and noise at the video detector is great enough to produce a noise-free picture.

The amount of noise which each stage in the receiver contributes to the total noise is determined by the amount of noise it actually generates and its position.

Assume that a certain receiver has five stages of amplification, as shown in Fig. 2-5A. The overall gain, the product of the individual gains of the stages, is 810,000. If the noise generated by the rf amplifier is equivalent to a 2 uV signal at its grid, this noise will be amplified by every stage, producing 1.62V at the detector. If the mixer has the same noise level as the rf amplifier, it will produce only 162 mV at the output, since the gain from mixer input to output is only 81,000. Similarly, if each of the three i-f amplifiers has the equivalent of 2 uV of noise generated at its grid, their contributions to the total noise will be 54 mV, 1.8 mV, and 60 uV, respectively.

It would seem that only the noise of the rf amplifier would be of any significance in determining the overall receiver noise and sensitivity. But this is not the case. Although the gains per stage shown in Fig. 2-5A are fairly accurate, the mixer noise might actually be 6 uV, three times as great as the rf amplifier noise. As a result, the mixer would actually produce nearly 500 mV of noise at the detector.

If the receiver in this explanation is designed for 3V output from the video detector and a 100 uV signal is present at the input, the agc circuit must reduce the overall gain from 810,000 to only 30,000. Depending on the setting of agc delay, this could be accomplished in either of the two ways shown in Figs. 2-5B and 2-5C. The example used in Fig. 2-5B is not typical since some gain in the rf amplifier always would be allowed with a 100 uV signal; nevertheless, it does illustrate the effect of too little rf gain. Obviously, a malfunction in the rf amplifier which reduces its gain has the same effect as improper agc delay; a snowy picture results.

MIXER AND OUTPUT

As shown in Fig. 2-1, the drain load of the rf amplifier is a tuned circuit resonated at the desired frequency by the electronic signal-selection system. Energy is capacitively coupled from this circuit to the input of the mixer. The mixer (Fig. 2-6), is a two-transistor cascode circuit. Both the rf signal and the local oscillator output are fed to the base of Q4; the bias network consisting of R7, R9, and R10, as well as the emitter resistor (R8), provide the correct bias for most efficient mixing and minimum noise. As with the KRK 142 tuner, the junction of the collector and emitter leads of the two transistors is used as a test-signal injection point for alignment and troubleshooting.

The tuned circuit in the collector lead of Q3 is resonant at the i-f and is part of the link coupling to the first i-f amplifier. The output matches the high-impedance collector to the low-impedance link cable.

LOCAL OSCILLATORS

Two local oscillators (one for channels 2-6 and one for channels 7-13) are used in the KRK 158 tuner. To understand the reason for two separate local oscillators, it is necessary first to examine how an AFT variable capacitance device (actually a transistor with no emitter connection) controls oscillator frequency.

The local oscillators are quite similar to each other. Figure 2-7 shows the basic circuit with circuit values. Wherever a component has different values for low and high channels, the high-channel value is shown in parentheses. A bandpass filter (R14 and L14), broadly tuned near channel 4, prevents harmonics of the low-band oscillator from reaching the mixer. The appropriate oscillator with the correct tuned circuit for the particular channel is connected into the circuit by the tuning circuits to be explained next.

SIGNAL SELECTION

Figure 2-8 shows, in simplified form, how the tuner is tuned either to channel 12 or 13. Assume that the channel 13 bus is connected to +16V. Since CR2313 is a "closed" switch in this case, the bottom of capacitor C2303 is placed at rf ground by the 1000 pF capacitor connected to the anode of CR2313. As the reactance of C2303 is very small at VHF, it has practically no effect other than to block dc, so the bottom of L2313 also is effectively at rf ground. L2313 is parallel-resonant at channel 13; thus, the input circuit of the rf amplifier is tuned to this channel.

CR2213 also is forward-biased by the channel 13 bus, completing the drain circuit of the MOSFET rf amplifier, via L2213. This inductance also is parallel-resonant at channel 13, and the rf energy developed at the drain of the amplifier is coupled to the mixer stage by C72.

The mixer circuit is tuned in much the same manner. The forward bias on CR2113 connects the bottom of L2113 (tuned to channel 13) to ground via a 1000 pF (0.001 uF) capacitor.

Operation of the oscillator-tuning circuit is slightly different. The forward bias on CR2013 places the bottom of L2013 at rf ground, tuning the oscillator; however, the top of L2013 is tuned and dc-isolated from the oscillator collector by C2006. The dc path from collector to the 16V supply is by way of L3 and the coils for channels 7-12, back to CR2013. The high reactance of L3 and the ferrite bead, coupled with the low reactance of C52 and C45, prevent any rf energy from flowing in this circuit.

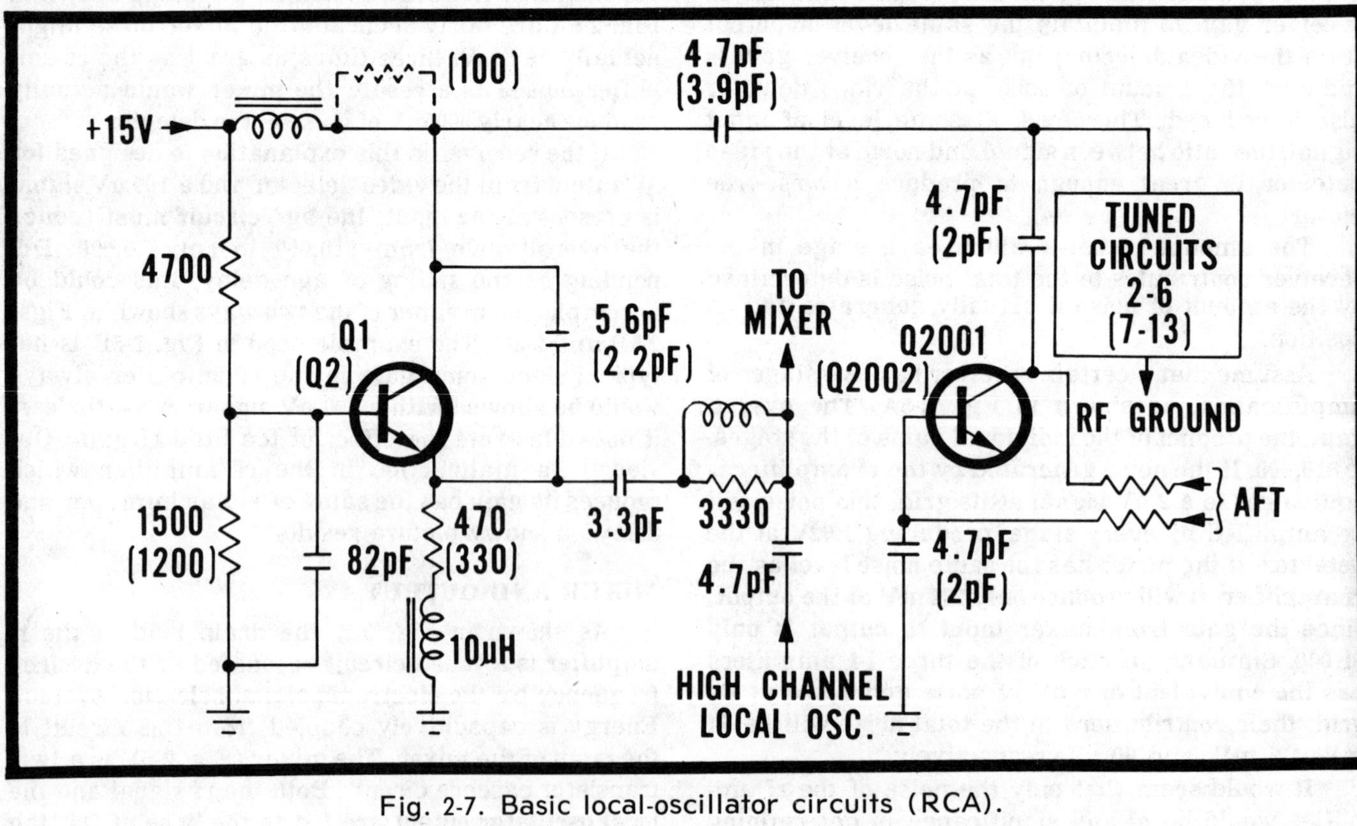

Fig. 2-7. Basic local-oscillator circuits (RCA).

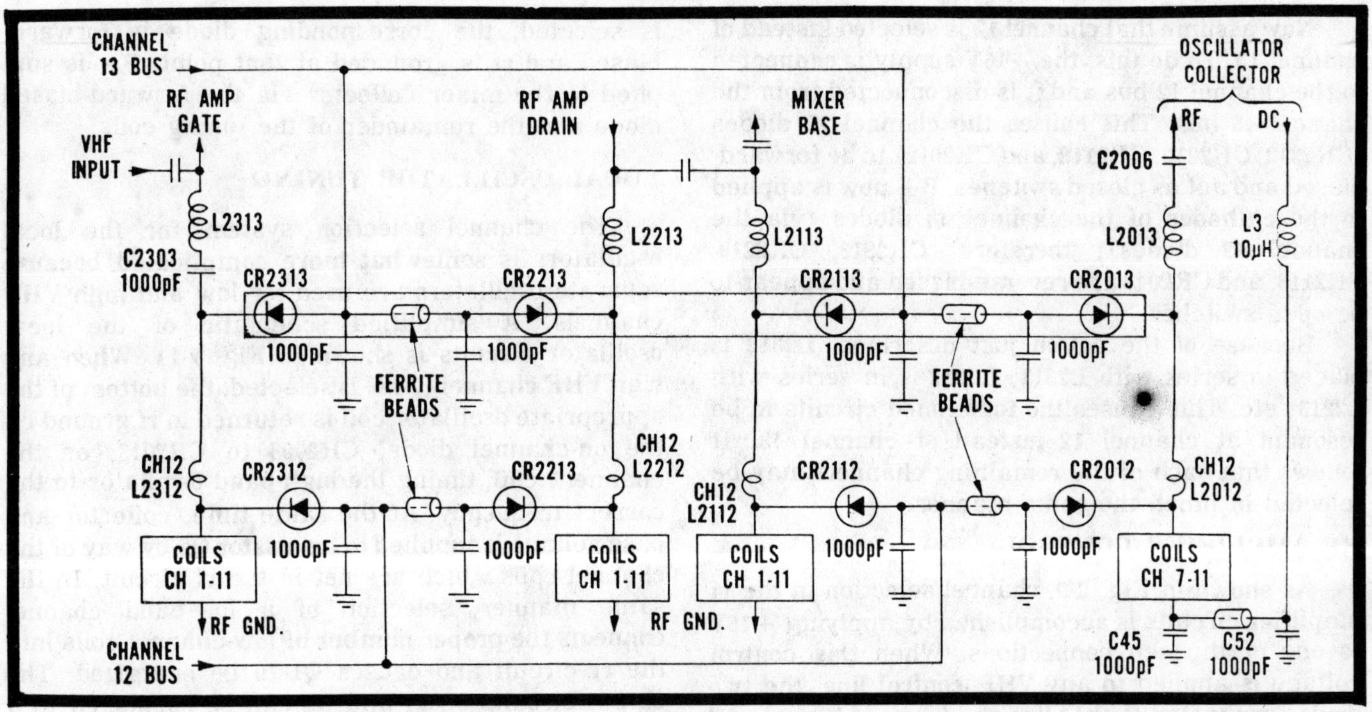

Fig. 2-8. Simplified channel-selection circuit (RCA).

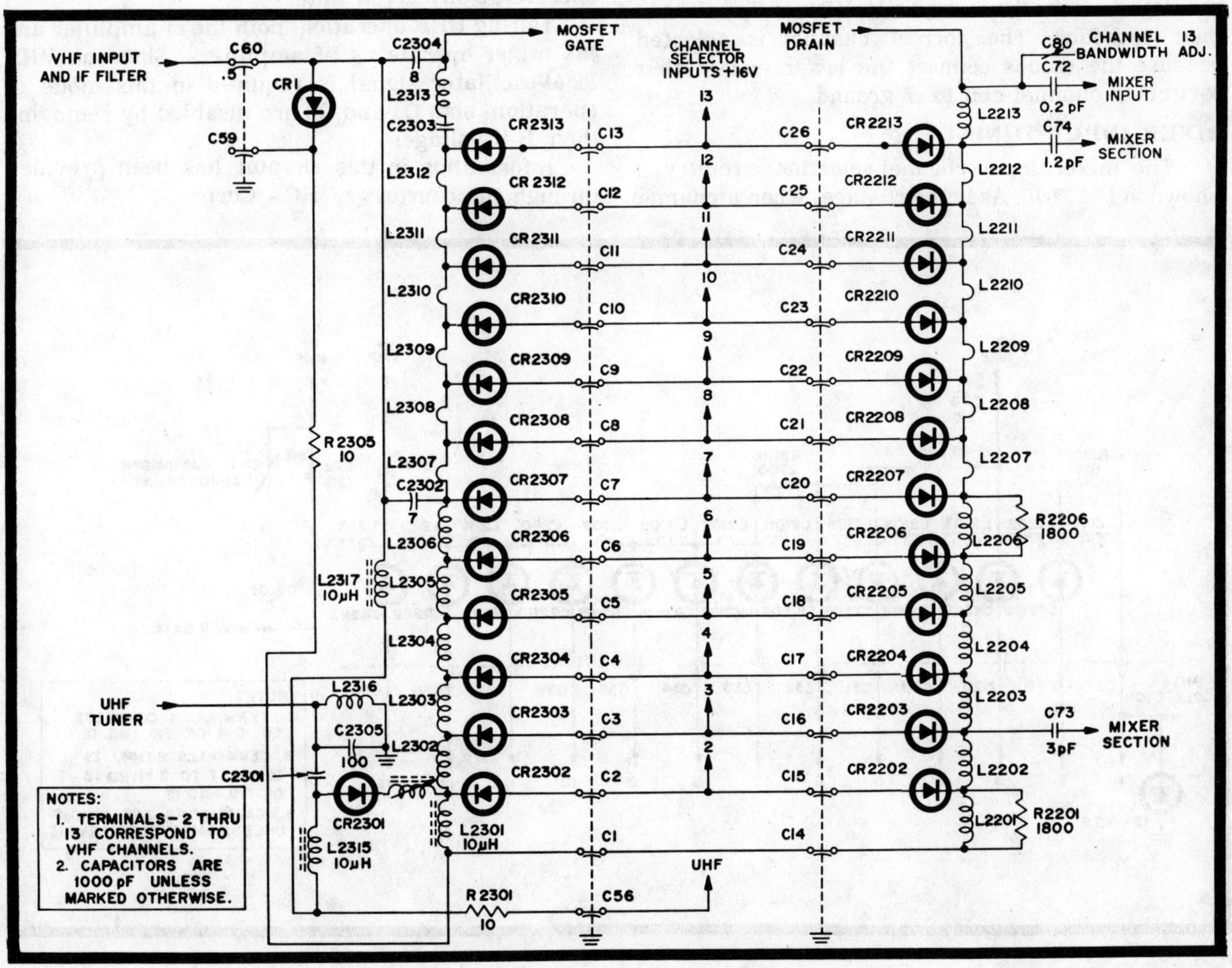

Fig. 2-9. Rf amplifier control (RCA).

Now assume that channel 12 is selected instead of channel 13. To do this, the +16V supply is connected to the channel 12 bus and it is disconnected from the channel 13 bus. This causes the channel 12 diodes (CR2312, CR2212, CR2112, and CR2012) to be forward-biased and act as closed switches. B+ now is applied to the cathodes of the channel 13 diodes (via the channel 12 diodes); therefore, CR2313, CR2213, CR2113, and CR2013 are reverse-biased and appear to be open switches.

Because of the action just described, L2312 is placed in series with L2313; L2212 is in series with L2213; etc. This causes the four tuned circuits to be resonant at channel 12 instead of channel 13. It follows that each of the remaining channels may be selected in much the same manner.

RF AMPLIFIER CONTROL

As shown in Fig. 2-9, channel selection in the rf amplifier circuits is accomplished by applying +16V to one of thirteen connections. When this control voltage is applied to any VHF control line, the two diodes connected to that line are forward-biased and all others are reverse-biased by virtue of the voltage developed across the zener diode, VR1, and applied to their cathodes. The correct channel is selected because the diodes connect the lower end of that particular channel coil to rf ground.

MIXER INPUT TUNING

The mixer input channel-selection circuitry is shown in Fig. 2-10. As in the rf stage, when a channel is selected, the corresponding diode is forward-biased and rf is grounded at that point. B+ is supplied to the mixer collector via the forward-biased diode and the remainder of the tuning coils.

LOCAL OSCILLATOR TUNING

The channel selection system for the local oscillators is somewhat more complicated because separate oscillators are used for low and high VHF channels. A simplified schematic of the local oscillator circuits is shown in Fig. 2-11. When any high VHF channel (7-13) is selected, the bottom of the appropriate oscillator coil is returned to rf ground by the on-channel diode, CR2008 to CR2013, or the channel 7 coil, tuning the high-band oscillator to the correct frequency. At the same time, collector and base voltage is supplied to transistor Q2 by way of the channel coils which are not in the rf circuit. In the same manner, selection of a low-band channel connects the proper number of low-channel coils into the rf circuit and causes Q1 to be energized. The output circuits of Q1 and Q2 can be connected to a common point, since only one oscillator is allowed to operate at any given time.

During UHF operation, both the rf amplifier and the mixer operate as i-f amplifiers. Since no VHF local-oscillator signal is required in this mode of operation, both Q1 and Q2 are disabled by removing their B+ voltage.

Information in this chapter has been provided through the courtesy of RCA Corp.

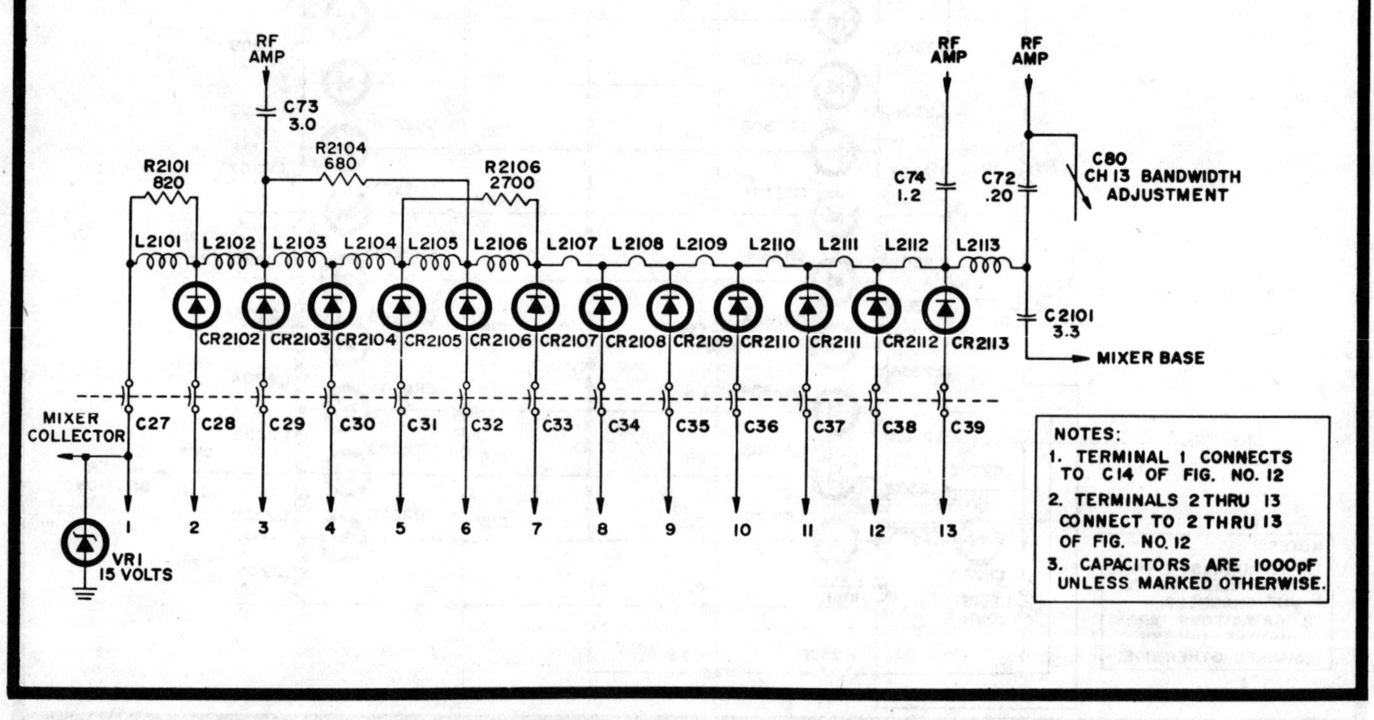

Fig. 2-10. Simplified mixer input control (RCA).

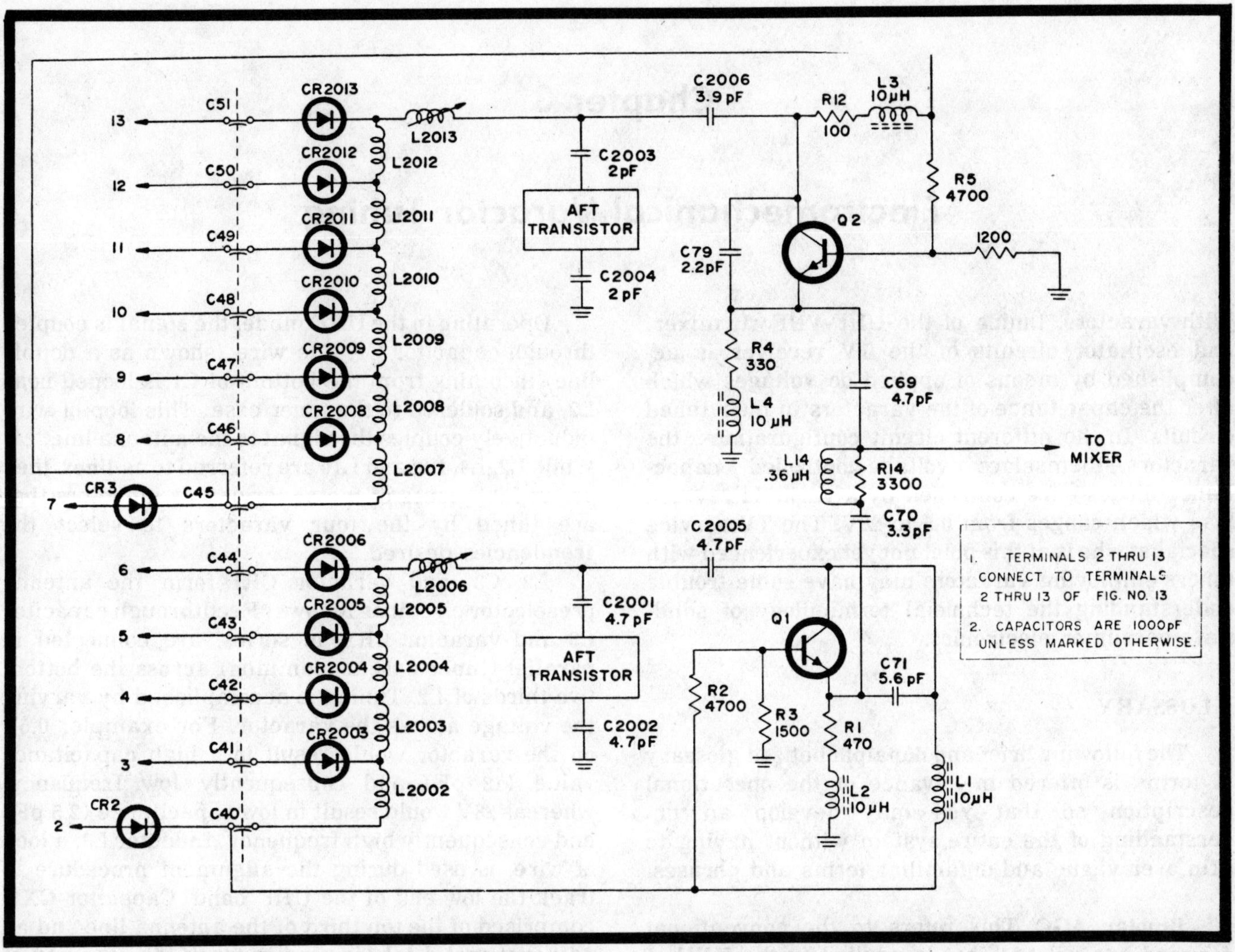

Fig. 2-11. Simplified local-oscillator control (RCA).

Chapter 3

Electromechanical Varactor Tuning

With varactors, tuning of the UHF-VHF rf, mixer, and oscillator circuits of the TV receiver is accomplished by means of applied dc voltages which alter the capacitance of the varactors in their tuned circuits. In the different circuit configurations, the varactors—themselves voltage-controlled capacitance diodes—are controlled by a variable reverse bias which ranges from 0.5 to 28V. The TV service specialist who is at this point not yet experienced with tuners employing varactors may have some trouble understanding the technicial terminology of solid-state consumer electronics.

GLOSSARY

The following brief and nonalphabetized glossary of terms is offered in advance of the operational description so that you may develop an understanding of the entire system without having to skip over vague and unfamiliar terms and phrases.

Bipolar AGC. This refers to the conventional **forward** agc voltage that is applied to the UHF rf amplifier transistor in a typical solid-state circuit.

FET AGC. This refers to a **negative-going** agc voltage developed by the receiver, and is used by the VHF rf stage. The rf stage uses a dual-gate FET. The FET gate used for gain control is biased for maximum gain. Under medium-strength signal conditions, the negative agc voltage is used to oppose the positive gate voltage, thereby reducing it to a value somewhat less positive.

Switch Diodes. Switch diodes may be used in VHF varactor tuners. For low-channel operation, such diodes are reverse biased, and for high channel operation they are forward biased.

UHF VARACTOR TUNER

The UHF tuner we'll be examining is shown in Fig. 3-1. The electronics here consist of a four-circuit, two-transistor, varactor-tuned front end. Unlike the previous species of tuner, it uses a transistor rf amplifier stage.

The receiver has a single antenna input. A balun mounted behind the antenna terminal board accommodates a 300-ohm balanced antenna line. The output to the tuner, of course, is the conventional unbalanced line.

Operating in the UHF mode, the signal is coupled through capacitor C1. The wire (shown as a double line) incoming from the bottom of C1 is looped near L2, and soldered to the tuner case. This loop of wire inductively couples the signal to the antenna line, L2. While L2, L4, L8, and L10 are referred to as **lines**, they are in fact pieces of metal serving as inductors that are tuned by the four varactors to select the frequencies desired.

L2, C3, and varactor CR1 form the antenna preselector circuit as follows: Feedthrough capacitor C3 and varactor CR1, in series, are connected in parallel (tuner case as common) across the bottom two-thirds of L2. Tuning is accomplished by varying the voltage across the varactor. For example: 0.5V on the varactor would result in a high capacitance value (12 pF) and consequently low frequency, whereas 28V would result in low capacitance (2.5 pF) and consequently high frequency. Inductor L3, a loop of wire, is used during the alignment procedure to track the low end of the UHF band. Capacitor CX1, comprised of the top third of the antenna line and an adjacent metal tab, is used to track the high end of the band. The two tuned circuits that follow (collector line and mixer line) tune and perform in a similar manner to the antenna line described above; therefore, in the discussion that follows, they will be referred to by their circuit name.

The rf signal is capacitively coupled from L2 by feedthrough C4 to the emitter of Q1. The capacitor taps the inductance at the proper point to provide rf impedance matching to the transistor junction. The rf amplifier (Q1) is a common-base amplifier operated in the forward agc mode. Maximum-gain base bias (+2.6V) is provided by a resistive voltage divider on the external component mounting board. Forward agc voltage (up to 7V) is applied to the rf base through R15 and the tuner control center.

The amplified rf signal appearing at the collector is capacitively coupled to the collector line by feedthrough capacitor C7. Collector line L4 and mixer line L8, along with varactors CR2 and CR3 and capacitors C6 and C10, form a mutually coupled rf-to-mixer interstage network. The coupling takes place due to the spacing between the collector and mixer lines.

The cathode side of the foward-biased hot carrier mixer diode is soldered to the mixer line at a point to

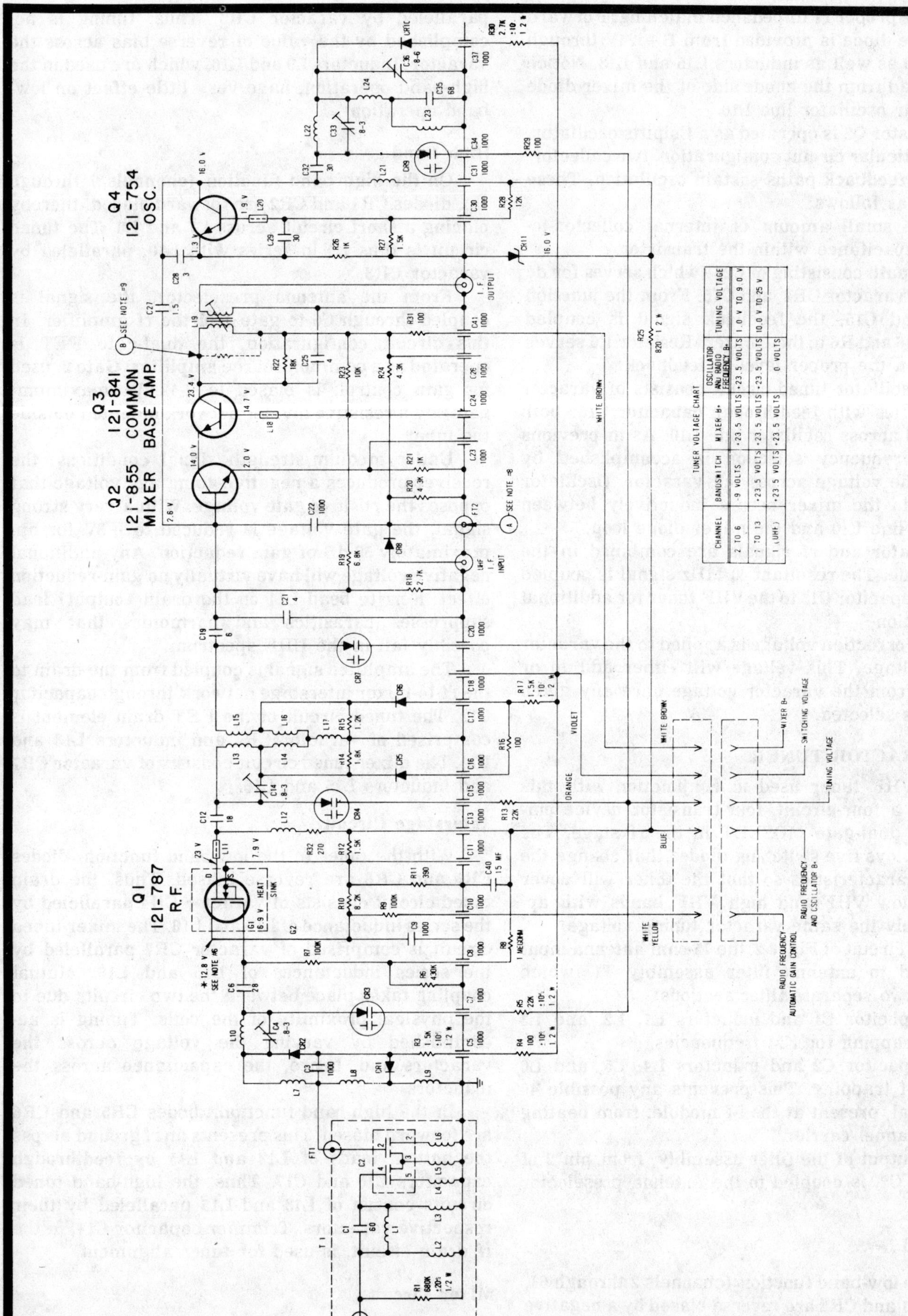

Fig. 3-1. UHF tuner.

provide the proper rf impedance matching. Forward bias for the diode is provided from B+ 24V through resistor R9 as well as inductors L15 and L13. Notice that the lead from the anode side of the mixer diode passes near oscillator line L10.

Transistor Q2 is operated as a Colpitts oscillator. In this particular circuit configuration, two collector-to-emitter feedback paths sustain oscillation. These paths are as follows:

1. The small amount of internal collector-to-emitter capacitance within the transistor.

2. The path consisting of C21 (which serves for dc blocking), varactor CR4, and C15. From the junction of CR4 and C15, the feedback signal is coupled through C16 and R6 to the emitter. Resistor R6 serves to maintain the proper level of feedback.

The oscillator tuned circuit consists of varactor CR4 in series with feedthrough capacitor C15; both paralleled across oscillator line L10. As in previous circuits, frequency selection is accomplished by varying the voltage across the varactor. Oscillator injection to the mixer occurs inductively between oscillator line L10 and the mixer diode loop.

Oscillator and rf signals are combined in the mixer diode. The resultant 40 MHz signal is coupled through capacitor C13 to the VHF tuner for additional amplification.

AFC correction voltage is applied to the varactor tuning voltage. This voltage will either add to or subtract from the varactor voltage once any given channel is selected.

VHF VARACTOR TUNER

The VHF tuner used in conjunction with this system is a four-circuit, four-transistor device employing a dual-gate MOSFET in the rf stage. The tuner employs five switching diodes that change the circuit characteristics so that the tuner will cover both the low VHF and high VHF bands with approximately the same varactor tuning voltage.

In the circuit of Fig. 3-2, the 75-ohm antenna input is coupled to antenna filter assembly T1, which contains two separate filter sections:

1. Capacitor C1 and inductors L1, L2, and L3 provide trapping for FM frequencies.

2. Capacitor C2 and inductors L4, L5, and L6 provide i-f trapping. This prevents any possible 40 MHz signal, present at the i-f module, from beating with a channel carrier.

The output of the filter assembly, from pin 2 of capacitor C2, is coupled to the antenna preselector circuit.

Low Band

On the low-band function (channels 2 through 6), diodes CR1 and CR2 are reverse biased by a negative voltage (—9V) coupled from the tuner control center. With diodes CR1 and CR2 cut off, the tuned circuit consists of inductors L7, L8, L9, L10 in series, paralleled by varactor CR3. Thus, tuning is accomplished by the value of reverse bias across the varactor. Inductors L9 and L10, which are used in the high-band operation, have very little effect on low-band operation.

High Band

On the high band function (channels 7 through 13), diodes CR1 and CR2 are forward biased, thereby placing a short circuit across L7 and L8. The tuned circuit is thus L9 in series with L10, paralleled by varactor CR3.

From the antenna preselector, the signal is coupled through C6 to gate 1 of the rf amplifier. In this circuit configuration, the dual-gate FET is operated as a common-source amplifier. Gate 2, used for gain control, is biased to +12.8V (maximum-gain) by a resistive divider network mounted outside the tuner.

Under medium-strength signal conditions, the receiver produces a negative-going agc voltage that opposes the positive gate voltage. With a very strong signal, the gate voltage is reduced to +3V for approximately 50 dB of gain reduction. Any additional negative voltage will have virtually no gain-reduction effect. Ferrite bead L11 on the drain (output) lead suppresses parasitics and harmonics that may possibly fall in the UHF spectrum.

The amplified signal is coupled from the drain to the rf-to-mixer interstage network through capacitor C12. The tuned circuit of the FET drain element is comprised of varactor CR4 and inductors L13 and L14. The mixer tuned circuit consists of varactor CR7 and inductors L15 and L16.

Interstage Circuit

With the tuner in the low-band function, diodes CR5 and CR6 are reverse biased. Thus, the drain tuned circuit consists of varactor CR4 paralleled by the series inductance of L15 and L16. The mixer tuned circuit is comprised of varactor CR7 paralleled by the series inductances of L15 and L16. Mutual coupling takes place between the two circuits due to the physical proximity of the coils. Tuning is accomplished by varying the voltage across the varactors and, hence, the capacitance across the inductors.

In the high-band function, diodes CR5 and CR6 are forward biased. This presents an rf ground across the bottom leads of L13 and L15 by feedthrough capacitors C16 and C17. Thus, the high-band tuned circuits consist of L13 and L15 paralleled by their respective varactors. Trimmer capacitor C14, in the rf drain circuit, is used for tuner alignment.

Mixer Circuit

The mixer circuit, which contains no trimmers, is the standard to which the other circuits are adjusted in the alignment procedure.

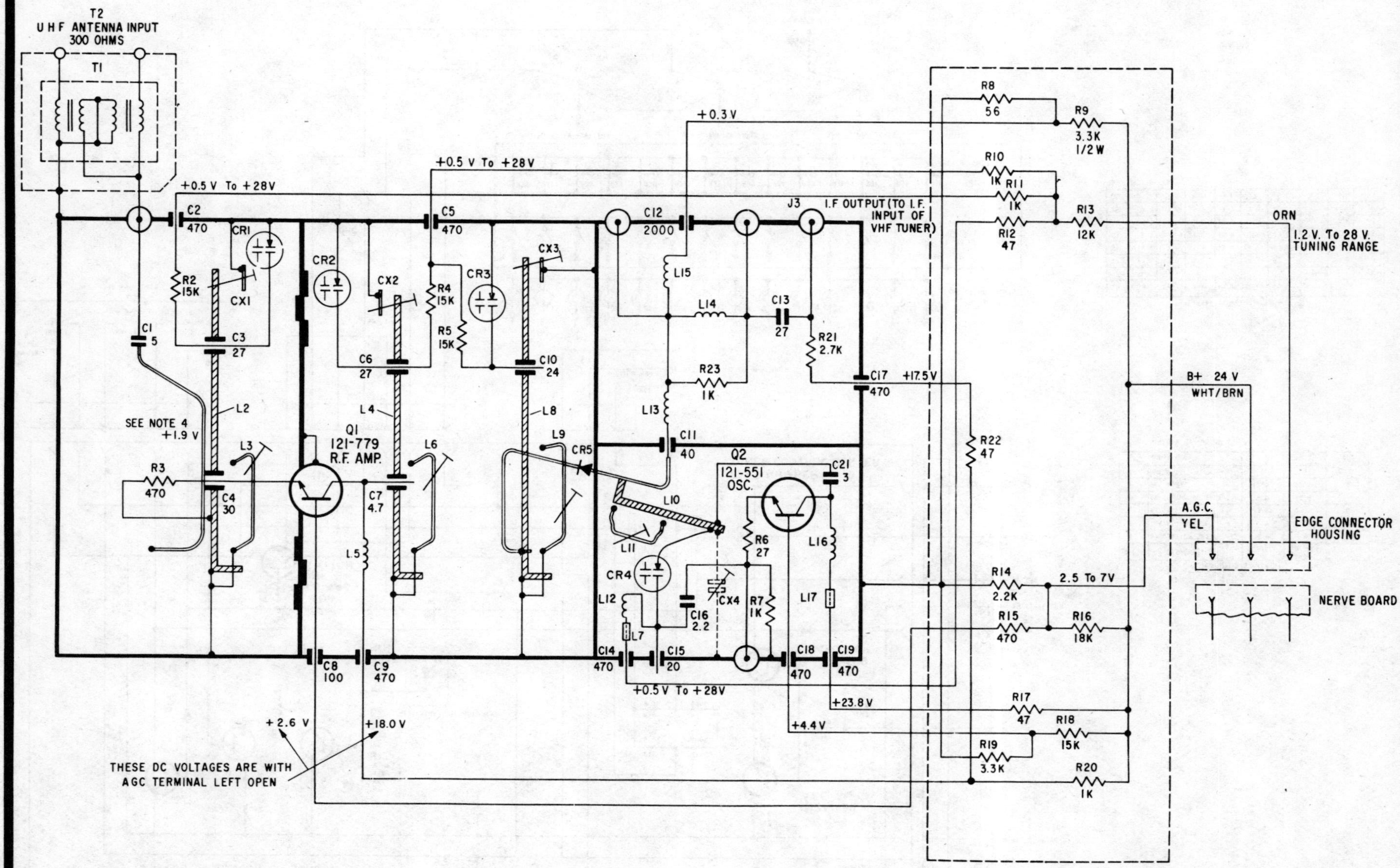

Fig. 3-2. VHF tuner.

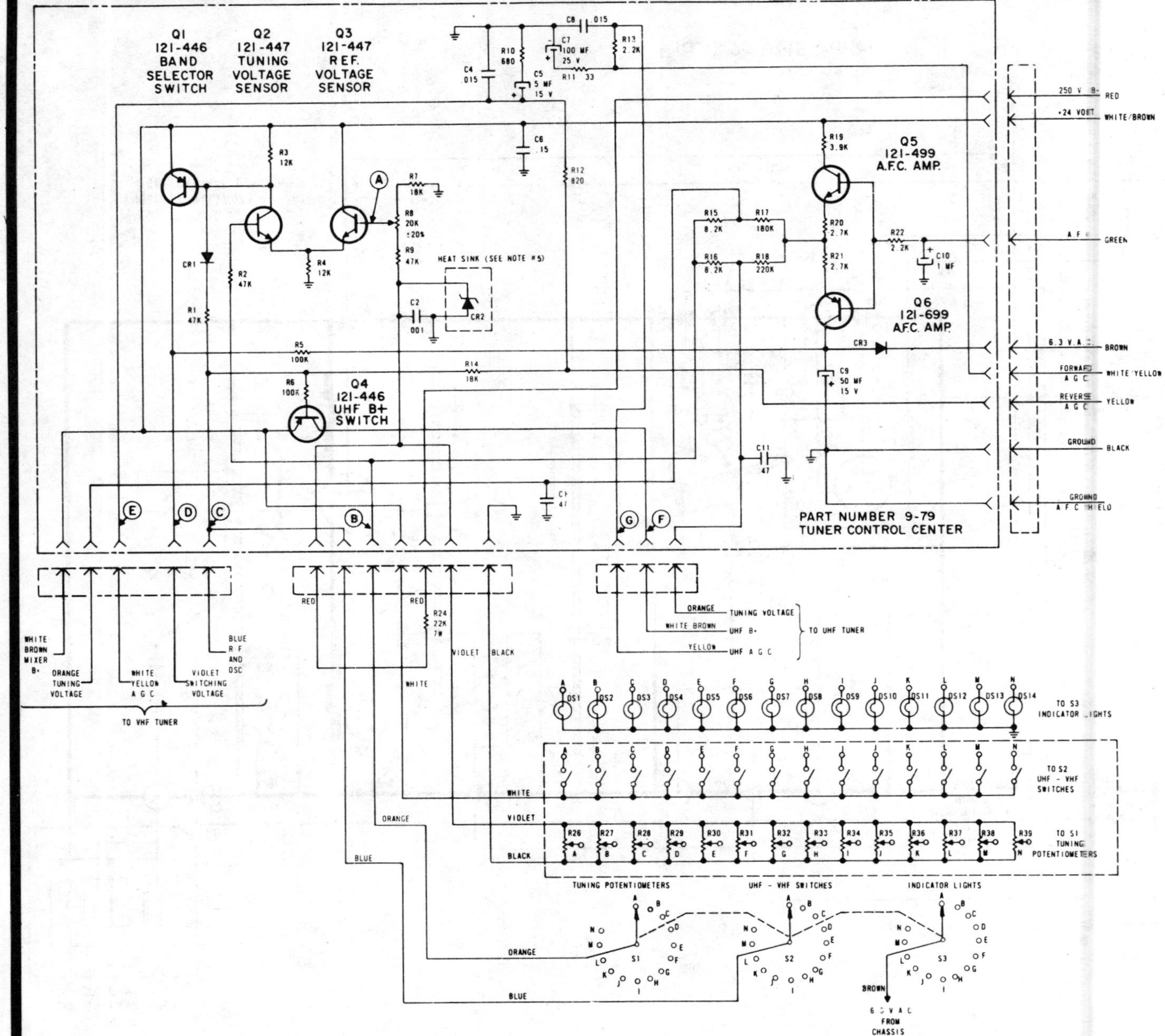

Fig. 3-3. Tuner control center.

TEST POINTS (T.P.)

T.P.	DESIGNATION	LO VOLTAGE	HI VOLTAGE	UHF
A	BANDSWITCH SET PT	9.8	*	*
B	TUNING VOLTAGE	0 - 9.8	9.8-25.0	0 28.0
C	VHF B+	23.5	23.5	0
D	SWITCHING VOLTAGE	-9.0	23.3	23.3
E	VHF A.G.C.	12.5	12.5	0
F	UHF B+	0	0	23.3
G	UHF A.G.C.	2.2	2.2	2.2

* = DEPENDS ON TUNING VOLTAGE

TRANSISTOR VOLTAGES

LO VOLTAGE (V_T = 2.0 V)				HI VOLTAGE (V_T 20.0 V)			UHF (V_T 20.0 V)		
	E	B	C	E	B	C	E	B	C
Q1	23.5	23.5	-9.0	23.5	22.8	23.3	23.5	22.8	23.3
Q2	8.9	2.2	23.5	18.9	19.4	22.8	18.9	19.4	22.8
Q3	8.9	9.8 *	23.5	18.9	11.9	23.5	18.9	11.9	[illegible]
Q4	23.5	23.3	0	23.5	23.3	0	23.5	22.8	23.3
Q5	2.3	3.0	22.0	4.2	3.0	23.5	4.2	[illegible]	23.5
Q6	-2.2	3.0	-9.0	-3.6	3.0	6.6	-3.6	3.0	6.5

* = SET BY R8

The signal is coupled from the interstage network to the mixer transistor base by C19. Oscillator injection, coupled through **gimmick** capacitor C27, is also applied to the mixer base. The resultant 40 MHz signal appearing at the collector (high-impedance) is directly coupled to the emitter (low impedance) of the common-base amplifier. (The output at the collector of the common-base amplifier is of high impedance to the requirements of i-f output coil L19.) A purposeful mismatch between the two stages provides loose coupling to the mixer, which offers circuit stability and isolation. This isolation allows the i-f output coil to be tuned during the i-f alignment procedure without affecting the mixer input circuit.

Transistor Q4 is operated as a Colpitts oscillator. The feedback circuit used to sustain oscillation consists of capacitor C28 (from the collector) in series with C30 to ground. The emitter is connected to the midpoint of this capacitive divider, thus providing the proper feedback level. The oscillator tuning circuit, coupled to the collector of Q4 by capacitor C32, operates as follows:

Oscillator Tuning Circuit

On low band, diode CR10 is reverse-biased. With the diode cut off, the tuned circuit is comprised of varactor CR9 paralleled by the series combination of inductors L22, L24, L23, and capacitor C35. It should be noted, however, that L22, used in high-band operation, has very little effect at the lower frequencies. Thus, the low-band tuned circuit actually consists of the varactor paralleled by L24 and C35. Inductor L23 provides a dc return path for CR9 and CR10.

On high band, diode CR10 is forward-biased. With the diode conducting, the junction of L22 and L24 is bypassed to ground through capacitor C37. The parallel combination of CR9 and L22 form the high-band tuning circuit. Trimmer capacitor C33 is used during tuner alignment. AFC correction voltage is applied to offset or increase the varactor tuning voltage, as necessary.

When the VHF tuner is utilized as a UHF i-f amplifier, the oscillator and rf B+ voltages are turned off by the tuner control center. The UHF output cable, coupled to the VHF tuner through the i-f input jack, has approximately +24V on the center conductor. This positive voltage forward biases diode CR8 so that the UHF i-f signal is coupled through C22 to the base of the mixer for amplification. Diode CR8 serves to isolate the mixer base from the agc system when the VHF tuner is being operated in the VHF mode. The UHF i-f signal receives some additional amplification through the common-base amplifier. The signal is then coupled to the receiver i-f system through L19.

TUNER CONTROL CENTER

The Zenith varactor tuning system is basically a 14-position electromechanical switching arrangement. For each of the 14 positions, a light bulb is turned on to indicate channel position, the voltage from a preset potentiometer is selected, and the output of a preset UHF-VHF switch is selected. These three functions are accomplished by three motor-driven rotary switches (Fig. 3-3).

The tuner control center uses these voltages to provide the correct voltages to the UHF and VHF varactor tuners.

The tuner control center provides the following actions:

A. VHF-UHF B+ switching
B. Electronic bandswitching for VHF
C. Agc filtering
D. Generation of a stable tuning supply voltage
E. Generation of a negative switching voltage
F. AFC amplification and equalization for UHF-VHF

B+ Switching

UHF B+ switching is accomplished by the setting of the individual UHF-VHF switches and the action of Q4, the UHF B+ transistor (Fig. 3-4). When the selected VHF-UHF switch is in the UHF position, the switch is open. The emitter of Q4 is connected to the +24V supply. The base sees a much less positive voltage and the transistor is biased into saturation. The transistor thus acts as a switch and provides B+ to the UHF tuner. When the switch is in the VHF or closed position, +24V is applied to the VHF tuner rf amplifier and mixer stage. This +24V is also applied to the base of Q4 through R6, which places the transistor into full cutoff. The cut-off transistor acts as an open switch and no B+ is applied to the UHF tuner. The mixer of the VHF tuner always receives B+ from the +24V supply.

VHF Electronic Bandswitch

In previous Zenith varactor tuning systems, UHF and VHF switching was accomplished by a mechanical bandswitch. The Zenith system now employs electronic bandswitching. The circuitry in Fig. 3-5, along with the preset two-position VHF-UHF switches, provides proper switching voltage (+24V on high band and —9V on low band) to the VHF tuner. Transistor Q1, controlled by transistor pair Q2 and Q3, functions as a basic electrical switch.

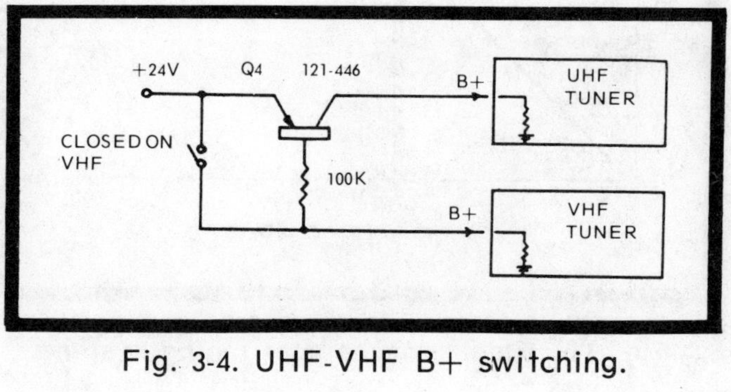

Fig. 3-4. UHF-VHF B+ switching.

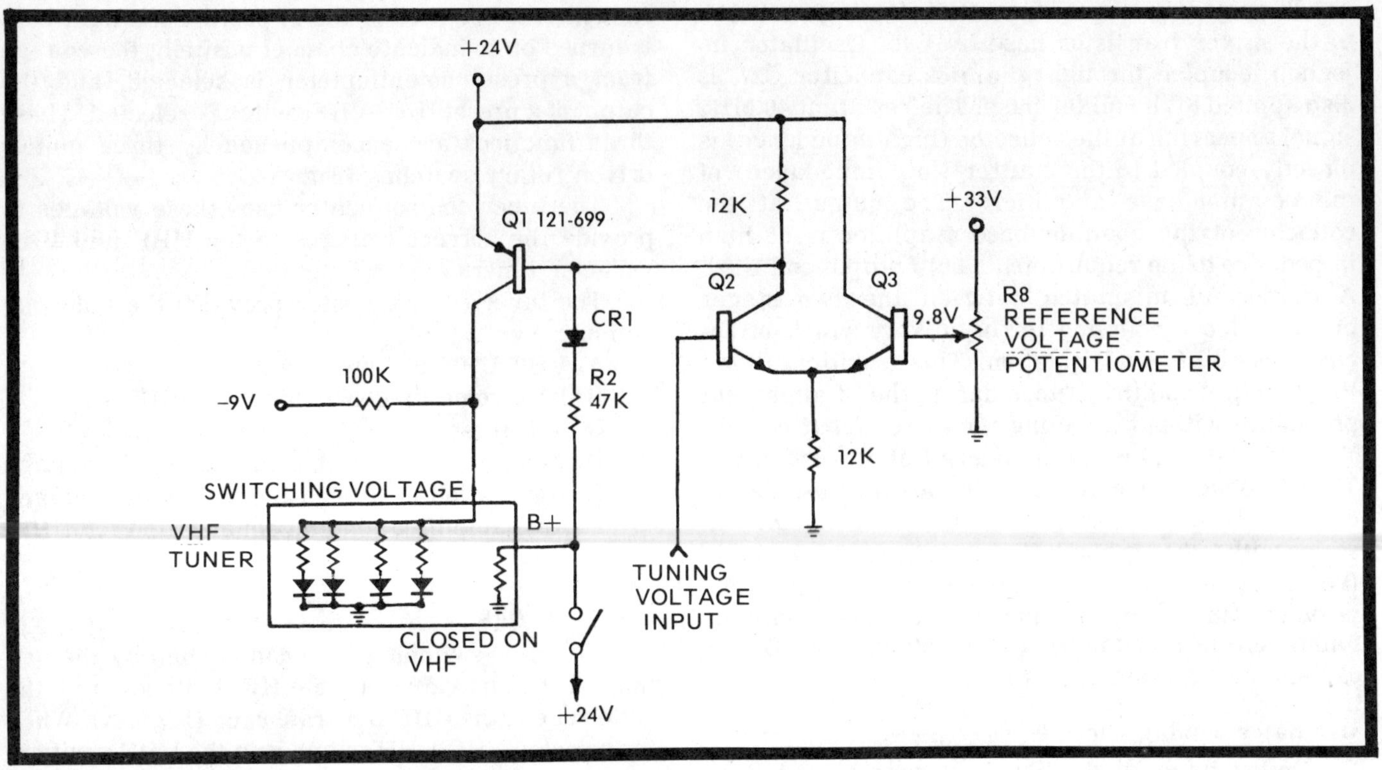

Fig. 3-5. Electronic bandswitching circuit.

When Q1 is off, the bandswitching voltage is —9V. This voltage, developed by CR3 and C9, is coupled to the collector of Q1 by R5. When Q1 conducts, the collector voltage is +24V. Conduction of Q1 is controlled directly by Q2, so it will conduct when Q2 conducts. Because of the differential action of Q2 and Q3, Q2 will conduct when the voltage on its base (tuning voltage) exceeds the voltage on the base of Q3. The voltage on the base of Q3 is manually set by potentiometer R8 to +9.8V. This adjustment is made when the tuning voltage is zero volts.

When the tuning voltage is below +9.8V (channels 2-6), the emitter voltage of Q2 (Fig. 3-6) will be

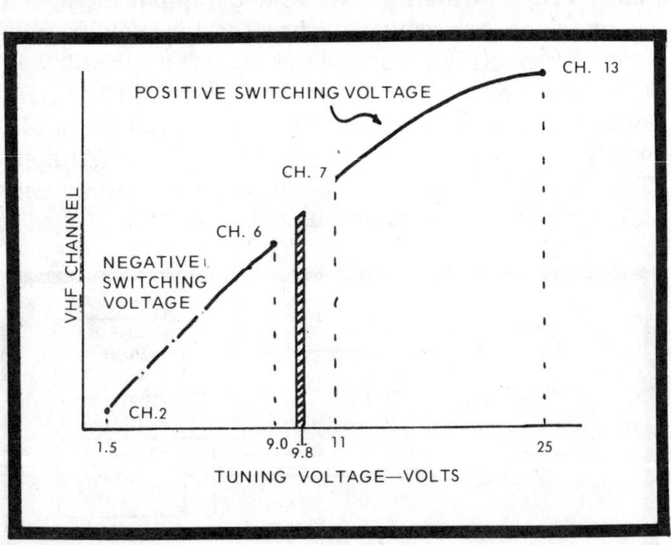

Fig. 3-6. Electromechanical tuning system.

higher than its base voltage and Q2 will be cut off. Since no current is flowing through Q2, its collector voltage and the base voltage of Q1 will be +24V. Q1 will be cut off since its base and emitter voltage will be near the same potential. The collector voltage of Q1 will be —9V (as described above).

When the tuning voltage increases above +9.8 volts (VHF channels 7-13) Q2 will begin to conduct. This causes its collector voltage to decrease. This in turn lowers the base voltage of Q1 sufficiently to cause it to conduct and reach saturation.

UHF Electronic Bandswitch

The collector voltage of Q1 becomes +24V, and this overrides the —9V supply to apply +24V to the switching line.

Diode CR1 and resistor R1 are used to insure that the switching voltage to the VHF tuner is +23.5V for UHF operation regardless of the tuning voltage. When the switch is in the UHF position (open) some current flows through the base-emitter junction of Q1 and through R3. Current flows from the +24V supply through diode CR1, resistor R1, and the VHF tuner load to ground. This forward biases the base-emitter junction of Q1 and causes the transistor to saturate.

When the VHF-UHF switch is closed (VHF position) diode CR1 is reverse biased and current is not allowed to flow through this path. Transistor Q2 now resumes complete control of Q1.

Diode CR2 is an integrated-circuit temperature-compensated voltage regulator circuit. It is used to provide an extremely stable +33V tuning supply

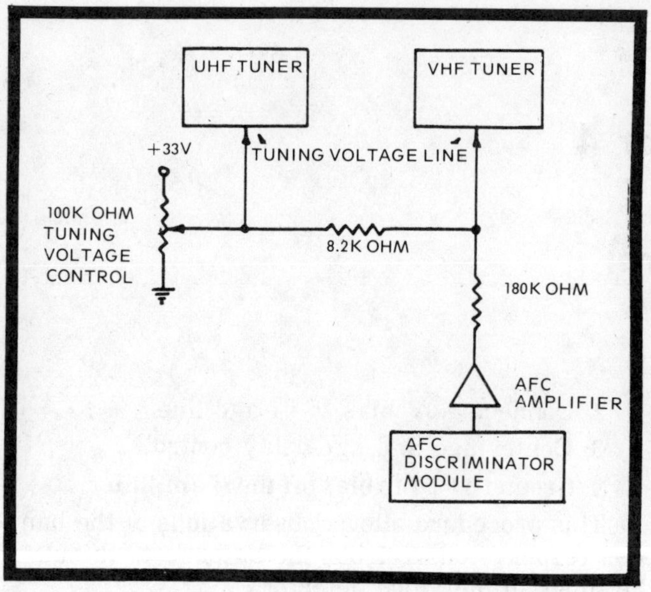

Fig. 3-7. AFC control system.

voltage needed for the varactors. This voltage is developed from the +250V supply in the receiver in conjunction with resistor R24. Capacitor C8 provides arc protection for semiconductors in the tuners and on the **tuner control center** module.

AFC correction voltage is applied in parallel with the tuning voltage. A conventional AFC system using an unbalanced discriminator is used. The AFC voltage from this circuit is used to drive a complementary-symmetry pair (Q5 and Q6). The conduction of Q5 and Q6 varies with tuning voltage and with output voltage from the AFC discriminator. The correction voltage is added to the VHF tuning voltage through R17 and to the UHF tuning voltage through R18. Values of these resistors are chosen so that the effective correction voltage is less on UHF than on VHF. (See Fig. 3-7.)

Capacitors C1 and C11 serve to remove unwanted transients and other "spurious" signals from the tuning voltage.

Chapter 4

Admiral

The service and alignment data contained in this chapter is applicable to the Admiral 99D1633D VHF tuner, as found in the T15K10 and T16K10 receiver chassis. (Information included in this chapter appears through the courtesy of Admiral Corporation.)

TUNER SWEEP ALIGNMENT

1. Set VHF tuner to channel 12 or 13—whichever channel is free of local signals and images.
2. Connect +6V bias to i-f agc line.
3. Center agc and agc-delay controls.
4. Ground TPB1 to bias off the rf amplifier.

This procedure allows observations of the bandpass response curve from the antenna terminals of the tuner to the video detector.

5. Apply 2.75V to the rf agc bias terminal of the tuner.

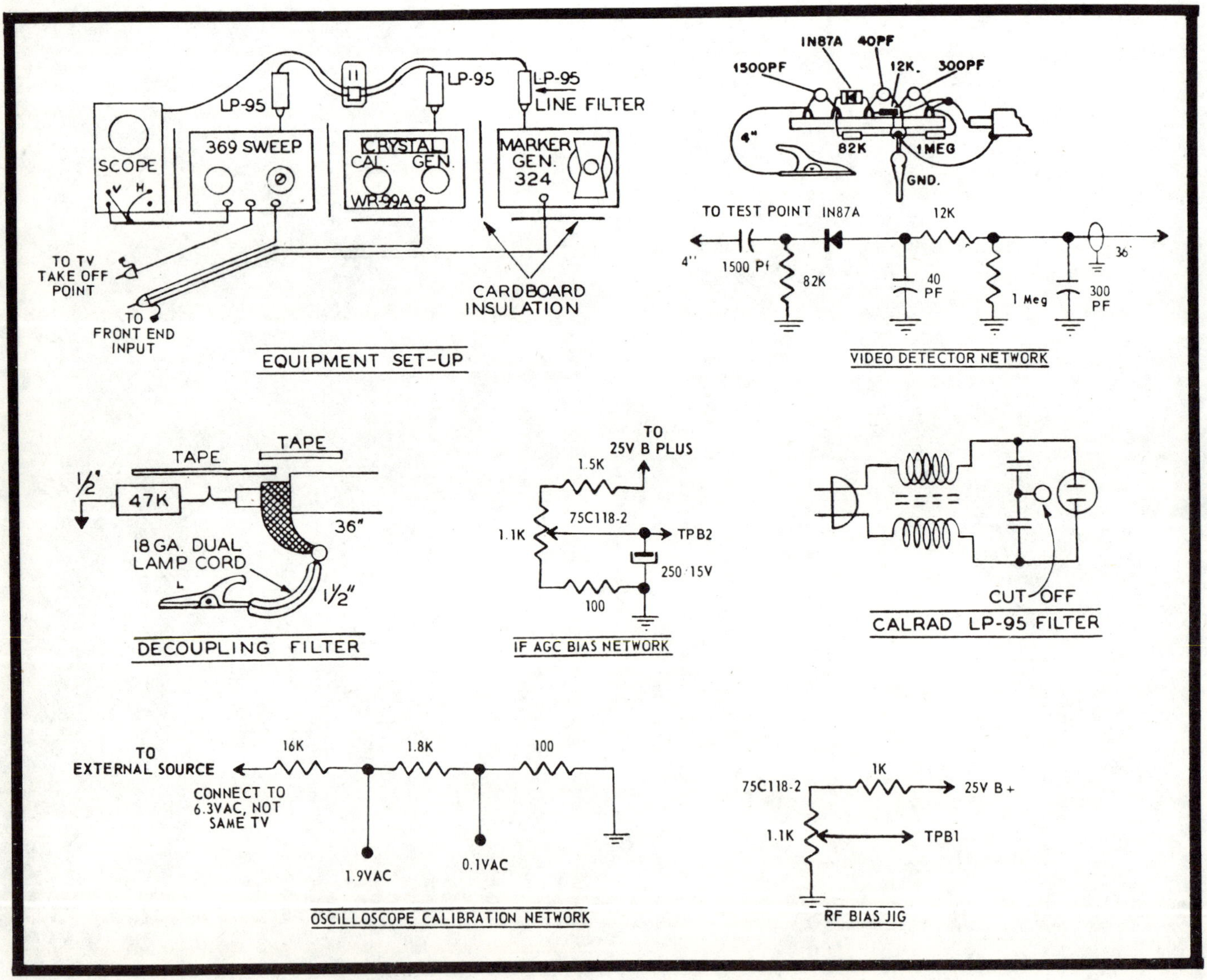

Fig. 4-1. Alignment jigs and networks (Admiral).

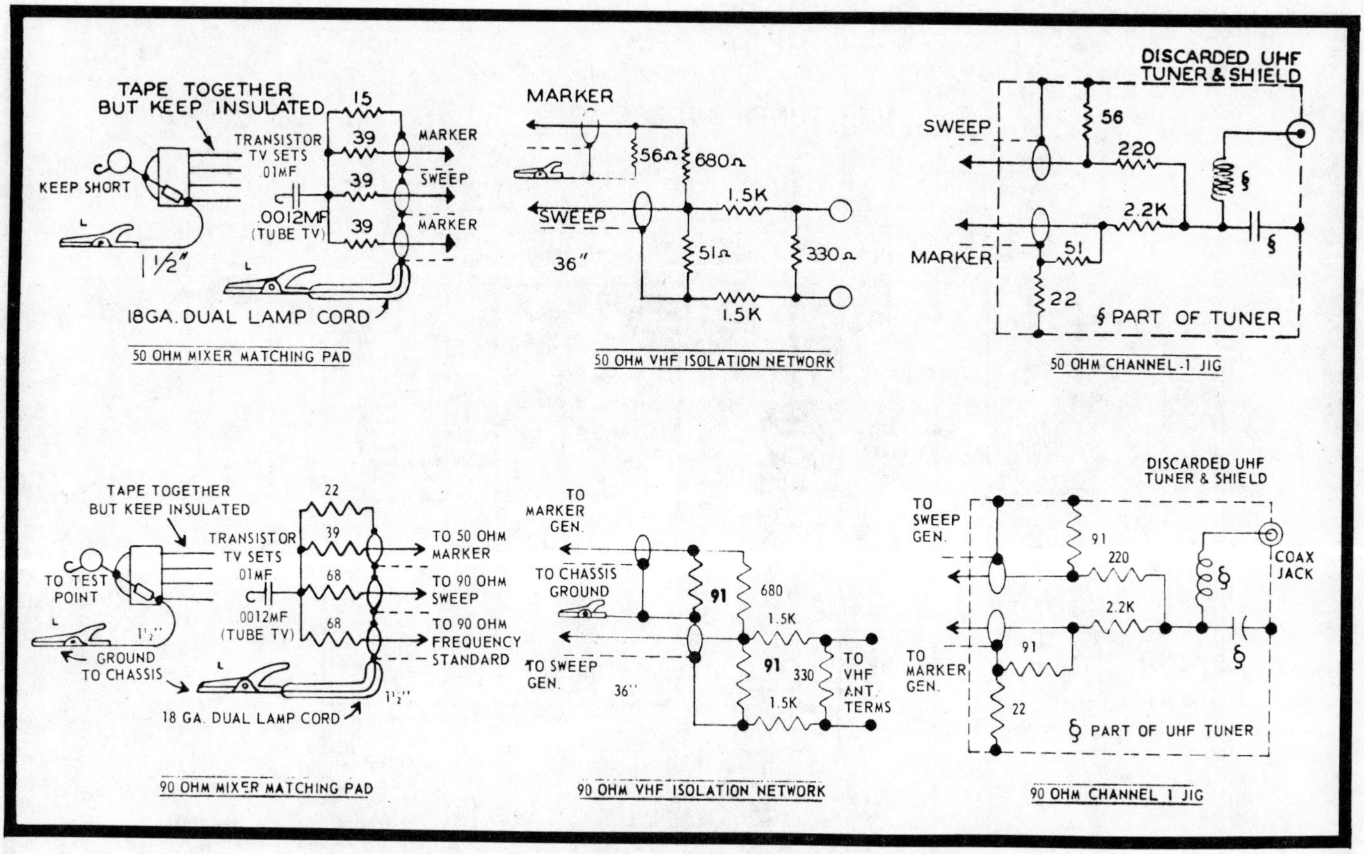

Fig. 4-2. Matching pads and networks (ADMIRAL).

8. Connect sweep and marker generators to the tuner antenna input terminals and set the rf outputs to minimum (**calibrate** switches set to off). When using the EICO 360 scope, always keep trace size at maximum.

9. Calibrate the oscilloscope vertical input to 2V peak-to-peak (p-p).

10. Connect VHF isolation network to VHF tuner by soldering directly to tuner terminal. All other leads should be removed from these terminals. Ground marker return to tuner frame.

11. Adjust sweep rf level to obtain 2V p-p output at TPB3.

12. Set the generator to the respective channel's picture carrier frequency. For channel 13, set the calibrating generator to produce an accurate 211.25 MHz marker point.

13. With the receiver's fine tuner control, set the picture carrier at 50 percent.

14. Observe the rest of the bandpass by moving the marker as required. Color will be 3.58 MHz above the picture carrier.

15. Continue this procedure on the other channels, checking for tilt, bandwidth, and gain. If tilt is slight and is on all channels, touch up tuner coil L9.

Figure 4-2 offers pictorial information for preparing various matching pads and isolation networks. The balance of the material in this chapter consists of diagrams, sketches, and schematics applicable to Admiral tuners.

6. Connect scope decoupling network directly to TPB3 (at the base of Q10 first video amplifier transistor in main chassis). Note the alignment instrument setup and test jig networks shown in Fig. 4-1. Clip network ground directly to chassis.

7. Set the i-f agc bias voltage to 6V with the set in operation.

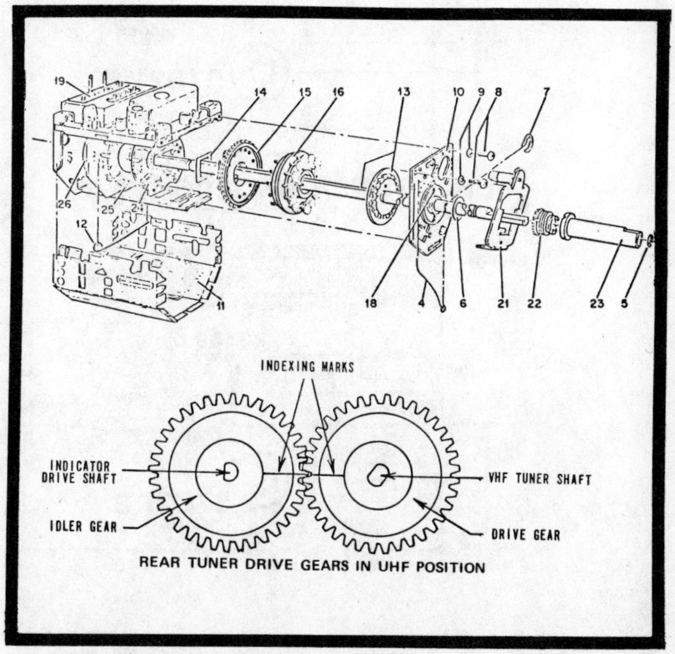

Fig. 4-3. VHF varactor tuner, exploded view (Admiral).

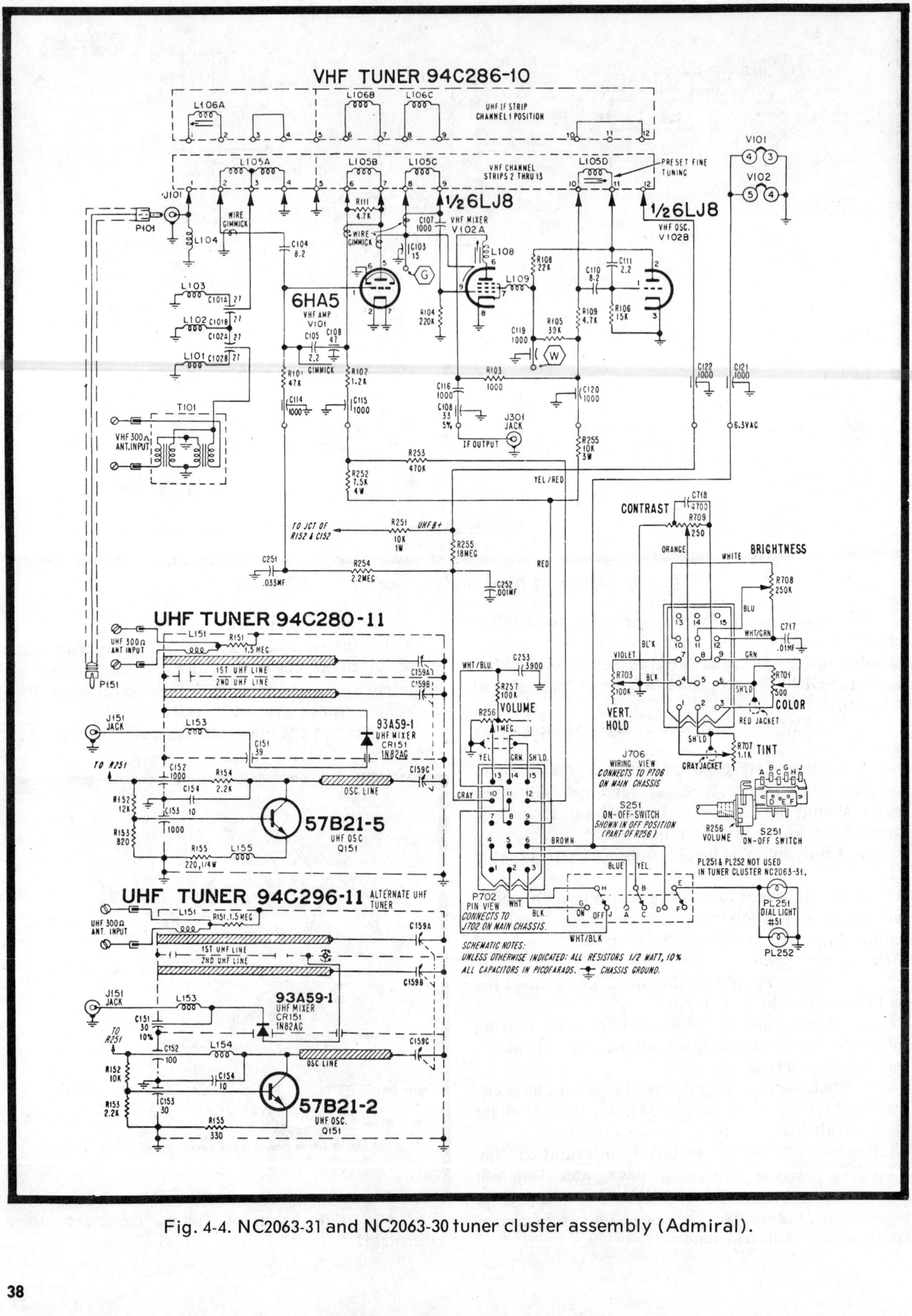

Fig. 4-4. NC2063-31 and NC2063-30 tuner cluster assembly (Admiral).

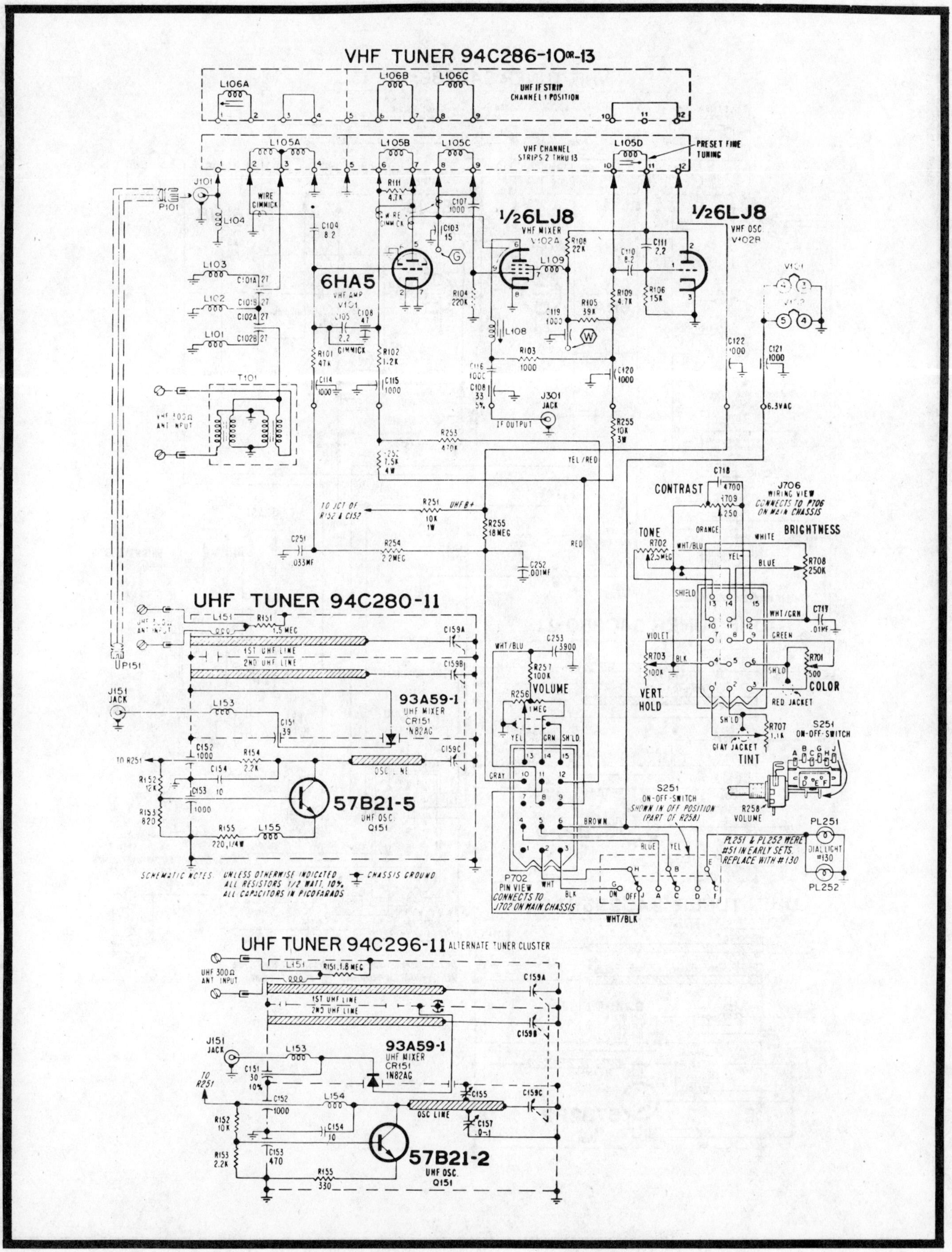

Fig. 4-5. NC2063-32 tuner cluster assembly (Admiral).

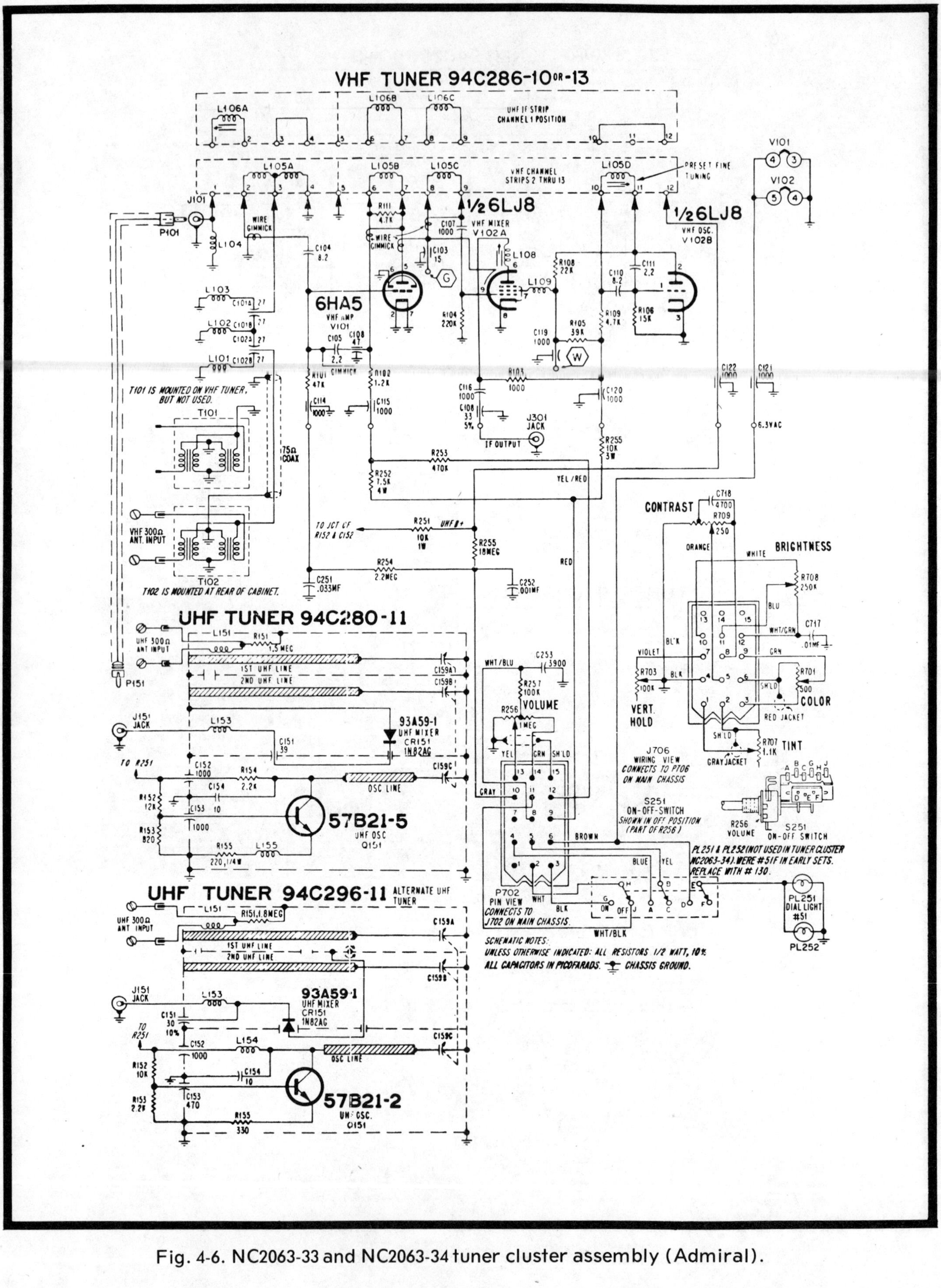

Fig. 4-6. NC2063-33 and NC2063-34 tuner cluster assembly (Admiral).

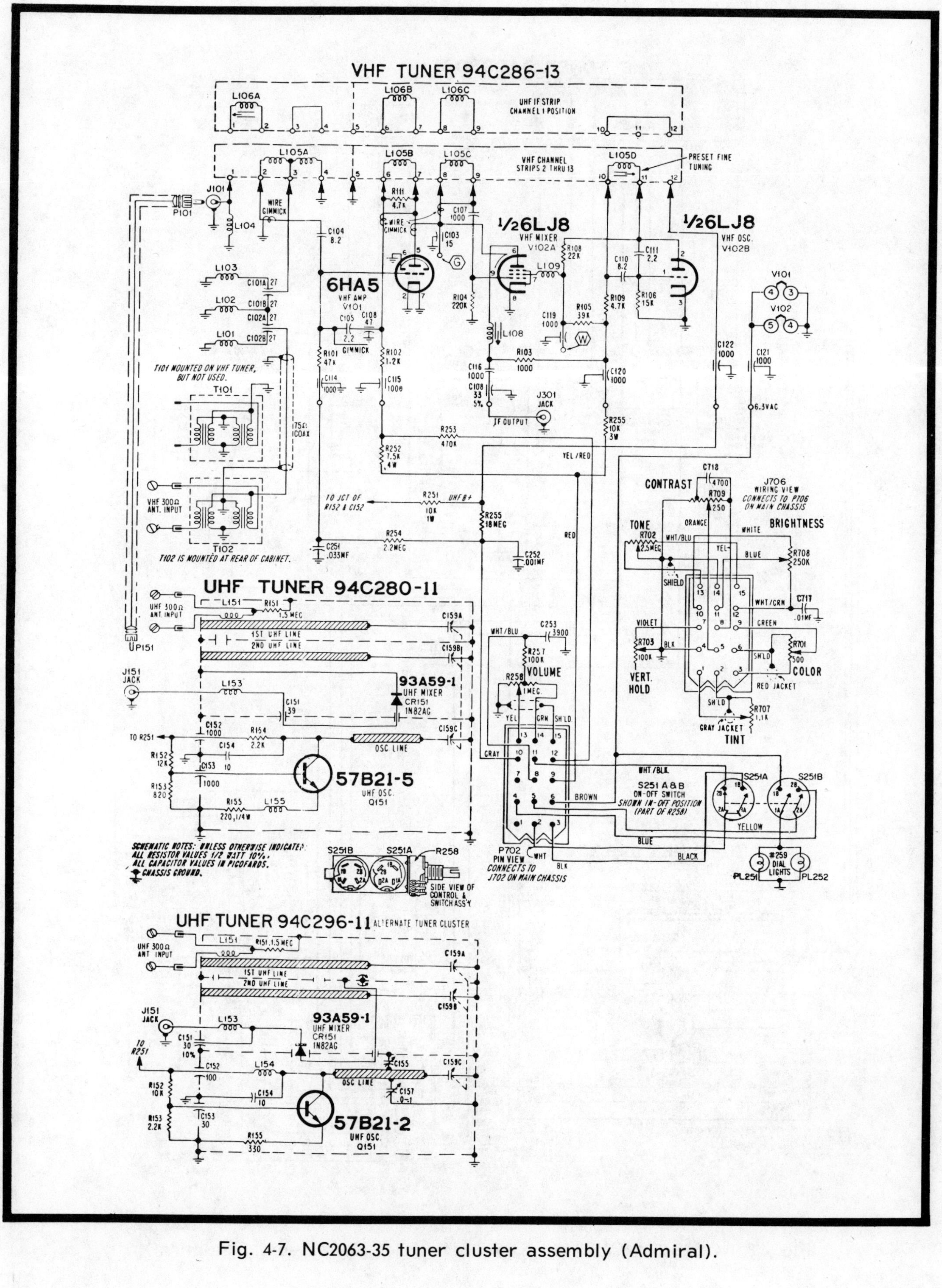

Fig. 4-7. NC2063-35 tuner cluster assembly (Admiral).

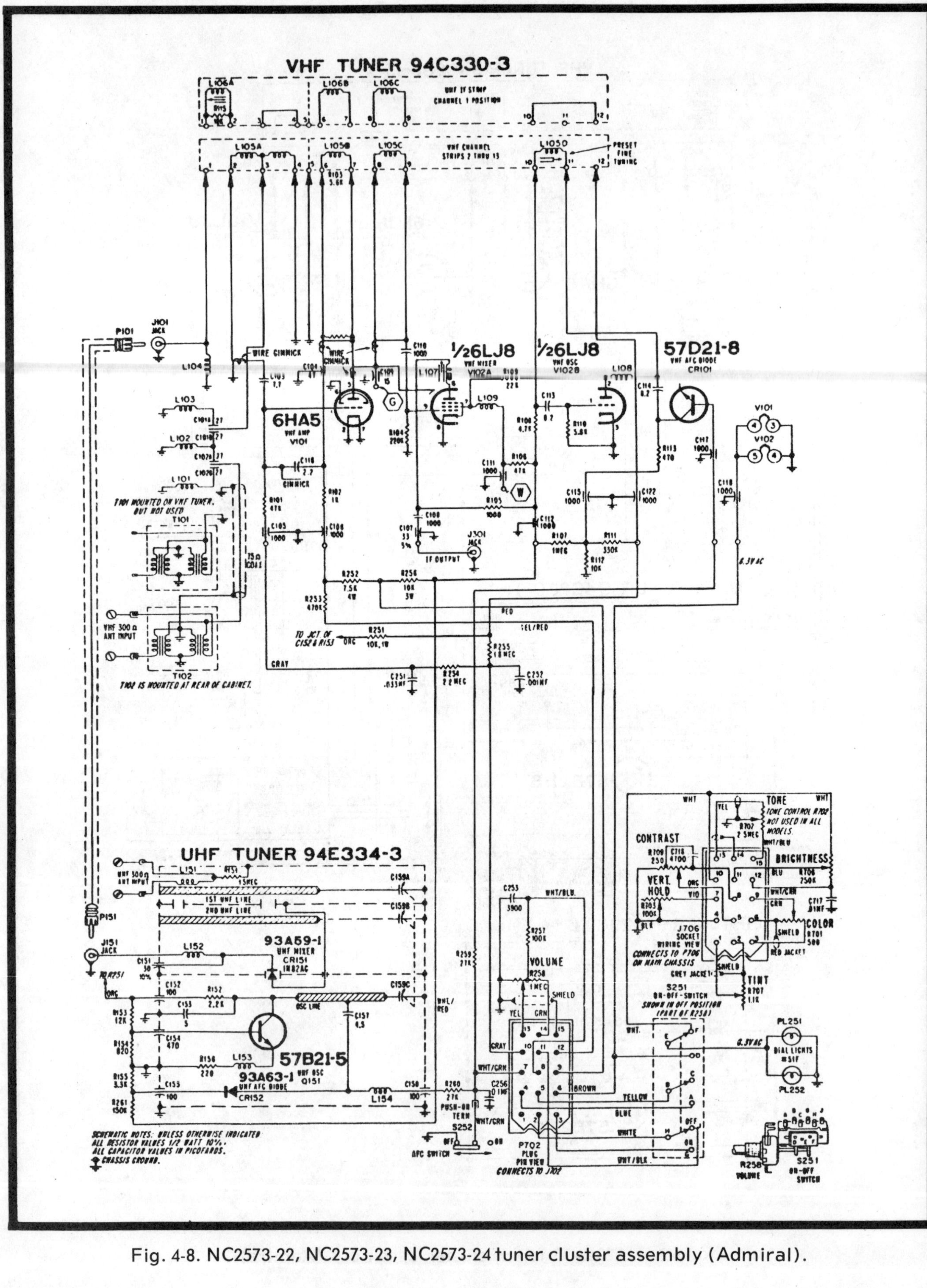

Fig. 4-8. NC2573-22, NC2573-23, NC2573-24 tuner cluster assembly (Admiral).

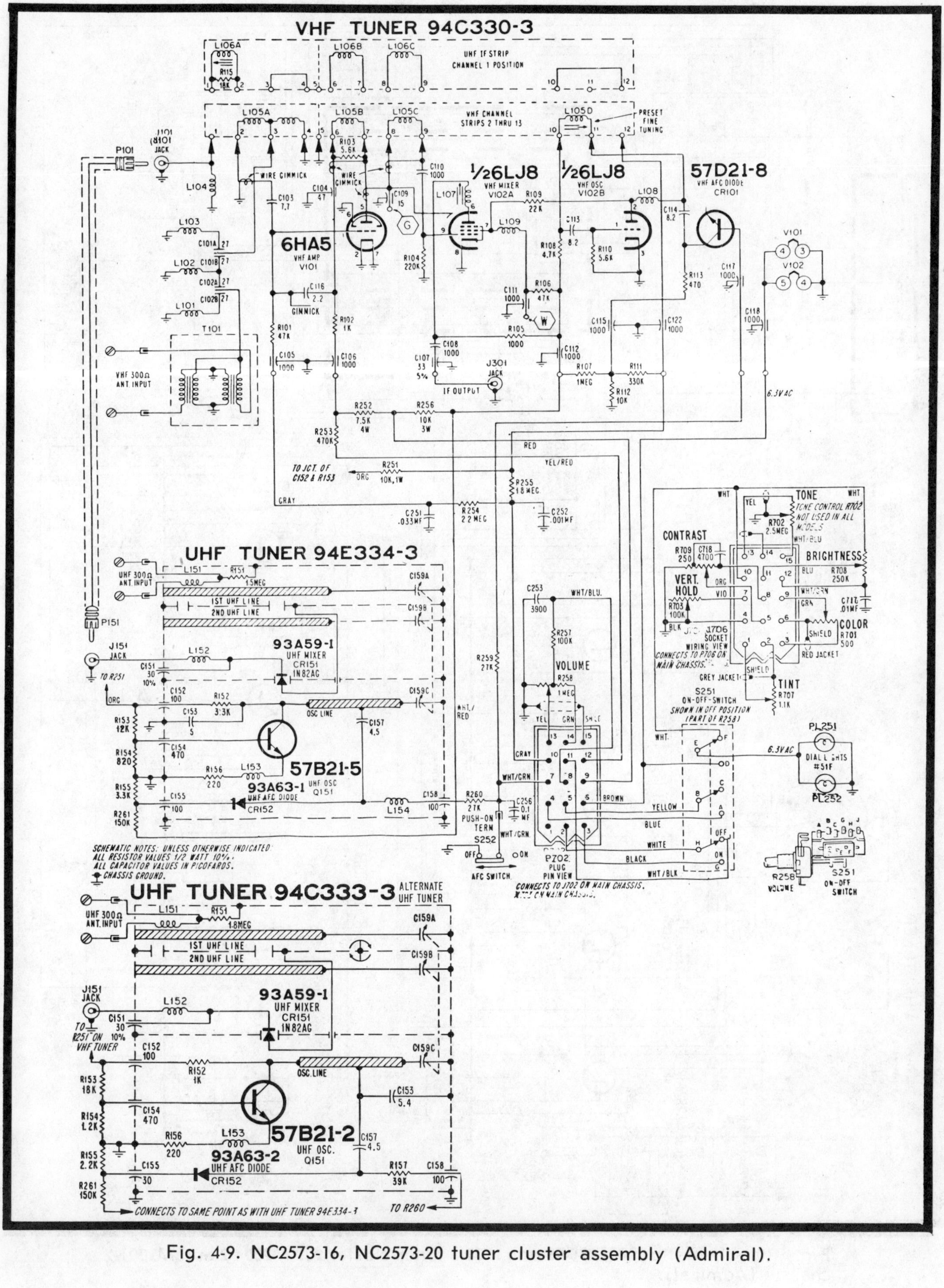

Fig. 4-9. NC2573-16, NC2573-20 tuner cluster assembly (Admiral).

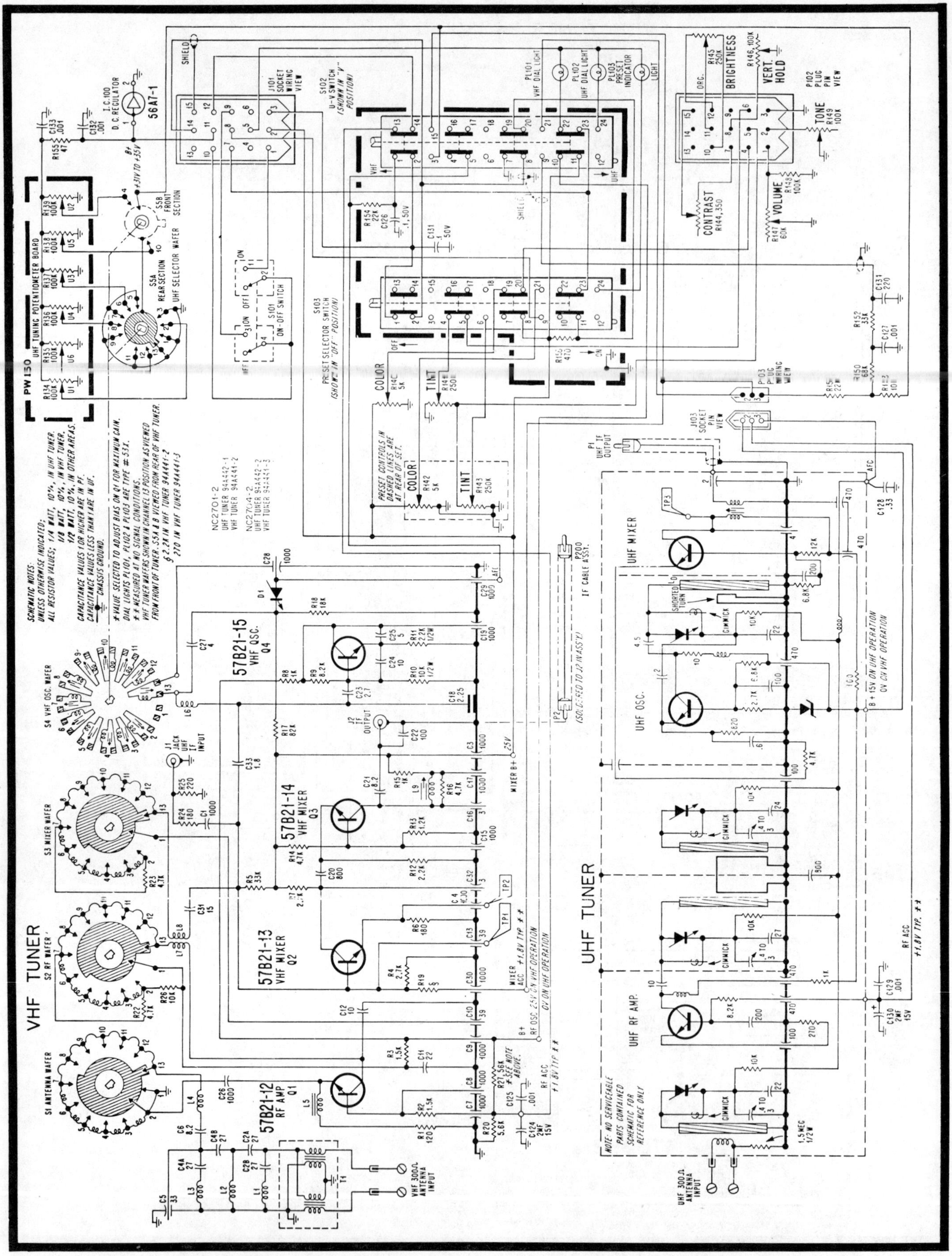

Fig. 4-10. Schematic of MC2704-2 and 2701-2 tuner cluster for M2004-2 and M2001-2 chassis (Admiral).

Chapter 5

General Electric

It is of course impossible to present all information applicable to all models of each manufacturer within the pages of a single volume of reasonable size. We have, therefore, selected a representative group of tuners from General Electric's model spectrum.

The models covered in this chapter are as follows:

1. EP85X2, 3, 6
2. EP86X2
3. EP86X265
4. EP86X4
5. EP86X5
6. ES85X1
7. ES85X7
8. ES86X3
9. ET85X61
10. EU85X58
11. EU85X59
12. EU86X1
13. EU86X3
14. EU86X304
15. EU86X305

UHF TUNERS EP85X2, 3, AND 6

These tuners are customarily used in TV receivers where the chassis is "hot"—that is, where the chassis is actually connected to one side of the 117V ac line. To preclude the possibility of shock hazard to the user, the tuning shafts are molded from plastic and the antenna circuit is isolated from chassis ground by a 2.2M half-watt resistor (R1 in Fig. 5-1). For continued protection, these items should be checked and defects corrected each time the tuner is serviced.

Before condemning a tuner to the junk bin, be sure to check the supply and control voltages as well as the UHF input portion of the VHF tuner. It is a simple matter to "patch in" a tuner known to be good as a comparative check for the tuner under suspicion. In general, alignment and repair are difficult unless you have the proper specialized equipment. Your best bet is to replace a defective tuner and repair it later under ideal bench conditions, where the equipment needed can be kept right at hand. The circuit shown in A is that of the EP85X2 tuner; that in B applies to both the EP85X3 and EP85X6 tuners.

VHF TUNERS EP86X2 AND EP86X265

The schematic shown in Fig. 5-2A is applicable to the EP86X2 and -X265 tuners. The differences between these two tuners are slight. A layout and exploded view of the X2 tuner is shown in Fig. 5-2B; you won't find any visible differences between this tuner and the X265. Figure 5-2C shows the channel strip layout (also applicable to both receivers).

Oscillator Adjustments

You'll need a sweep generator with a marker adder (or a sweep-marker generator) for this adjustment, but the process is simple and doesn't take long.

1. Connect a properly terminated sweep generator (sweep width of 10-15 MHz) to the tuner's VHF antenna input, and loosely couple the marker unit to the same point.
2. Connect a negative 2V dc bias to the tuner's agc terminal (with positive lead connected to chassis of tuner).
3. Set fine-tuning control and all oscillator cores to centers of their range, and connect your oscilloscope through a 10K load resistor to test point III (video detector load).
4. Use only enough generator output signal to provide a usable scope pattern and adjust the oscillator coil winding for each channel to place the picture carrier marker at the 50-percent point on the high-frequency slope of the response. The table of Fig. 5-3 shows the sweep and receiver settings and marker generator frequencies. Be sure to set each oscillator coil to the channel it is intended to represent.

40 MHz Channel Alignment

With receiver and sweep generator set for channel 13 (marker frequencies 211.25 and 215.75), adjust C104 first to place sound marker on peak, then adjust C105 to place picture marker on peak. To observe the waveform properly, connect the scope to the mixer test point through the RC network shown in Fig. 5-4A. Once this has been accomplished, connect the scope to test point III through a 10K load and, with —18V bias, adjust rf neutralization capacitor C109 for minimum response on scope. (You may have to adjust the signal input level a bit for this.) Now, if the oscillator is in good adjustment, you can begin the business end of the alignment procedure:

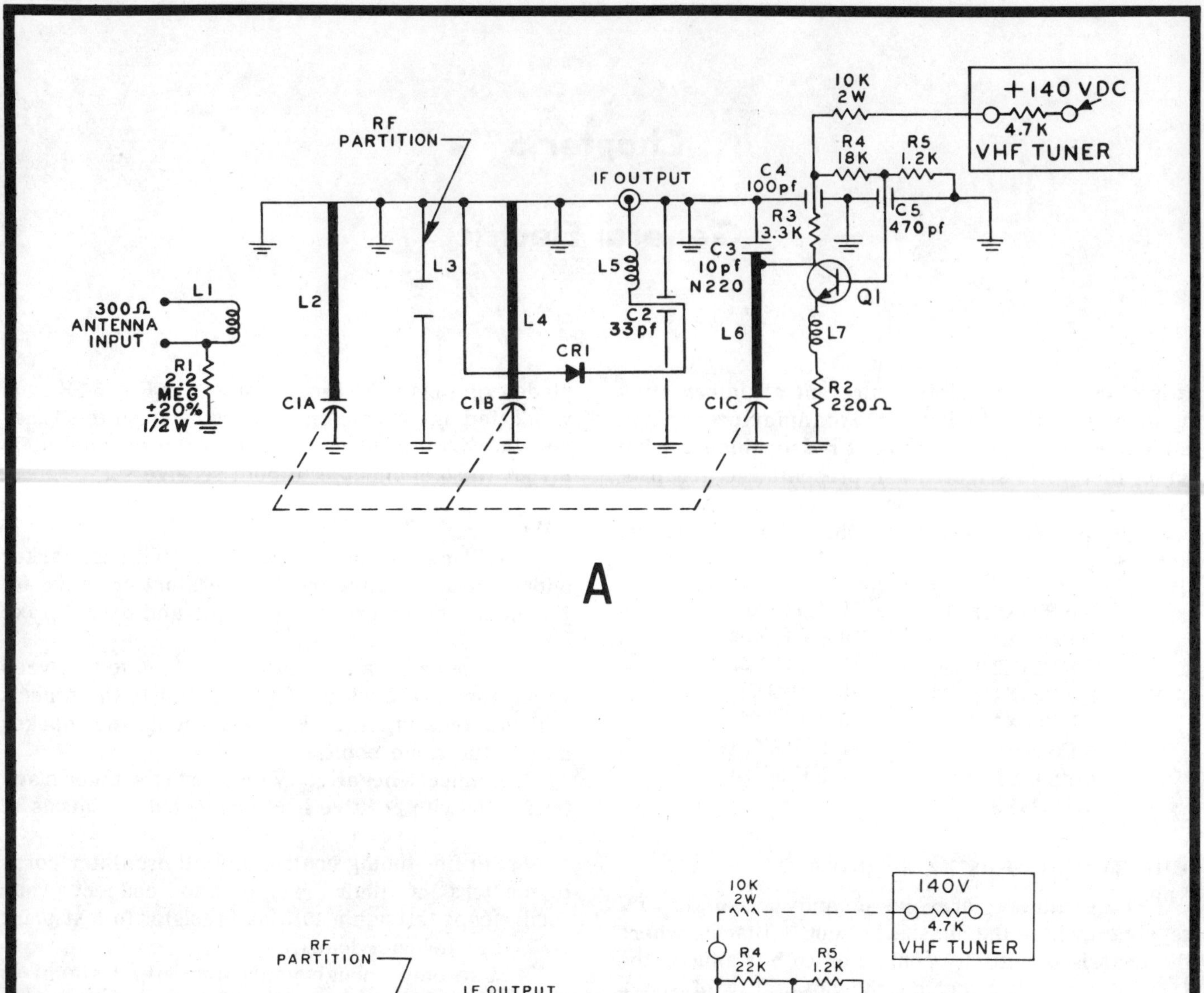

Fig. 5-1. GE UHF tuners; A applies to EP85X2, and B is applicable to EP85X3 and EP85X6.

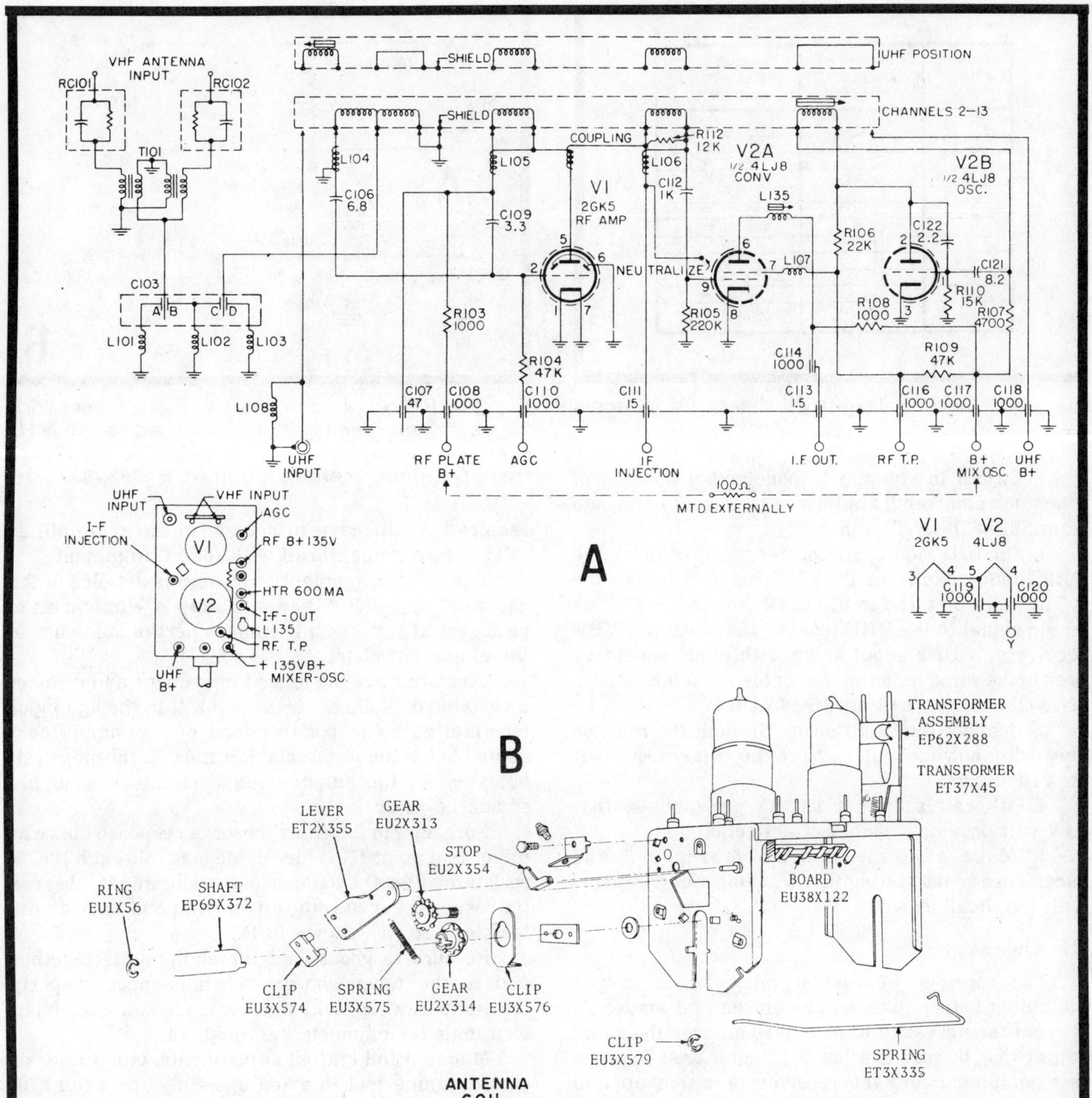

Fig. 5-2. GE UHF tuners EP86X2 and EP86X265; A is schematic applicable to both models: B is exploded view of X2 (which differs insignificantly from X2); C is the channel strip layout for these models.

STEP	SWEEP & RECEIVER POSITION	MARKER GENERATOR FREQUENCY
1	CH. 13	211.25 MC
2	CH. 12	205.25 MC
3	CH. 11	199.25 MC
4	CH. 10	193.25 MC
5	CH. 9	187.25 MC
6	CH. 8	181.25 MC
7	CH. 7	175.25 MC
8	CH. 6	83.25 MC
9	CH. 5	77.25 MC
10	CH. 4	67.25 MC
11	CH. 3	61.25 MC
12	CH. 2	55.25 MC

Fig. 5-3. Marker frequencies for TV channel settings.

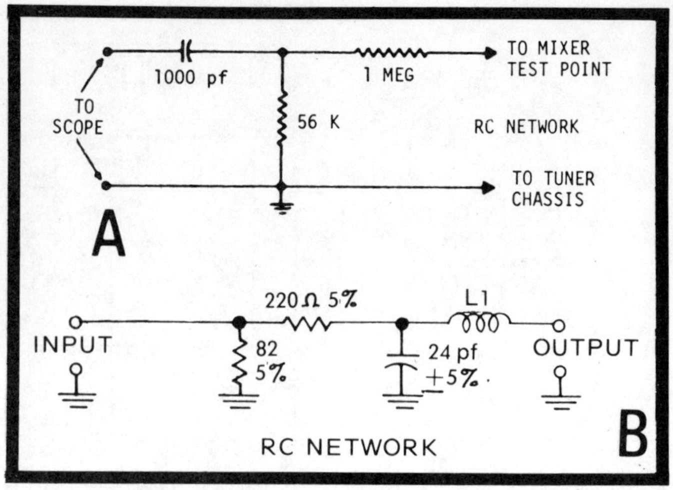

Fig. 5-4. RC coupling networks; A shows network for use in alignment; B network used for 40 MHz channel alignment. Latter network available from GE under catalog number ET36X783.

1. Switch to channel 1 (nonchannel position of tuner selector) and apply a —2V bias to the agc terminal of the VHF tuner.

2. On UHF-equipped receivers, disconnect the UHF input cable from the UHF tuner. Through the RC network pictured in Fig. 5-4B, connect a 40 MHz sweep signal to the UHF input of the tuner. On VHF receivers, a UHF input cable with plugs should be used in the same manner. The cable is a 9-inch length of 50-ohm coax, such as RG-58A / U.

3. Connect the oscilloscope through the resistor network (shown in Fig. 5-4A) to the **mixer** test point on VHF tuner.

4. Calibrate scope for 100 mV per inch—so that 0.3V will provide 3 inches of deflection.

5. Make adjustments noted in Fig. 5-5 and ascertain at each setting that waveform corresponds with the ideal shown.

RF Alignment

The receiver i-f system must be in proper alignment before tuner alignment can be made. To prevent the necessity of having to go over the same steps twice, be sure to allow 20 minutes or so for your test equipment and the receiver to warm up and stabilize before alignment. If a complete alignment is required, you'll have to neutralize the rf amplifier (V1) before going ahead with the rf alignment.

The actual alignment procedure is detailed in the chart of Fig. 5-6. But there are a few preliminaries to be observed and a couple of notes pertaining to use of the alignment chart.

First, remove the tuner shield cover and connect a variable dc voltage (bias) of 0-20V to the agc input terminal with the positive lead of the supply connected to the tuner chassis. For most of the steps (all but step 2), the output voltage of the bias supply should be at 2V.

For step 2 in the chart, observe response curve at mixer test point III (video diode load) through 10K of resistance; for the balance of the alignment, observe the response curve at the mixer test point through the resistor network shown in Fig. 5-4A.

Be sure to ground all the equipments together and to the tuner chassis. Then connect properly terminated sweep (with marker) to the antenna input terminals for channels 2 through 13.

Make all coil knifing adjustments with a fiber or plastic tuning tool to avoid upsetting the actual indications you'll be observing. And to observe proper

TUNE	TO SET	IDEAL CURVE LIMITS
RF Plate Coil	For Proper Frequency	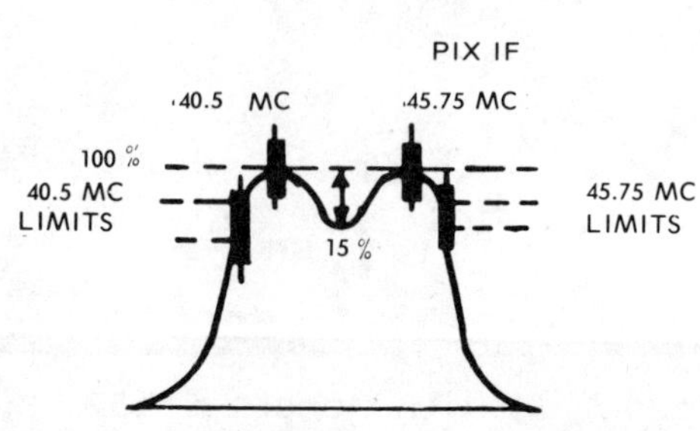
Converter Grid Coil	For Tilt	
RF Grid Coil	For Max. Midband Gain & Zero Tilt	
RF Plate & Converter Grid "Coupling"	For Bandwidth	

Fig. 5-5. Tuning procedure and waveform sample for 40 MHz channel alignment.

STEP	RECEIVER & SWEEP GENERATOR CHANNEL	MARKER GENERATOR FREQUENCY MC	ADJUST	REMARKS
1.	No. 13	211.25 215.75	Adjust C104 to place sound marker on peak then adjust C105 to place picture marker on peak.	Scope connected to mixer test point thru network.
2.	No. 13	211.25 215.75	With -18 volts bias, adjust R-F neut. on C109 for minimum response on scope. Adjust signal input level as necessary.	Scope connected to test point III thru 10,000 ohms.
3.			Recheck Step One & Oscillator alignment.	
4.	No. 13	211.25 215.75	With the scope connected to the mixer T.P. thru the resistor network & the bias voltage set at 2 volts. Knife or compress corresponding channel RF grid, plate & converter grid Coils starting with Channel 13 and going in sequence to Channel 2.	Adjust R-F plate coils for proper frequency, converter grid, coils for tilt of curve, and R-F grid coils for Min. saddle in center of band. Bandwidth for chan 7-13 is varied by spacing between RF plate & converter grid windings. Channels 2-6 by coupling winding
5.	No. 12	205.25 209.75		
6.	No. 11	199.25 203.75		
7.	No. 10	193.25 197.75		
8.	No. 9	187.25 191.75		
9.	No. 8	181.25 185.75		
10.	No. 7	175.25 179.25		
11.	No. 6	83.25 87.75		
12.	No. 5	77.25 81.75		
13.	No. 4	67.25 71.75		
14.	No. 3	61.25 65.75		
15.	No. 2	55.25 59.75		

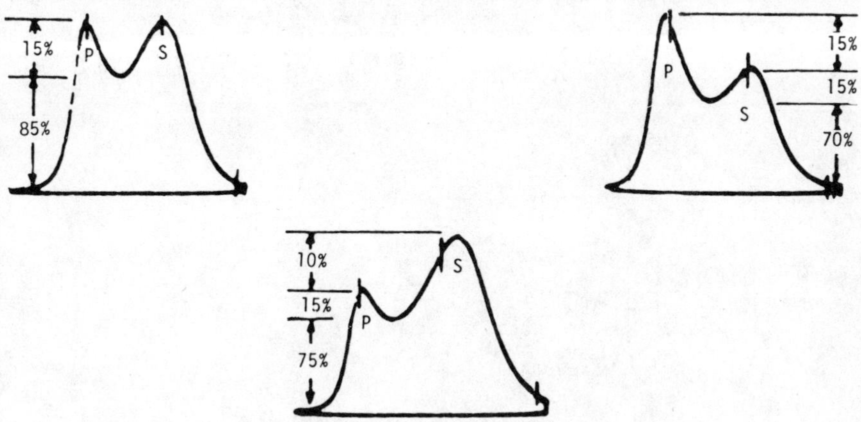

SADDLE, TILT, AND BANDWIDTH LIMITS

Fig. 5-6. This rf alignment chart is applicable to several GE models; refer to chart as required in text.

frequency response, you'll have to replace the tuner shield cover.

Remember, the alignment shown in Fig. 5-6 is applicable to the basic VHF TV channels; the proper alignment procedure for the channel 1 (13th detent) is described above under the heading of **40 MHz Channel Alignment**.

VHF TUNERS EP86X4 AND EP86X5

While the tuners here are very similar, there are a few subtle circuit differences and part differences. The circuit diagram of Fig. 5-7A applies to the EP86X4 tuner; Fig. 5-7B is the schematic for EP86X5. The exploded view shown in Fig. 5-7C applies to both; the part number differences are shown on the diagram.

Alignments and Adjustments

The oscillator adjustments, 40 MHz channel alignment, and rf alignment are exactly the same as those procedures under **VHF TUNERS EP86X2 AND EP86X265**. Similarly, all figures referenced in those procedures are fully applicable to the X4 and X5 tuners.

UHF TUNER ES85X1

This tuner is used in television receivers that have "hot" chassis (connected to one terminal of the incoming 117V ac line). It is important to observe certain precautions when repairing such sets to prevent safety hazards for the user. The antenna circuit, for example, is isolated from the chassis by a 2.2M half-watt resistor (R4 in Fig. 5-8). Both the wattage value and the resistance value are quite important to preserve the factory's built-in protection—be sure to replace this resistor with the same value.

Your GE distributor will not have parts for this tuner in stock. In the event of component failure, the entire tuner here was intended to be replaced. Still, you'll probably find that you will be able to effect some repairs, if only because of the inherent simplicity of the circuit.

An exploded view of the UHF tuner is pictured in Fig. 5-9A. Here the various compartmentalized sections of the tuner are identified with the cover removed. A top view of the tuner, which also shows major terminals, is provided in Fig. 5-9B.

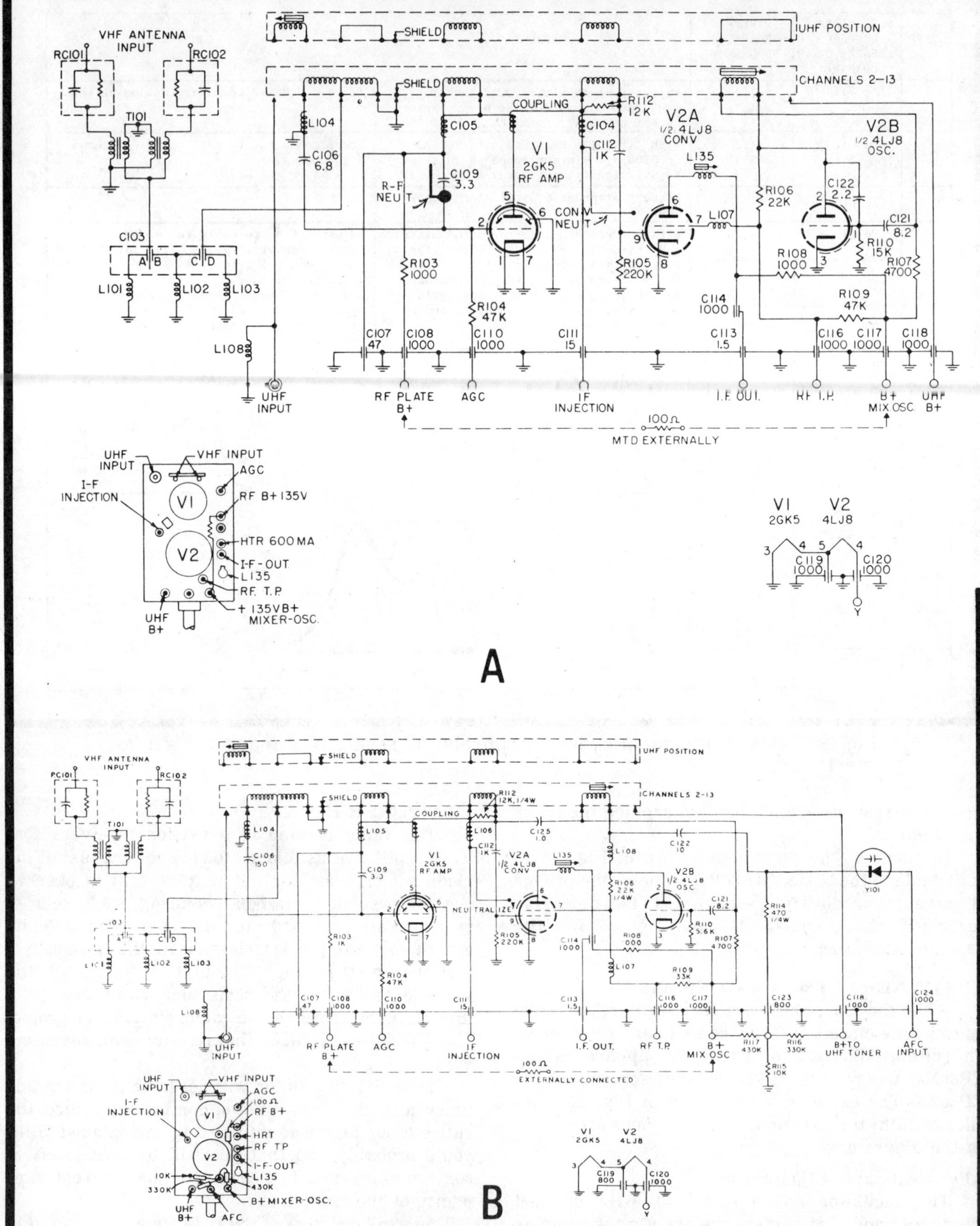

Fig. 5-7. GE VHF tuners; A is schematic for EP86X4; B is schematic for EP86X5; the exploded view in C applies to both models, the only differences being the part numbers of shaft and shaft assembly.

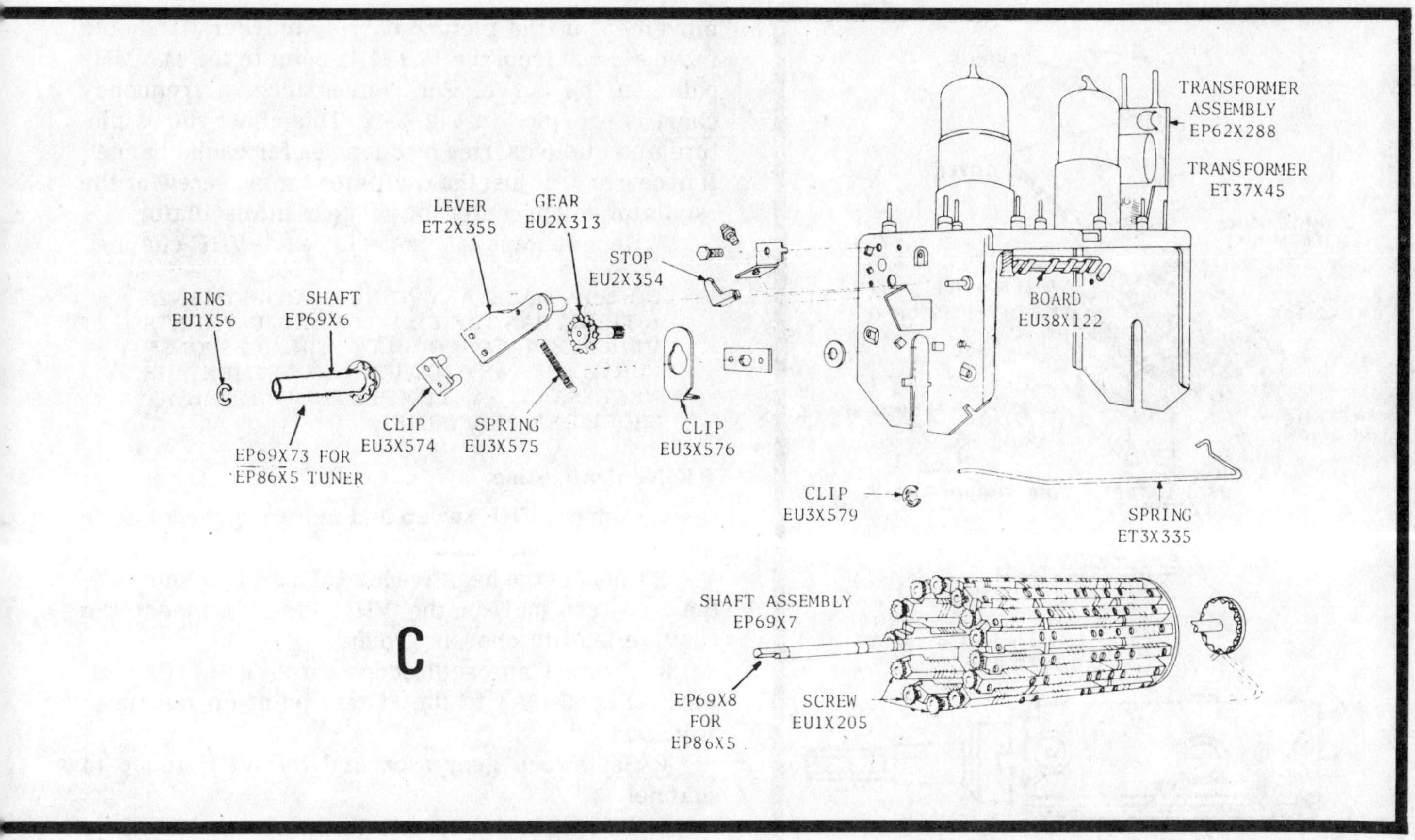

This tuner is designed to operate from a B+ input voltage of 22V at 6.2 mA of current drain. Its tuning range is from 470 MHz through 890 MHz when supplied with a 300-ohm signal from a balanced line. The unit's i-f output frequencies are 45.75 MHz for the picture carrier and 41.25 MHz for the FM audio carrier.

VHF TUNERS ES86X3 AND ES86X5

The ES series VHF tuners described here are designed for use in receivers in which the chassis is connected to one side of the ac line. The tuning shafts and the antenna terminals are electrically isolated from the tuner chassis. Replacement tuners or replacement parts installed in the original tuners must provide identical isolation in order to afford adequate protection for the user.

A schematic diagram of the ES86X3 tuner is shown in Fig. 5-10A; Fig. 5-10B is an exploded view.

Complete alignment of a tuner is seldom necessary. Most faults are due to defective tubes or mechanical problems. Inspect the tuner for loose or shorted connections and clean the coil-strip and stator contacts with a good grade of contact cleaner.

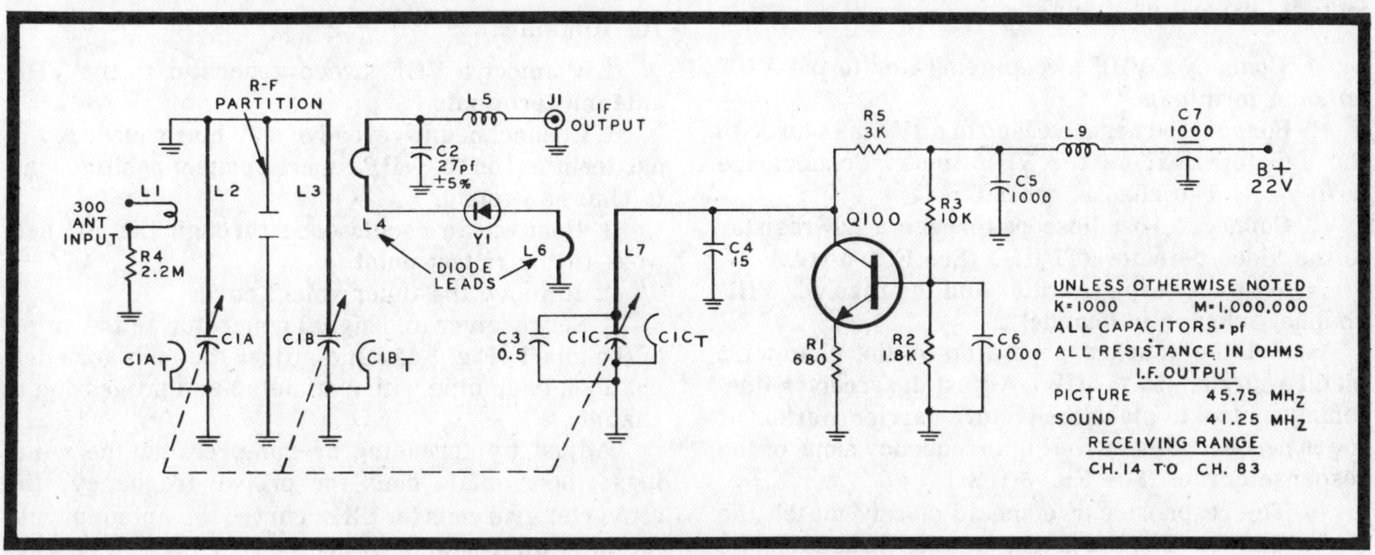

Fig. 5-8. UHF tuner ES85X1 schematic diagram.

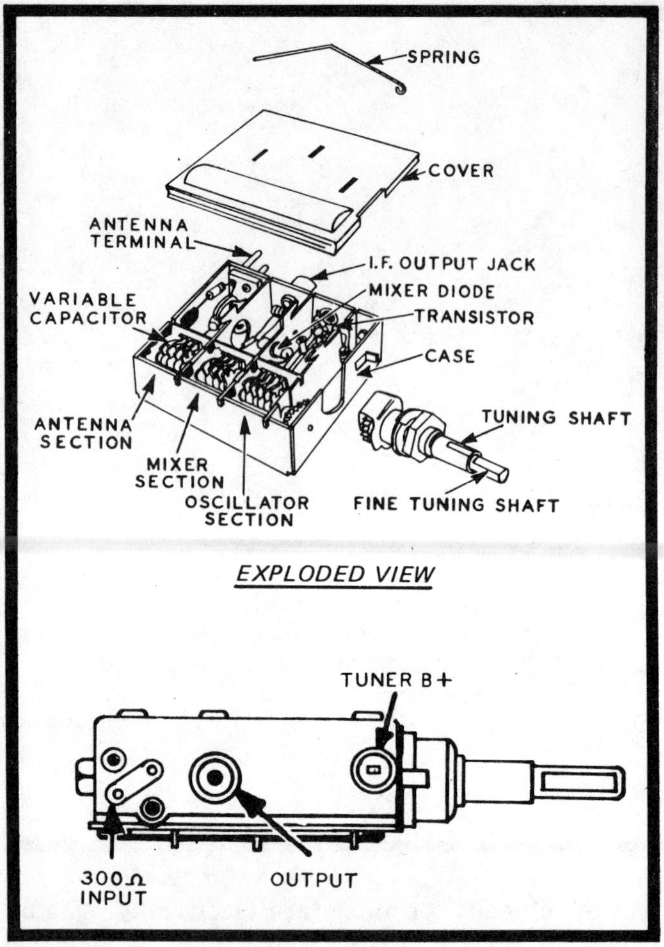

Fig. 5-9. Exploded view (A) and top-view sketch (B) of GE UHF tuner ES85X1.

If necessary, re-form the stator contacts so that they will make firm contact with the strip. After cleaning the tuner, lubricate the contacts with Rykon EP2 lubricant or its equivalent.

RF Response Check and Oscillator Adjustments

The overall response of the tuner and receiver can be checked as follows:

1. Connect a VHF sweep generator to the VHF antenna terminals.
2. Connect the negative lead of a 2V bias source to the agc terminal on the VHF tuner. Connect the positive lead to chassis ground.
3. Connect an oscilloscope through a 22K resistor to the video detector (TPIII). (See Fig. 5-11A.)
4. Set the sweep generator and the receiver VHF channel selector to channel 2.
5. Set the marker generator to the channel 2 picture carrier (55.25 MHz). Adjust the receiver fine-tuning control to place the picture carrier marker at the 50 percent point on the high-frequency slope of the response curve. (See Fig. 5-11B.)
6. The response curve should closely match the curve shown. To check the tuning range of the oscillator, rotate the fine-tuning control and note the movement of the picture carrier marker. It should move at least from the 46.5 MHz point to the 44.5 MHz point on the curve. For convenience, a frequency chart is provided in Fig. 5-12. This chart shows picture and audio carrier frequencies for each channel. If necessary, adjust the oscillator tuning screw or the oscillator coil to center the range of the oscillator.
7. Repeat steps 4, 5, and 6 for each VHF channel.

DO NOT MAKE ANY TUNING ADJUSTMENTS (OTHER THAN THE FINE-TUNE ADJUSTMENTS INDICATED) TO CORRECT THE RESPONSE CURVE. IF ANY OTHER ADJUSTMENT IS NECESSARY, A COMPLETE ALIGNMENT SHOULD BE PERFORMED.

RF Neutralization

1. Connect VHF sweep and marker generators to the VHF antenna terminals.
2. Connect the negative lead of a 2V bias source to the agc terminal on the VHF tuner. Connect the positive lead to chassis ground.
3. Connect an oscilloscope through the RC network (Fig. 5-13A) to the rf test point on the tuner (Fig. 5-11A).
4. Set sweep generator and the VHF tuner to channel 13.
5. Set marker generator to the channel 13 audio carrier frequency, and adjust L106 to place the audio marker on the peak (Fig. 5-13B).
6. Set marker generator to the channel 13 picture carrier frequency, and adjust L105 to place the picture carrier on peak. (Fig. 5-13B.)
7. Reset the agc bias to —18V; connect oscilloscope through a 22K resistor to the video detector test point (TPIII) on receiver.
8. Adjust gimmick wire on C109 for minimum response on scope.
9. Reset the agc bias to —2V, connect oscilloscope to rf test point through RC network, and recheck steps 5 and 6.

RF Alignment

1. Connect a VHF sweep generator to the VHF antenna terminals.
2. Connect negative lead of a 2V bias source to the agc terminal on the VHF tuner. Connect positive lead to chassis ground.
3. Connect an oscilloscope through the RC network to the rf test point.
4. Remove the tuner shield cover.
5. Set receiver and signal generator as indicated in the chart (Fig. 5-12) and adjust the coils for each channel, beginning with channel 13 and progessing to channel 2.

Adjust by spreading or compressing the windings: the rf plate coils for proper frequency; the converter grid coils for tilt of curve; the antenna coils for minimum saddle in center of curve. (See Fig. 5-14A and B.)

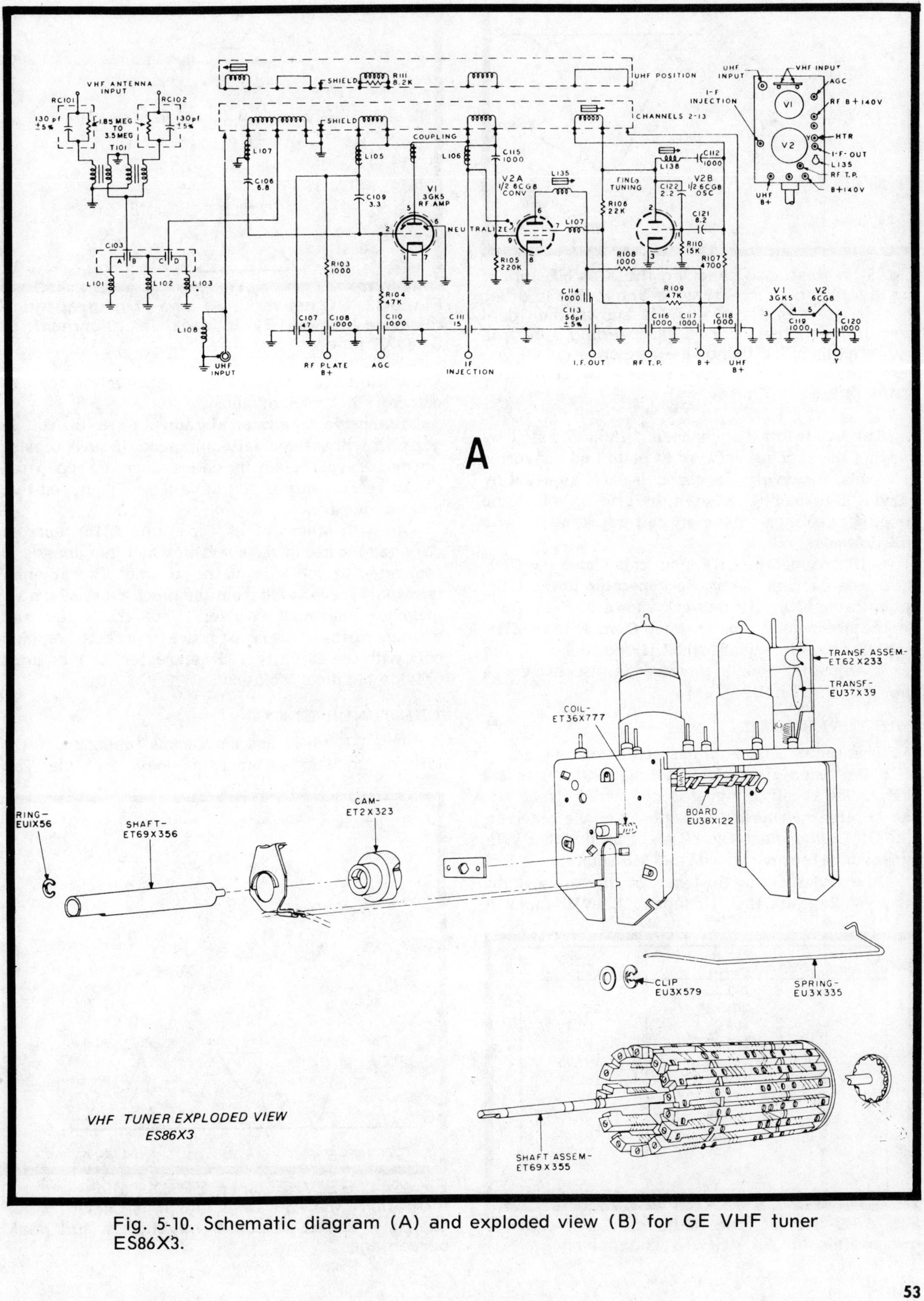

Fig. 5-10. Schematic diagram (A) and exploded view (B) for GE VHF tuner ES86X3.

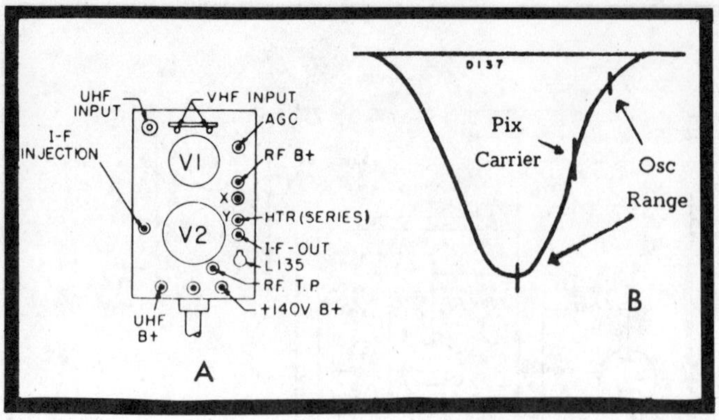

Fig. 5-11. Response checking the X3 VHF tuner; the layout shown in A may be applicable to other models (see text); sketch B shows the ideal overall response curve. Depending on the equipment you're using, the response curve you view may be reversed or inverted when compared to the one shown here.

The bandwidth for channels 7-13 is adjusted by varying the spacing between rf plate and converter grid coils. Bandwidth for channels 2-6 is adjusted by varying the spacing between the grid coils and the coupling windings. **Make all coil adjustments with plastic tools!**

6. Disconnect the UHF input cable from the UHF tuner and connect the sweep generator to the UHF cable through the RC network shown in Fig. 5-15A. Set the sweep generator to sweep from 39 to 46 MHz.

7. Set the VHF tuner to the UHF position.

8. Adjust the coils to produce the response curve shown in Fig. 5-15B.

UHF TUNER ES85X7

This tuner (Fig. 5-16) has continuous tuning for UHF television channels 14 through 83 (470 to 890 MHz). The received signal is converted directly to the 45 MHz intermediate frequency of the receiver. The UHF tuner must, therefore, be used with a VHF tuner which is provided with a 45 MHz input.

In troubleshooting the tuner, be sure to check the supply voltage and the UHF input of the VHF tuner. A known tuner can be quickly "patched in" to confirm or contradict your diagnosis.

Channel	Picture Carrier	Audio Carrier
13	211.25	215.75
12	205.25	209.75
11	199.25	203.75
10	193.25	197.75
9	187.25	191.75
8	181.25	185.75
7	175.25	179.75
6	83.25	87.75
5	77.25	81.75
4	67.25	71.75
3	61.25	65.75
2	55.25	59.75

Fig. 5-12. Picture and audio carrier frequency information for all VHF TV channels.

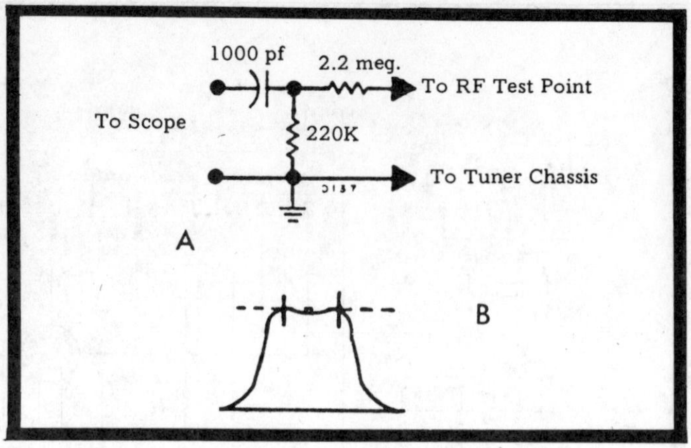

Fig. 5-13. RC network (A) and rf neutralization response curve (B) to seek in alignment of EP86X5 tuner.

Extensive repairs or alignments are difficult to perform without special equipment. If such repairs are necessary, replace the entire tuner and repair the defective unit only if you have a well-equipped test and service bench.

As with others of its type, this UHF tuner is designed for use in receivers in which the chassis is connected to one side of the ac line. The antenna terminals are isolated from the tuner chassis by a 3.6 megohm, half-watt resistor. Both the value and wattage rating of the resistor are important. Replace only with the same type. Replacement tuners **must** provide identical isolation.

UHF TUNER ET85X61

This GE tuner has continuous tuning from 470 MHz to 890 MHz and operates converter-style. The

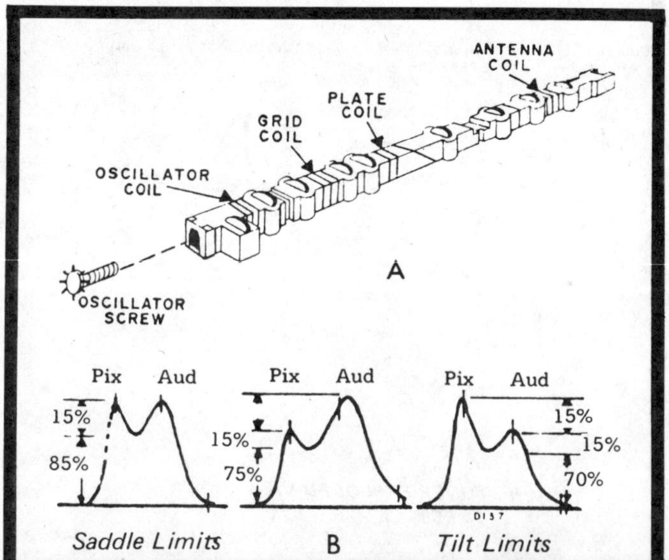

Fig. 5-14. GE VHF tuner EP86X5 alignment information; the coil strip layout is shown in A; sketch B shows response waveforms and peak percentages.

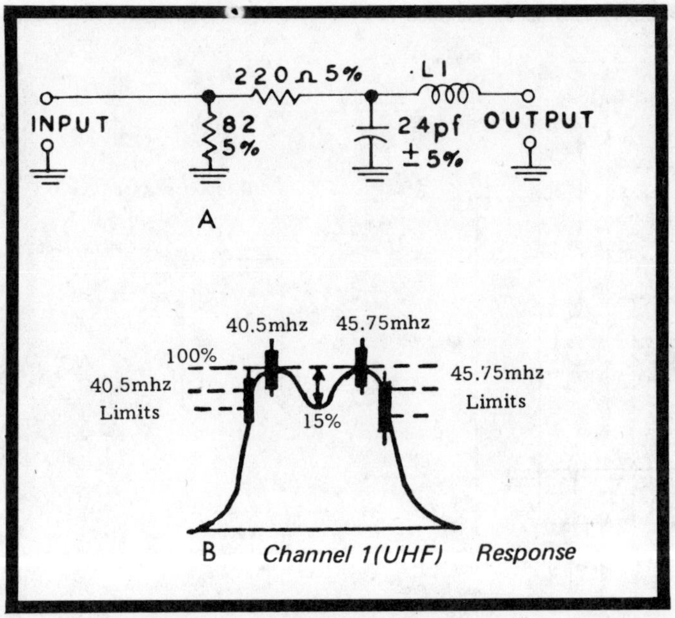

Fig. 5-15. Recommended RC network (A) and off-channel response curve for GE VHF tuner EP86X5.

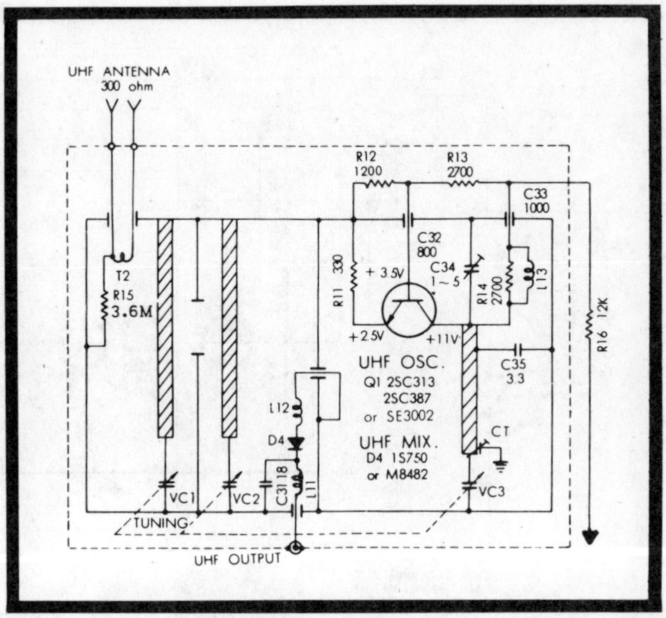

Fig. 5-16. GE UHF tuner ES85X7 schematic diagram.

received signal is converted directly to the 45 MHz intermediate frequency. The schematic of this simple one-transistor UHF tuner is pictured in Fig. 5-17. As a safety precaution, when you service this model, be sure to reconnect the 22M resistor on the antenna input coil (L1). It connects to chassis ground and provides current limiting to protect the user from electrical shock when the line cord is plugged into the receptacle incorrectly. The resistor is a half-watt type; if it's replaced, be sure to use a replacement of like value—in both resistance and wattage.

VHF TUNERS EU86X1 AND EU86X3

Although these tuners are quite similar, there are distinct physical and electrical differences between units. Figure 5-18A is the schematic diagram of the X1 unit; Fig. 5-18B shows the general layout in an exploded view. The circled numbers on the latter view may be referenced against like numbers appearing on the parts list at the back of this book. Figure 5-18B applies to both X1 and X3 tuners. Figure 5-19 is the schematic for the X3 unit.

Alignment and Adjustments

The alignment and adjustment procedures described for the ET series may be used for these EU tuners; however, be sure to note test point differences and tuning adjustment identification differences. As you perform the procedures, make repeated references to the layout (top view) drawing insert of Fig. 5-18A.

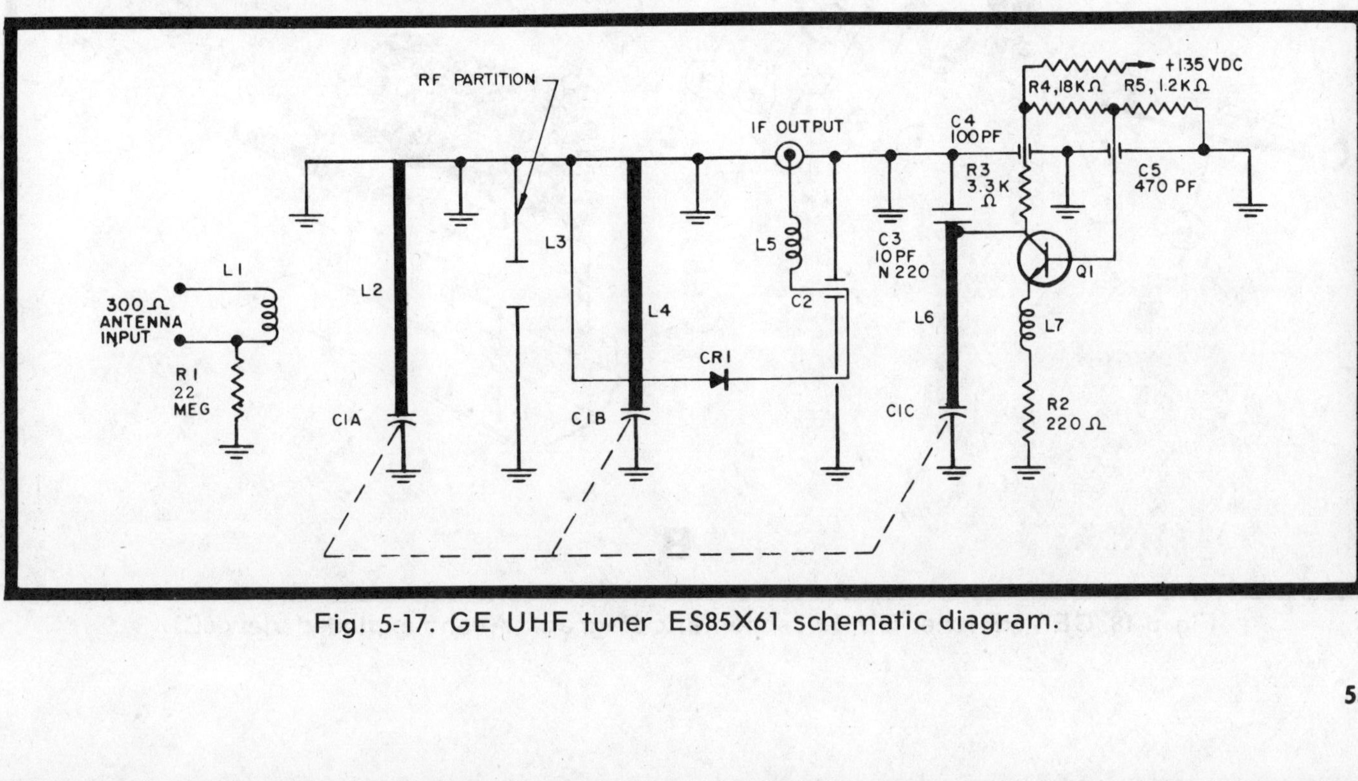

Fig. 5-17. GE UHF tuner ES85X61 schematic diagram.

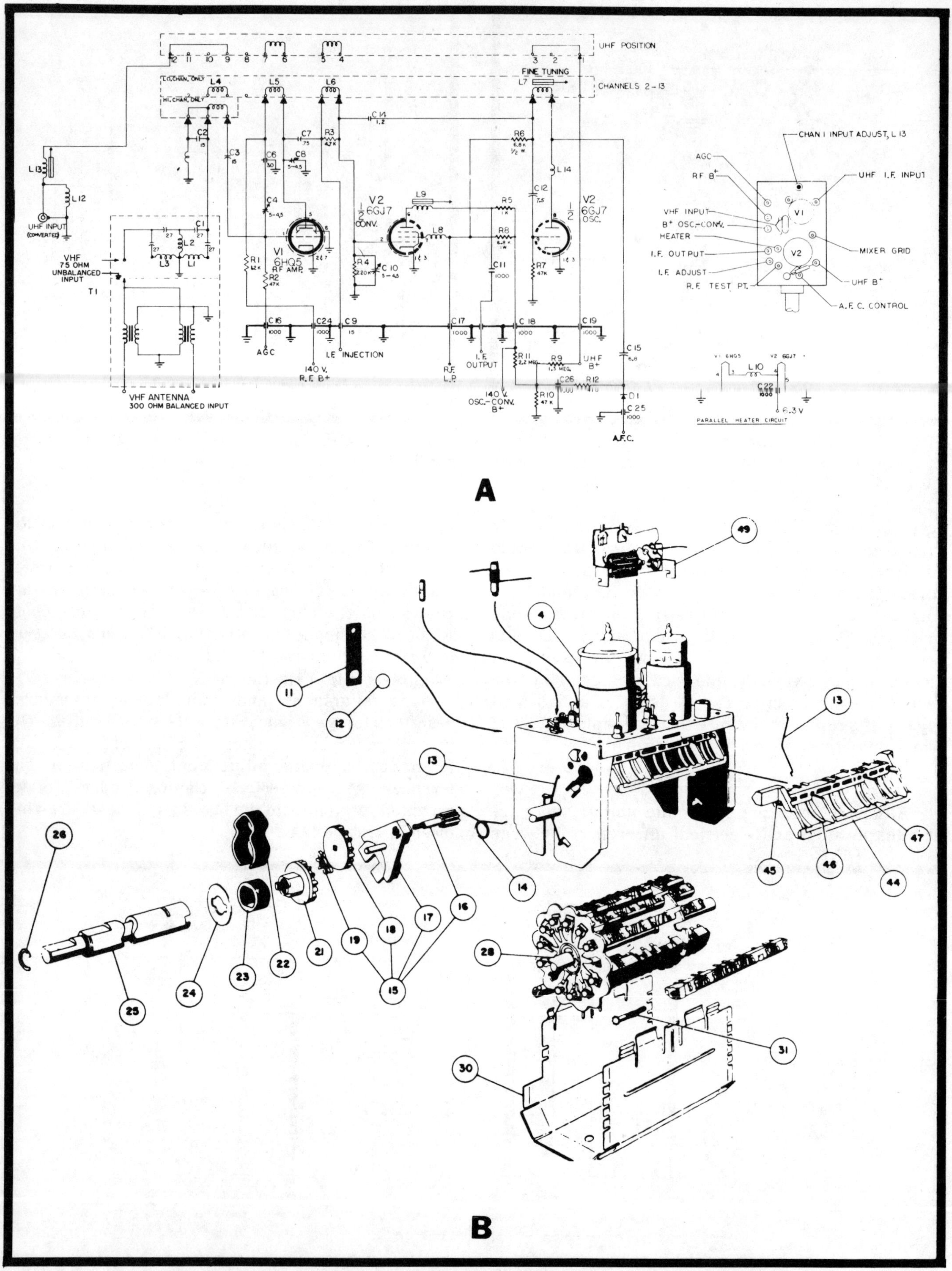

Fig. 5-18. GE VHF tuner EU86X1 schematic diagram (A) and exploded view (B).

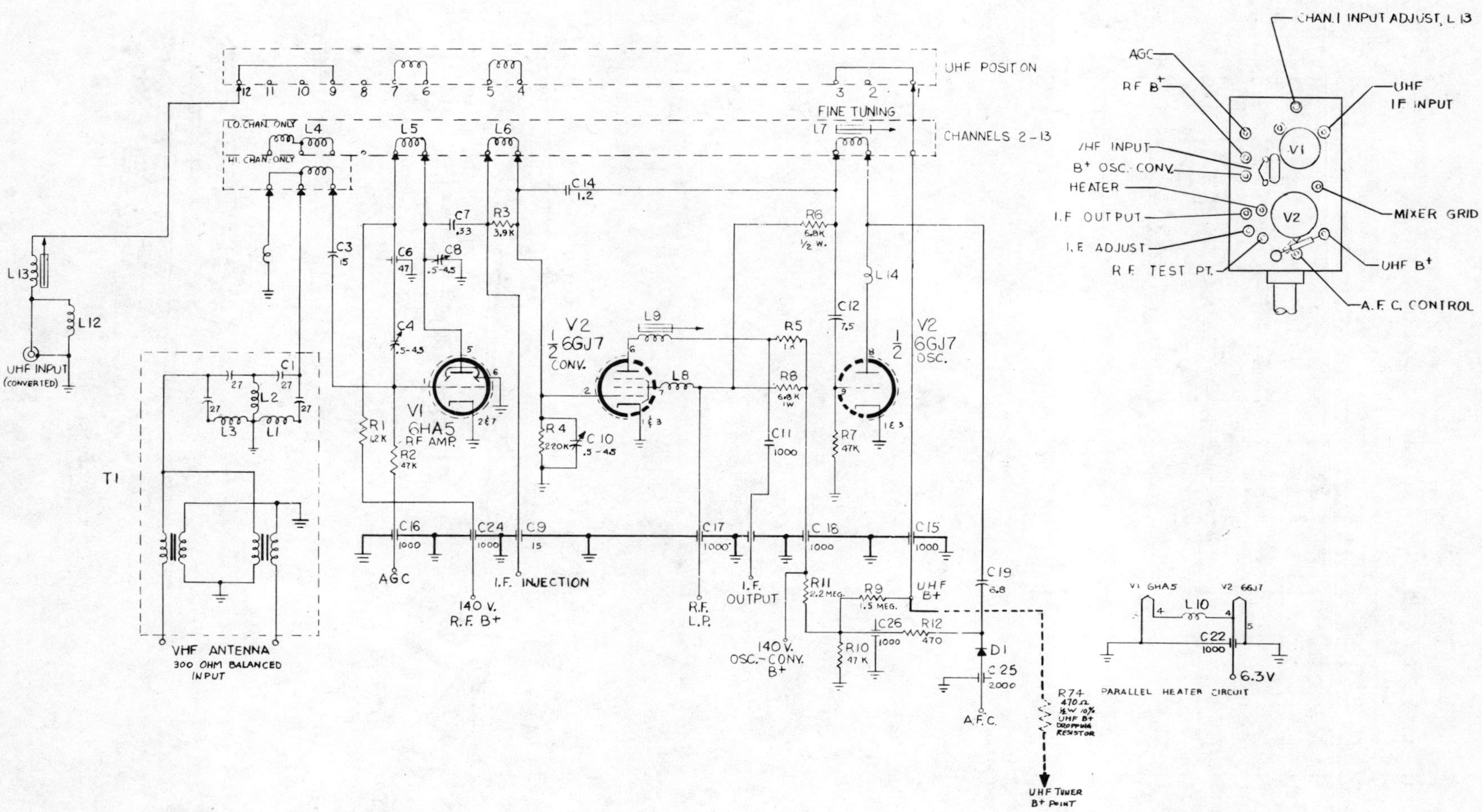

Fig. 5-19. GE VHF tuner EU86X3 schematic diagram; exploded view is same as that of EU86X1 (Fig. 5-18B).

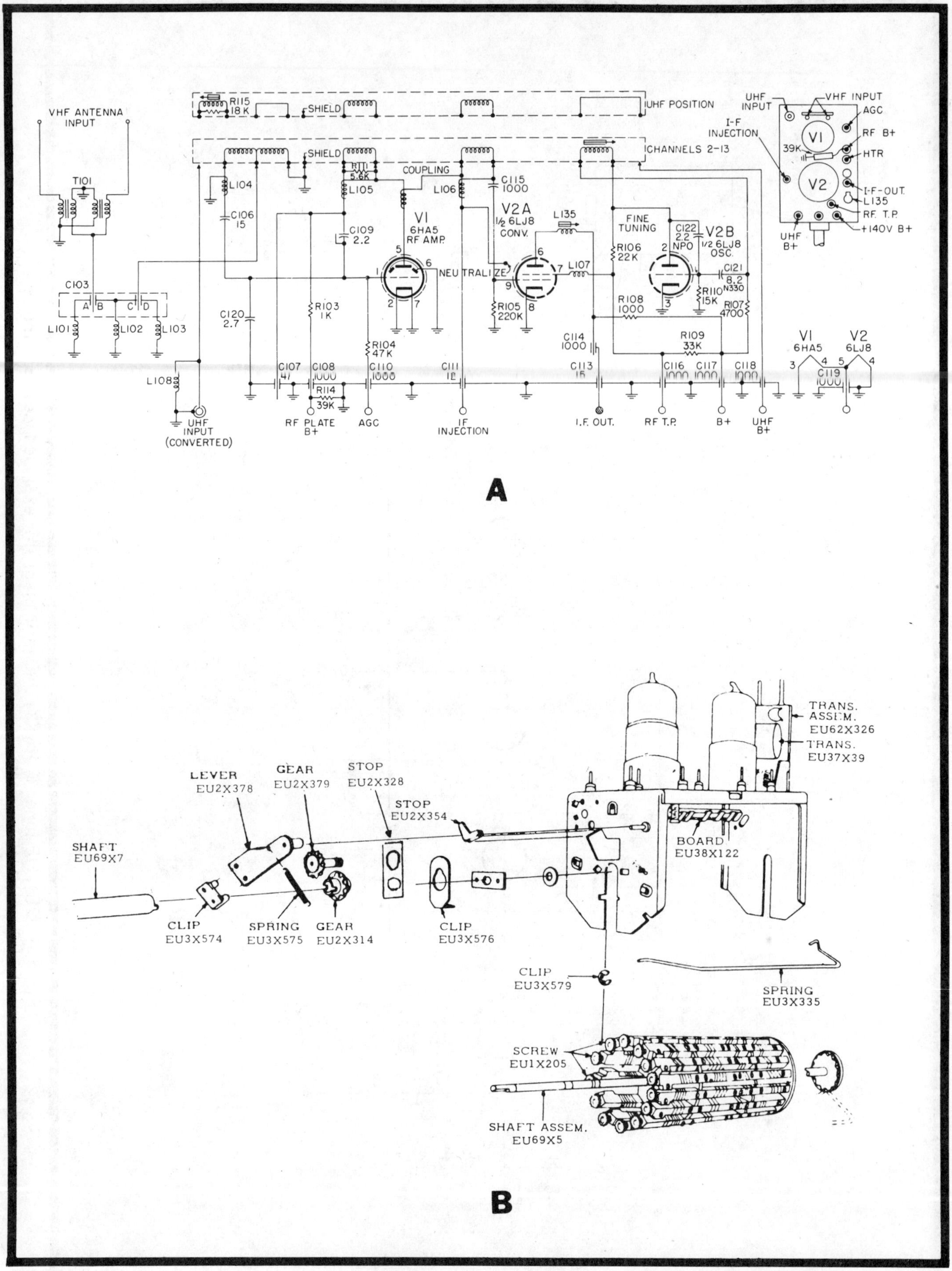

Fig. 5-20. GE VHF tuner EU86X304 schematic diagram (A) and exploded view (B).

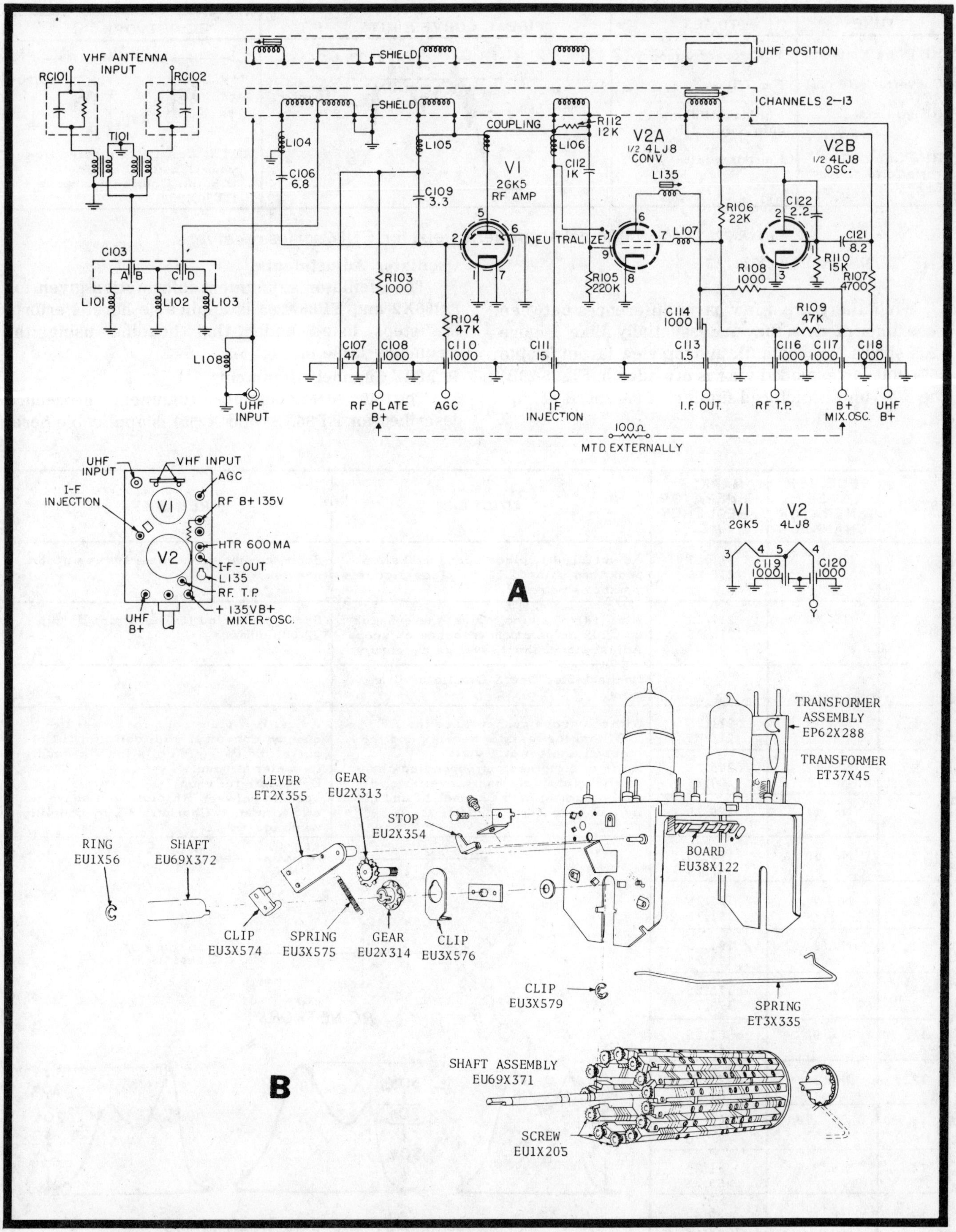

Fig. 5-21. GE VHF tuner EU86X305 schematic diagram (A) and exploded view (B). Unit is very similar to EP86X265, and alignment procedure for EP unit applies to this tuner.

TUNE	TO SET	IDEAL CURVE LIMITS	RC NETWORK
RF Plate Coil	For proper frequency		
Converter grid coil	For Tilt		
RF grid coil	For Max. Midband gain & zero tilt		
RF Plate & Converter grid "coupling"	For bandwidth		NOTE: Coil L1 is ordered from your General Electric distributor. Catalog number is ET36X733.

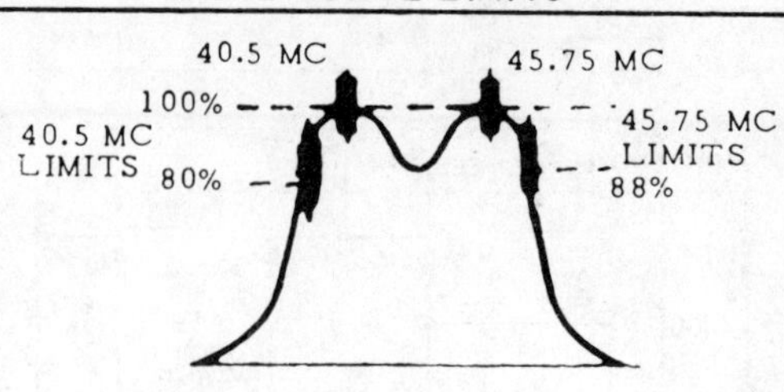

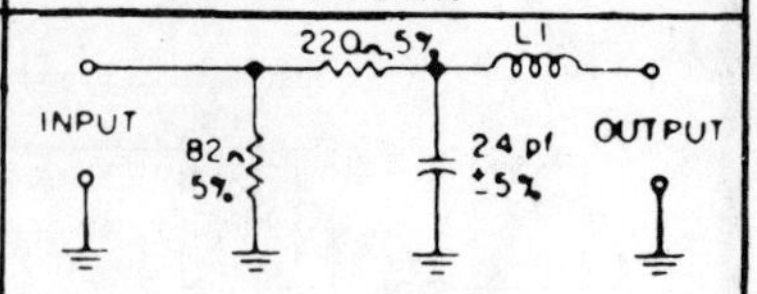

Fig. 5-22. 40 MHz channel alignment steps for EU86 series receivers.

VHF TUNERS EU86X304, 5

While there are a few basic differences between these tuner models, they are essentially alike. Figure 5-20A shows the schematic and top-view layout for the X304 unit; an exploded view is provided in Fig. 5-20B. The X305 unit is pictured similarly in A and B of Fig. 5-21.

Oscillator Adjustments

The oscillator adjustment information given for EP86X2 and EP86X265 is applicable here. Perform the steps listed under that heading using the frequency table of Fig. 5-3.

40 MHz Channel Alignment

The 40 MHz channel alignment procedure described for EP86X2 (and X265) is applicable here.

STEP	RECEIVER & SWEEP GENERATOR CHANNEL	MARKER GENERATOR FREQUENCY MC	ADJUST	REMARKS
1.	No. 13	211.25 215.75	Adjust L106 to place sound marker on peak then adjust L105 to place picture marker on peak.	Scope connected to mixer test point thru network.
2.	No. 10	211.25 215.75	With -18 volts bias, adjust the gimmick on C109 for minimum response on scope. Adjust signal input level as necessary.	Scope connected to Test Point III thru 22,000 ohms.
3.			Recheck Step One & Oscillator alignment.	
4.	No. 13	211.25 215.75	With the scope connected to the RF T.P. thru the resistor network and the bias voltage set at 2 volts.	Adjust R-F plate coils for proper frequency, convert er grid, coils for tilt of curve, and R-F grid coils for Min. saddle in center of band.
5.	No. 12	205.25 209.75	Knife or compress corresponding channel RF grid, plate and converter grid Coils starting with Channel 13 and going in sequence to Channel 2.	Bandwidth for chan 7-13 is varied by spacing between RF plate and converter grid windings. Channels 2-6 by coupling winding.
6.	No. 11	199.25 203.75		
7.	No. 10	193.25 197.75		
8.	No. 9	187.25 191.75		
9.	No. 8	181.25 185.75		
10.	No. 7	175.25 179.75		
11.	No. 6	83.25 87.75		
12.	No. 5	77.25 81.75		
13.	No. 4	67.25 71.75		
14.	No. 3	61.25 65.75		
15.	No. 2	55.25 59.75		

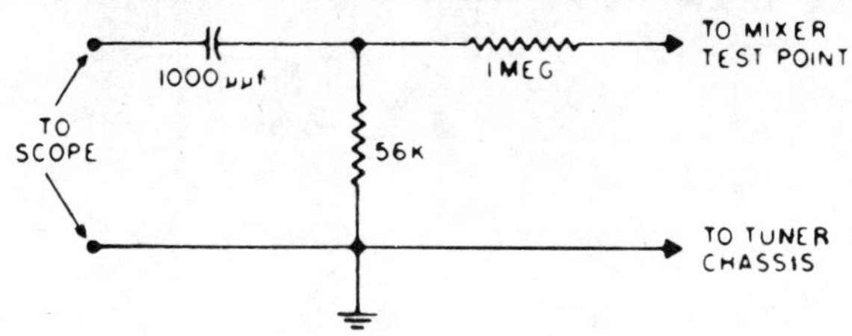

RC NETWORK

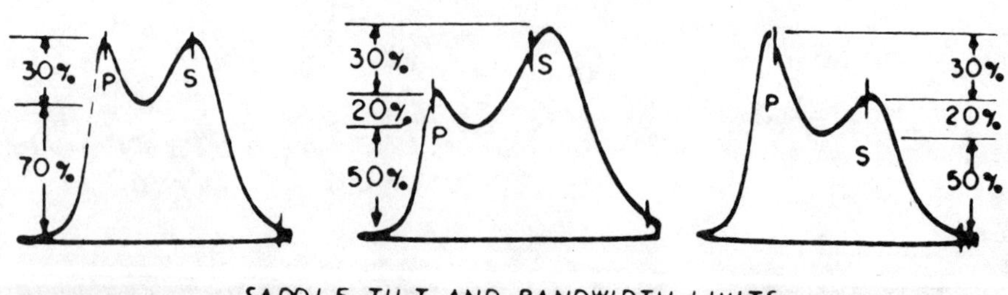

SADDLE TILT AND BANDWIDTH LIMITS

Fig. 5-23. Rf alignment procedure for EU86X304, 5 tuners.

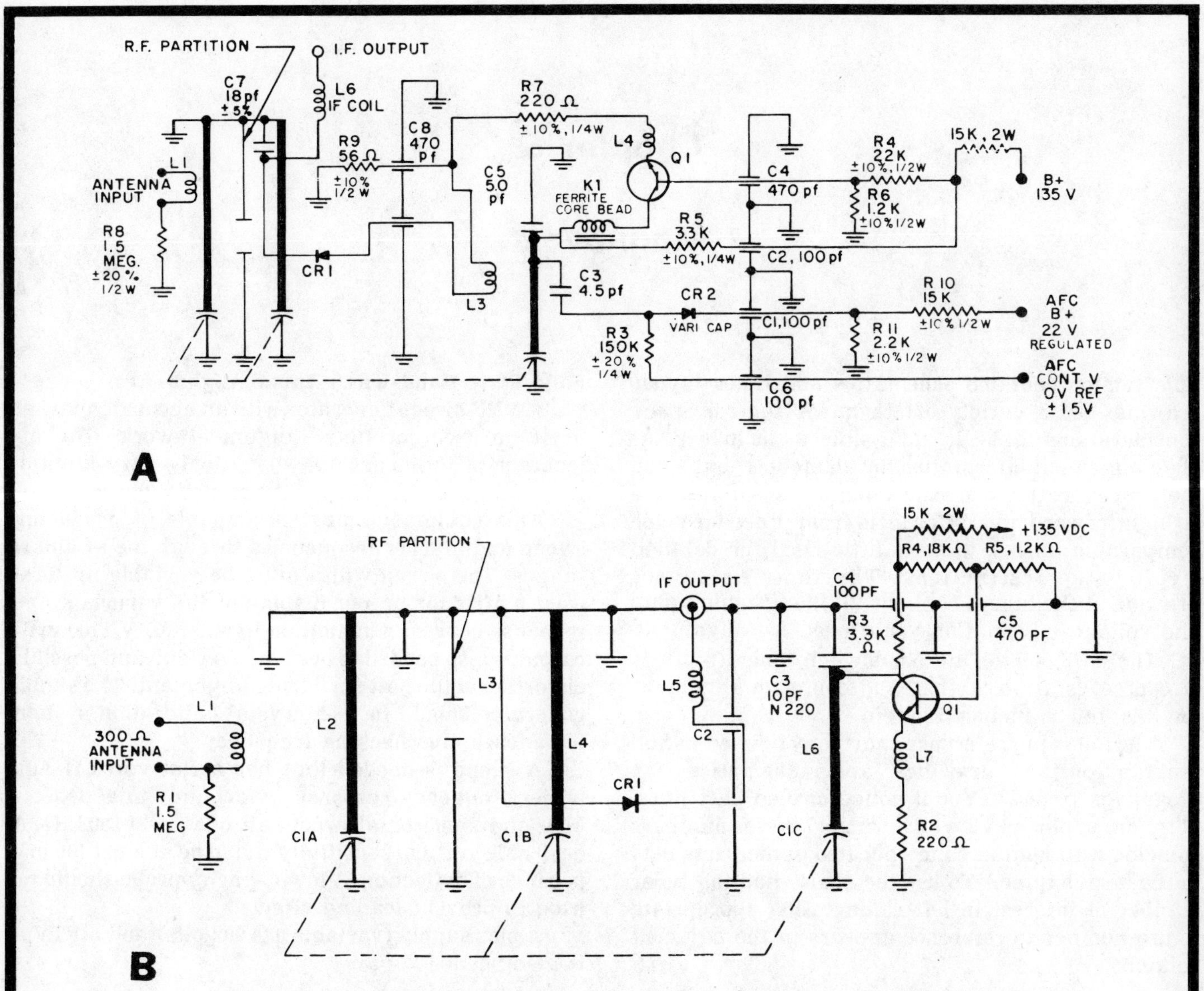

Fig. 5-24. (A) Schematic of tuner EU85X58. (B) Schematic of EU85X59.

The only exception is step 5, which refers to Fig. 5-5. Instead, when performing this procedure for the EU86 series receivers, refer to Fig. 5-22.

RF Alignment

1. Remove the tuner shield cover and connect variable bias voltage (0-20V) to the agc input terminal with the positive lead connected to the tuner chassis. For step 2 in Fig. 5-23, bias voltage is set for 18-20V and for the balance of the rf alignment 2V bias is used.

2. For step 2 in the Fig. 5-23 observe response curve at test point III (video diode load) through 22K; for the balance of the alignment procedure observe response curve at mixer test point through network shown.

3. Ground test equipment to the tuner chassis and connect the properly terminated sweep (with marker) to the antenna input terminals for channels 2 through 13.

4. Make all coil knifing adjustments with plastic or fiber alignment tools.

5. Replace tuner shield cover after adjustment to observe proper frequency.

6. See 40 MHz alignment for channel 1 (or 13th position) alignment.

UHF TUNERS EU85X58 AND 59

Like other units of its kind, this UHF tuner has continuous tuning and covers the range of 470 to 890 MHz. The received signal is converted directly to the 45 MHz intermediate frequency of the receiver. The UHF tuner must, therefore, be used with a VHF tuner which is provided with a 45 MHz input.

Should it become necessary to completely remove the UHF tuner, first remove the knobs from the front of the receiver and then remove the cabinet back as described. Remove the complete control cluster from the cabinet front. In Fig. 5-24 are the applicable tuner diagrams; the schematic in A represents the EU85X58, and that in B the EU85X59.

Chapter 6

Magnavox

This section contains schematics and tuner layout drawings for a variety of Magnavox tuners. Space considerations make it impossible to include complete alignment information for all models; but since the procedure for the conventional assemblies are straightforward and differ little from procedures for comparable tuners, there is little need for detailed step-by-step instructions. The top-view layout drawing of the tuners contains all the test-point data and voltage information you'll need for alignment.

The VHF solid-state bandswitch tuner (340152), of course, is an exception, and complete alignment for this unit is included herein.

The following reference chart may be used to find the appropriate drawings and schematics for Magnavox tuners. (You'll notice circled part numbers on exploded-view drawings; these numbers coincide with numbers incorporated in the parts lists in the last chapter.) To use the chart, find the tuner number of interest in left column; the appropriate figure number to reference appears in the adjacent columns.

Solid-State Bandswitch Tuner Alignment

A VHF sweep generator with an accurate marker must be used for tuner alignment work. The attenuator of the generator should only vary the amplitude and not the shape of the response curve.

This equipment must be capable of producing sweep and marker frequencies through the i-f and rf ranges. The sweep width must be variable up to at least 8 MHz for proper display of the various scope response curves. Attenuation to about 0.1V (100 mV) output is also needed to prevent overload and possible distortion of the pattern during alignment. The signal generator should have a crystal calibration system for accurately checking frequency.

A scope is needed that has a flat vertical amplifier frequency response (wideband) to at least 4 MHz and a horizontal sweep rate of at least 100 kHz. A desirable vertical sensitivity would be at least 100 mV per inch of deflection. A low-capacity probe should be used to prevent loading effects.

A bias supply (variable 0 to 20 volts) and a VTVM is also needed.

Note: All graphic material in this chapter appears through the courtesy of Magnavox Co.

MAGNAVOX NUMBER	EXPLODED VIEW	PHOTO & PART I.D.	LAYOUT TOP VIEW	SCHEMATIC DIAGRAM
340131	Fig. 6-1		Fig. 6-2	Fig. 6-3
340137		Fig. 6-6	Fig. 6-7	Fig. 6-8
340145		Fig. 6-9		Fig. 6-10
340146	Fig. 6-11		Fig. 6-4	Fig. 6-5
340152		Fig. 6-11	Fig. 6-2	Fig. 6-12
340156	Fig. 6-13		Fig. 6-14	Fig. 6-15
340163	Fig. 6-1		Fig. 6-4	Fig. 6-5
340183		Fig. 6-16	Fig. 6-17	Fig. 6-18
340184			Fig. 6-19	Fig. 6-20
340185		Fig. 6-21	Fig. 6-22	Fig. 6-23
340186			Fig. 6-24	Fig. 6-25
340187		Fig. 6-26		Fig. 6-27
340192				Fig. 6-28
340199			Fig. 6-29	Fig. 6-30
340200			Fig. 6-30	Fig. 6-31
340202				Fig. 6-33

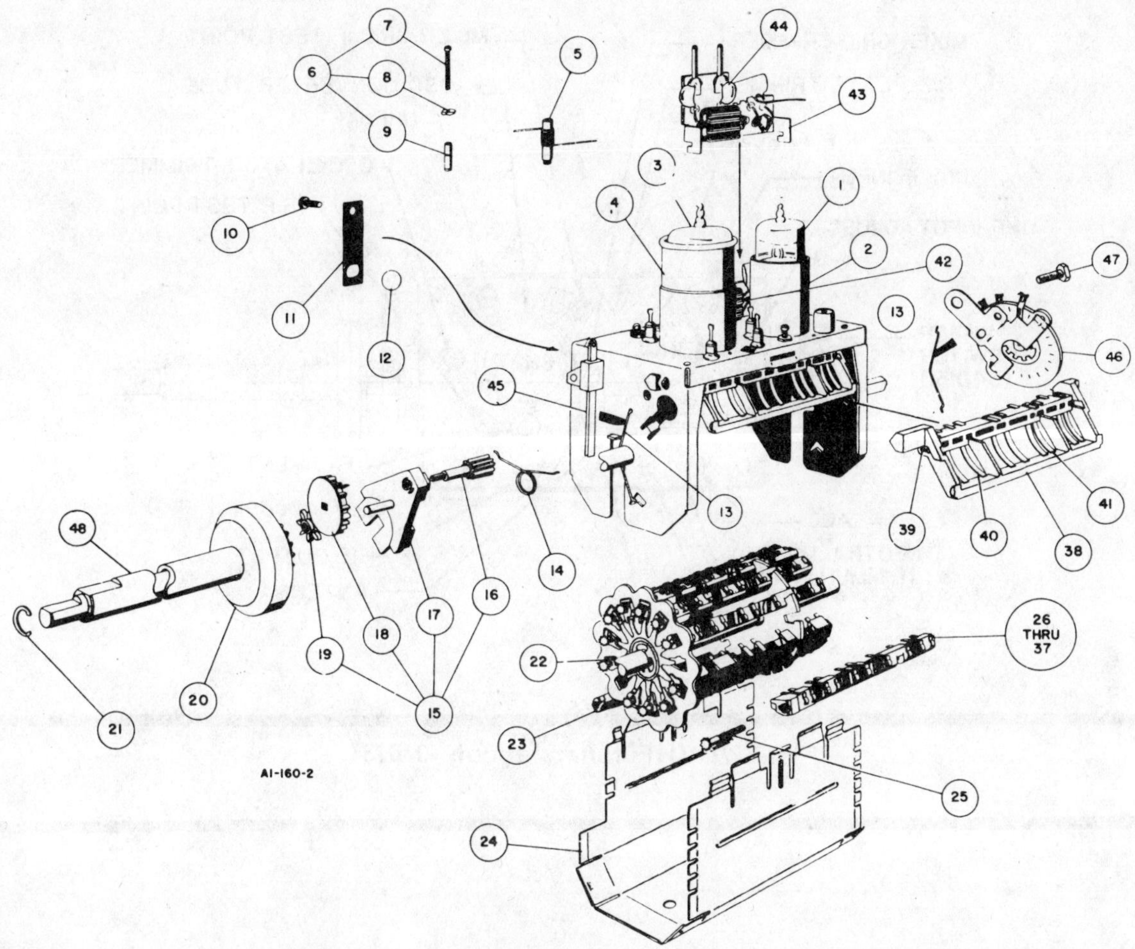

Fig. 6-1. Exploded view for 340131, 340146, and 340163.

Alignment Procedure

1. Connect the scope to i-f output test point on the VHF tuner through an rf bandpass isolation network. Set scope to maximum sensitivity.

Note: The 40 MHz i-f output jack on top of tuner must be terminated with a 75-ohm carbon resistor to prevent distortion of patterns.

2. Connect sweep generator to the antenna terminals (300-ohm input). Alignment inside the VHF tuner consists of spreading or squeezing the proper coils; this must be done with the tuner cover in place. The coils are accessible through holes in the cover. Use only a nonmetallic tool for coil adjustments. Also, the channel 5 tuning adjustment should be near midposition.

3. Set VHF tuner and adjust the sweep generator to channel 13. Then adjust channel 13 master oscillator coil L39 for correct frequency (257.0 MHz).

4. For alignment purposes, apply a dc bias of +2V from a low-impedance supply to the agc terminal (negative lead to tuner chassis). This rf transistor is designed for forward (positive) bias only.

5. To obtain proper high-band alignment response (location of markers and maximum gain on channel 13), spread or squeeze the rf collector coil for maximum gain between the markers.

6. For low-band alignment, set tuner and sweep generator to channel 6. Adjust channel 6 antenna coil for resonance between picture and sound markers.

7. Adjust all other low-band antenna coils (channels 5, 4, 3, and 2) for proper resonance and center bandpass.

8. Adjust channel 6 interstage coils for proper bandpass.

9. Note that the i-f output coil L26 is adjusted as part of the i-f alignment procedure.

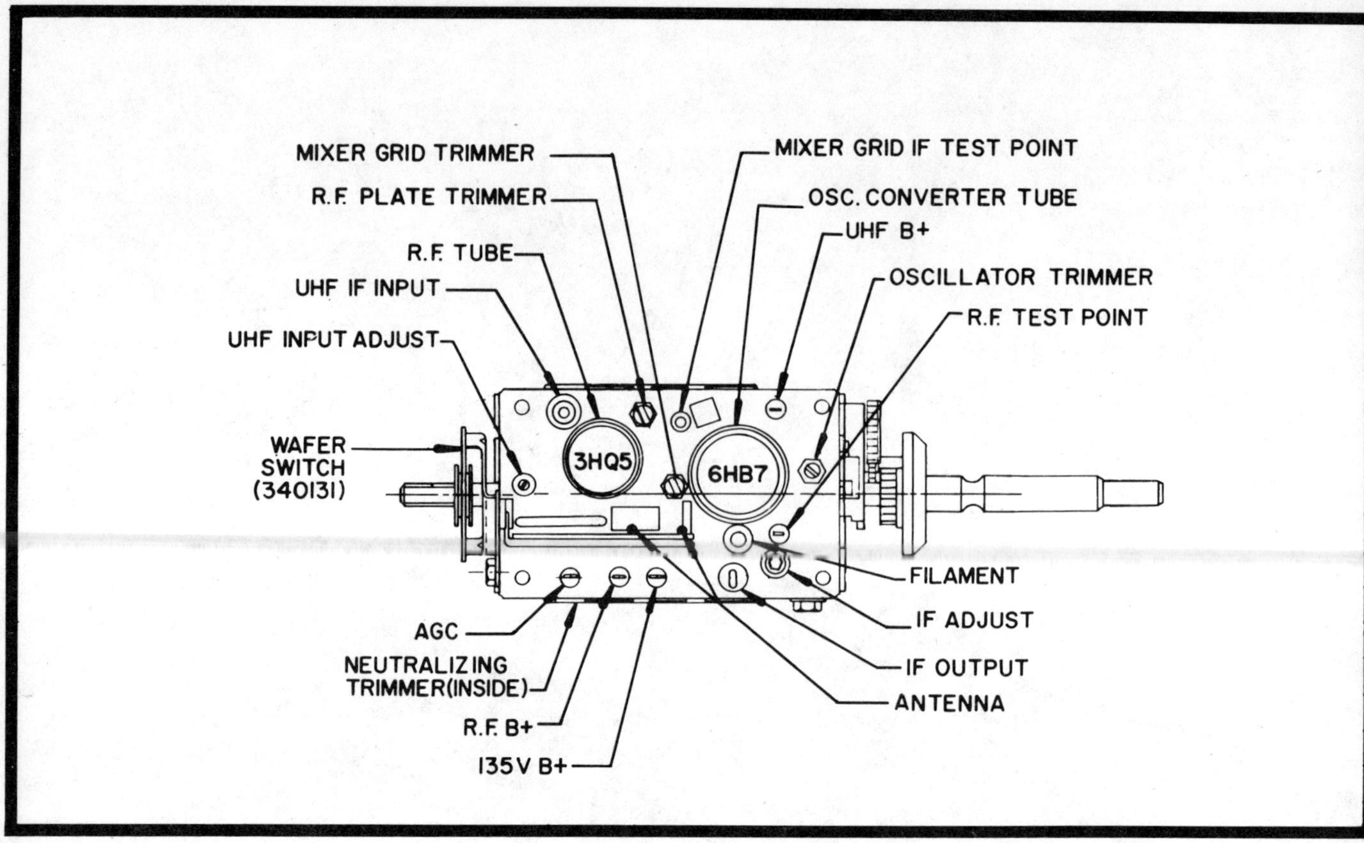

Fig. 6-2. VHF tuner layout—340131.

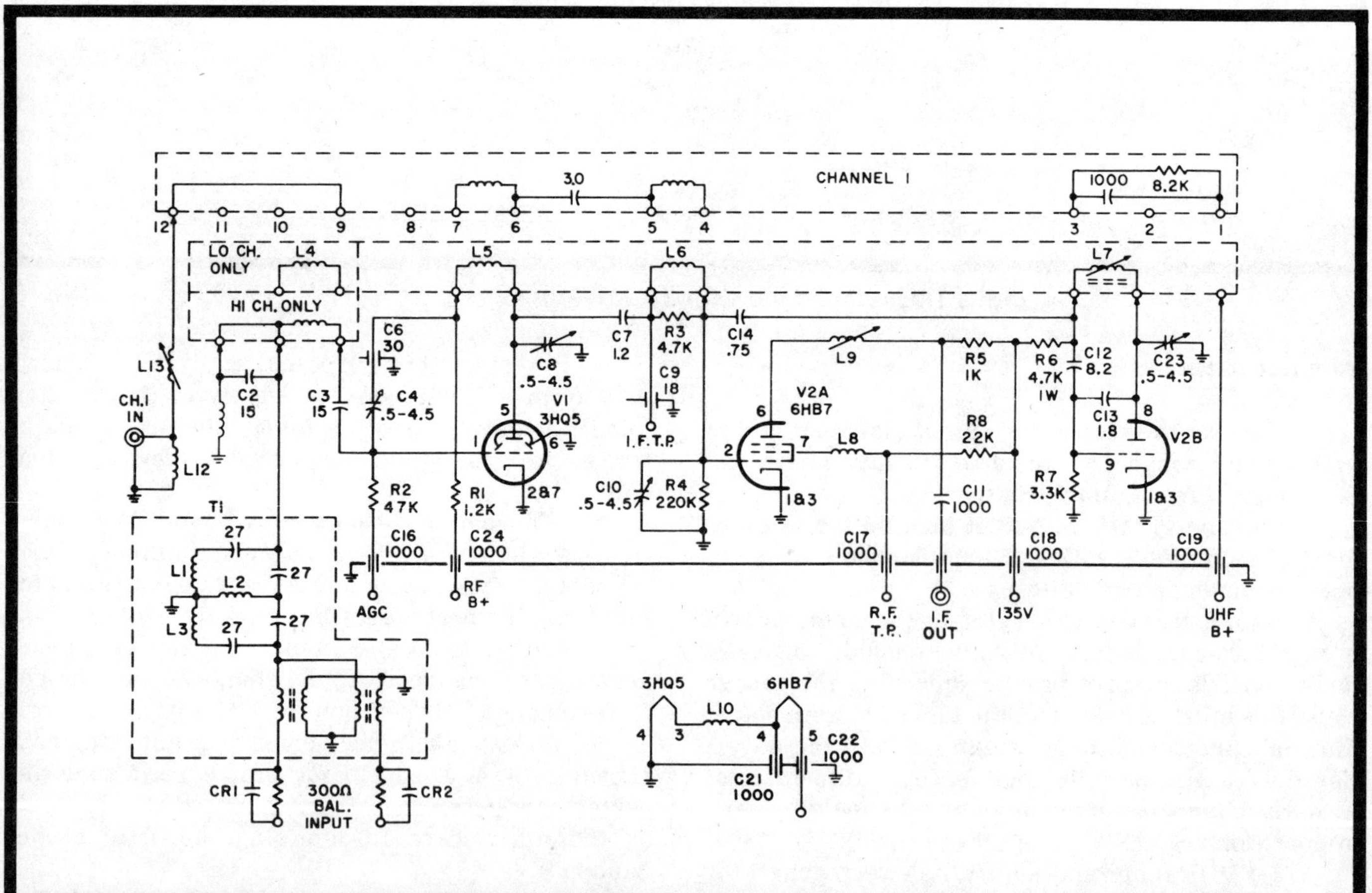

Fig. 6-3. VHF tuner schematic—340131.

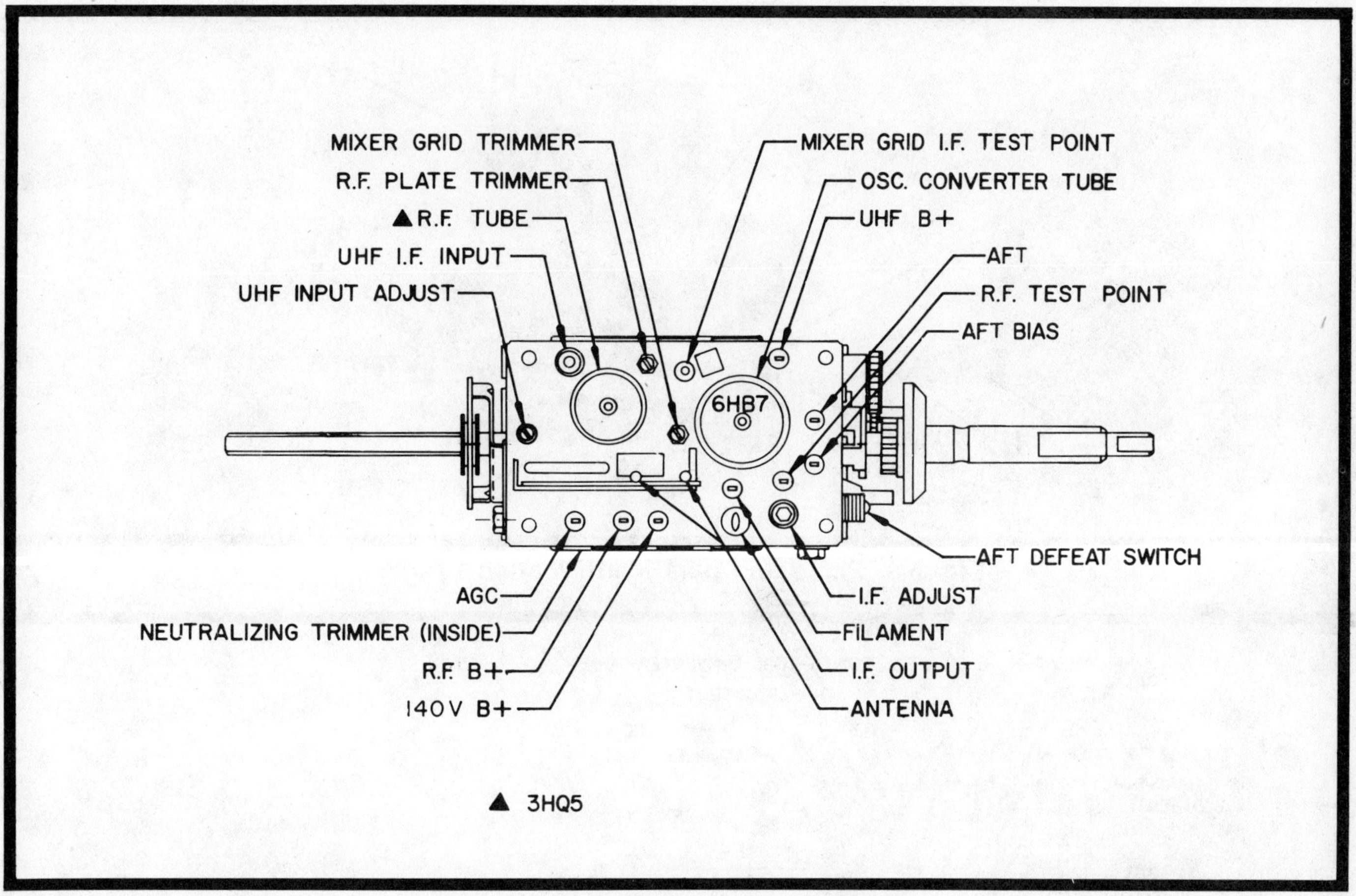

Fig. 6-4. VHF tuner layout—340146.

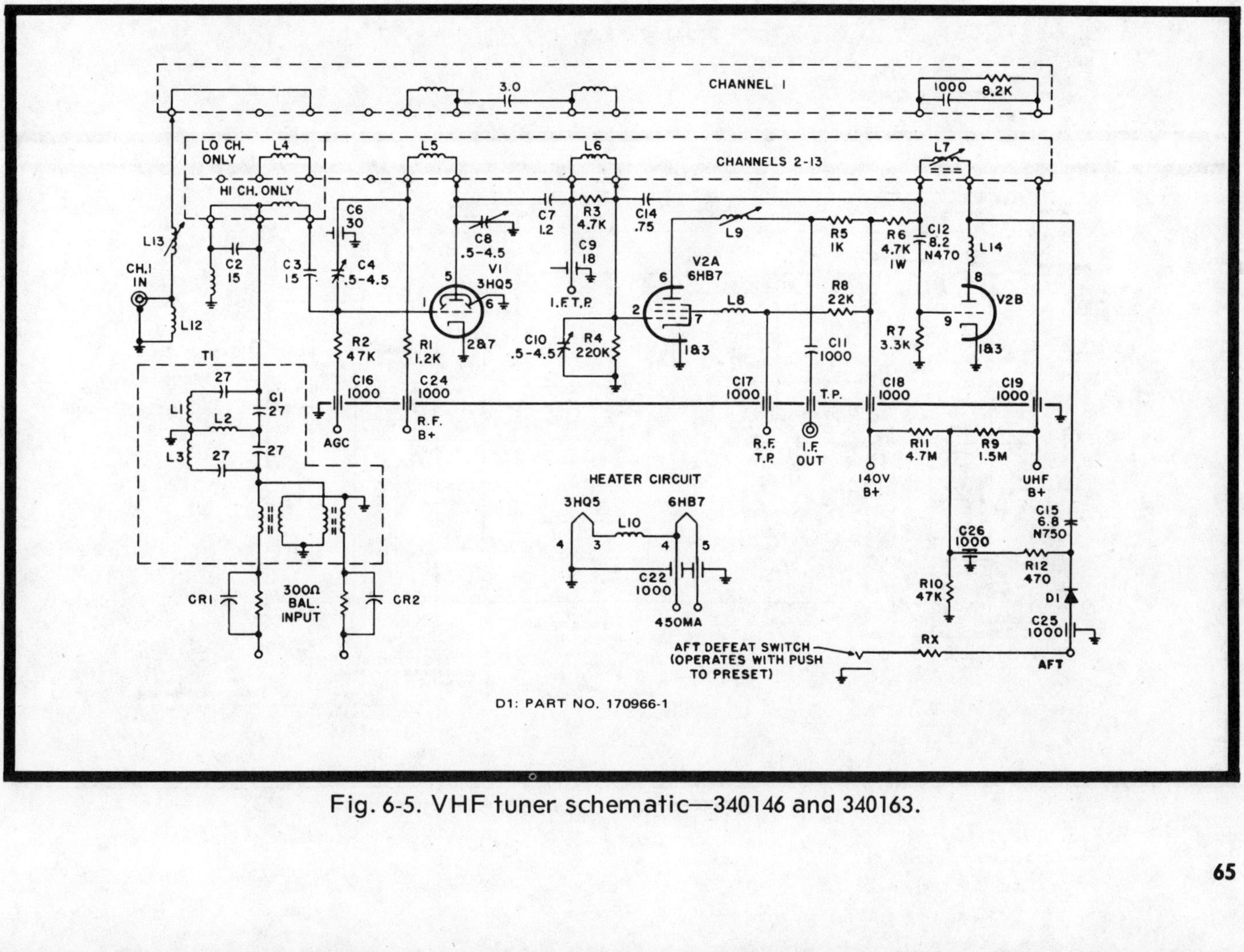

Fig. 6-5. VHF tuner schematic—340146 and 340163.

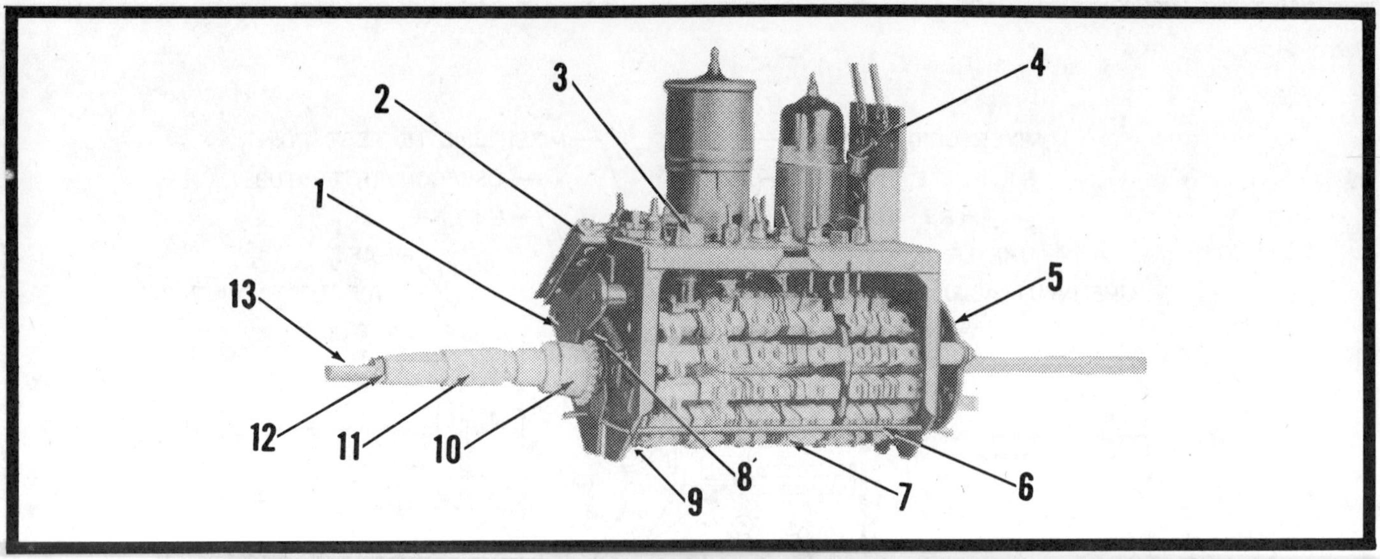

Fig. 6-6. Photo and part identification—340137.

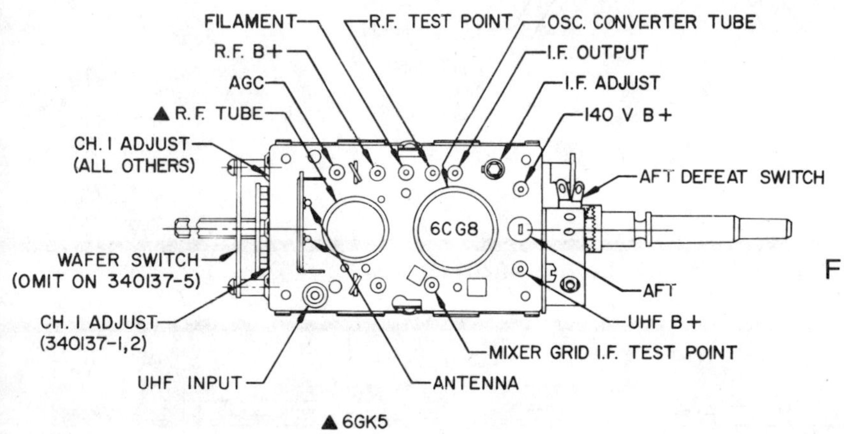

Fig. 6-7. VHF tuner layout—340137.

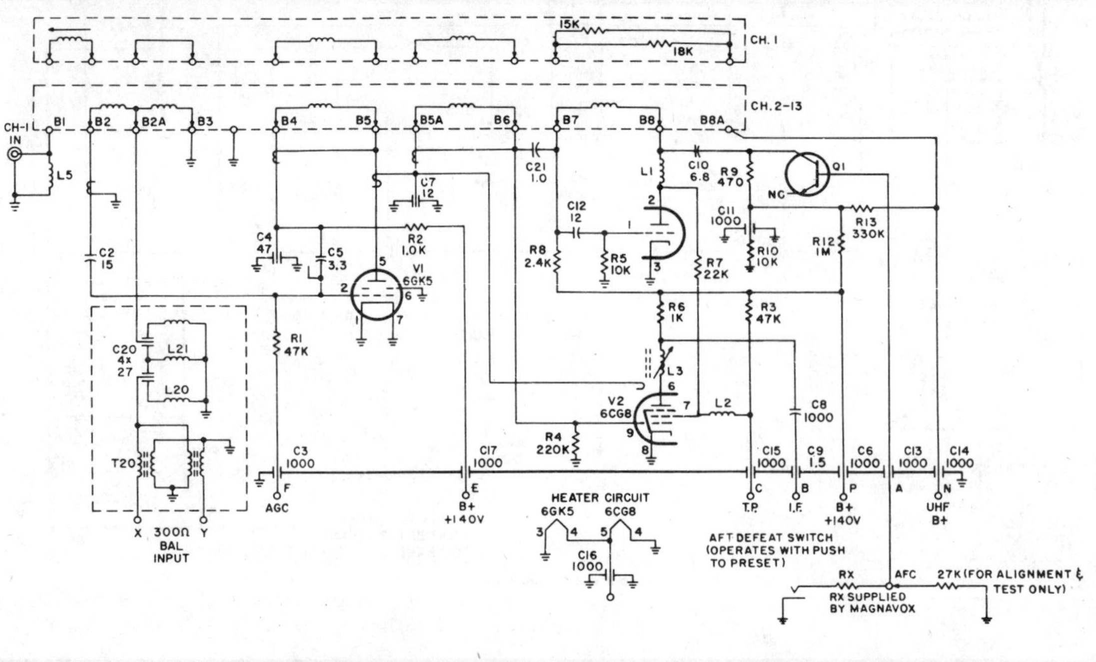

Fig. 6-8. VHF tuner schematic—340137.

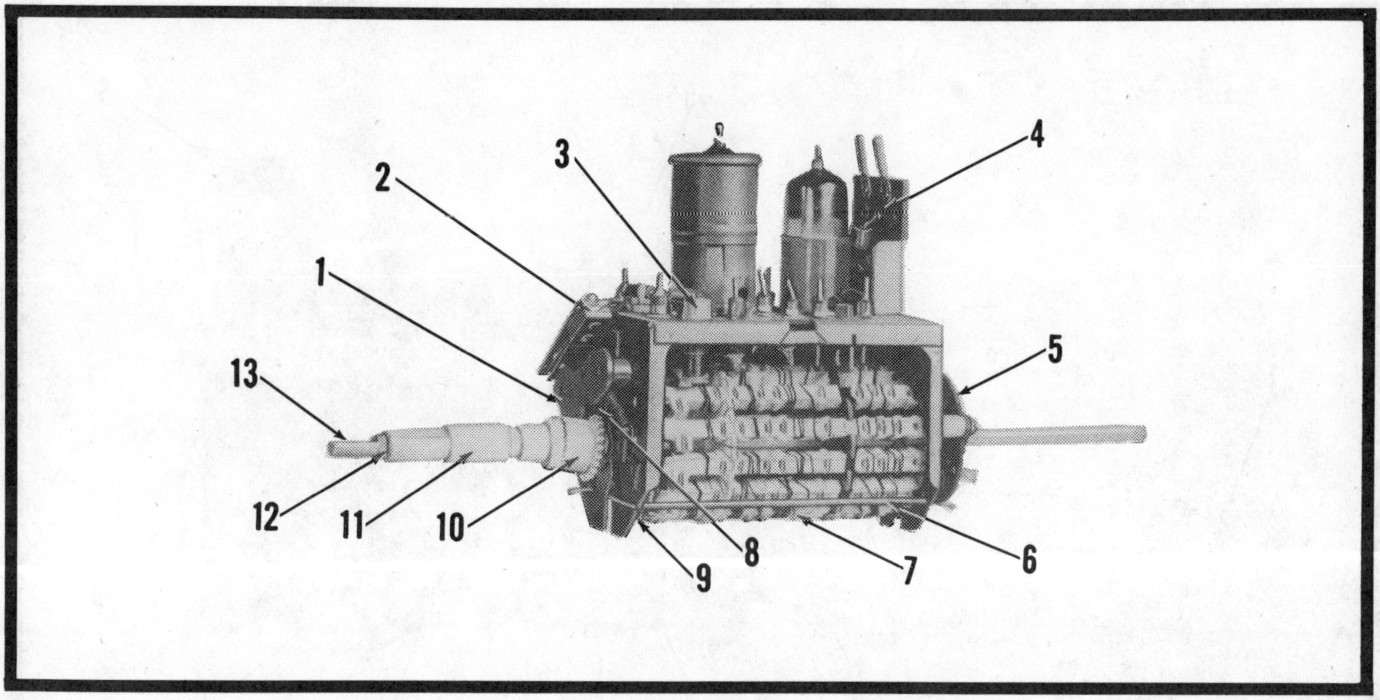

Fig. 6-9. Photo and part identification—340145.

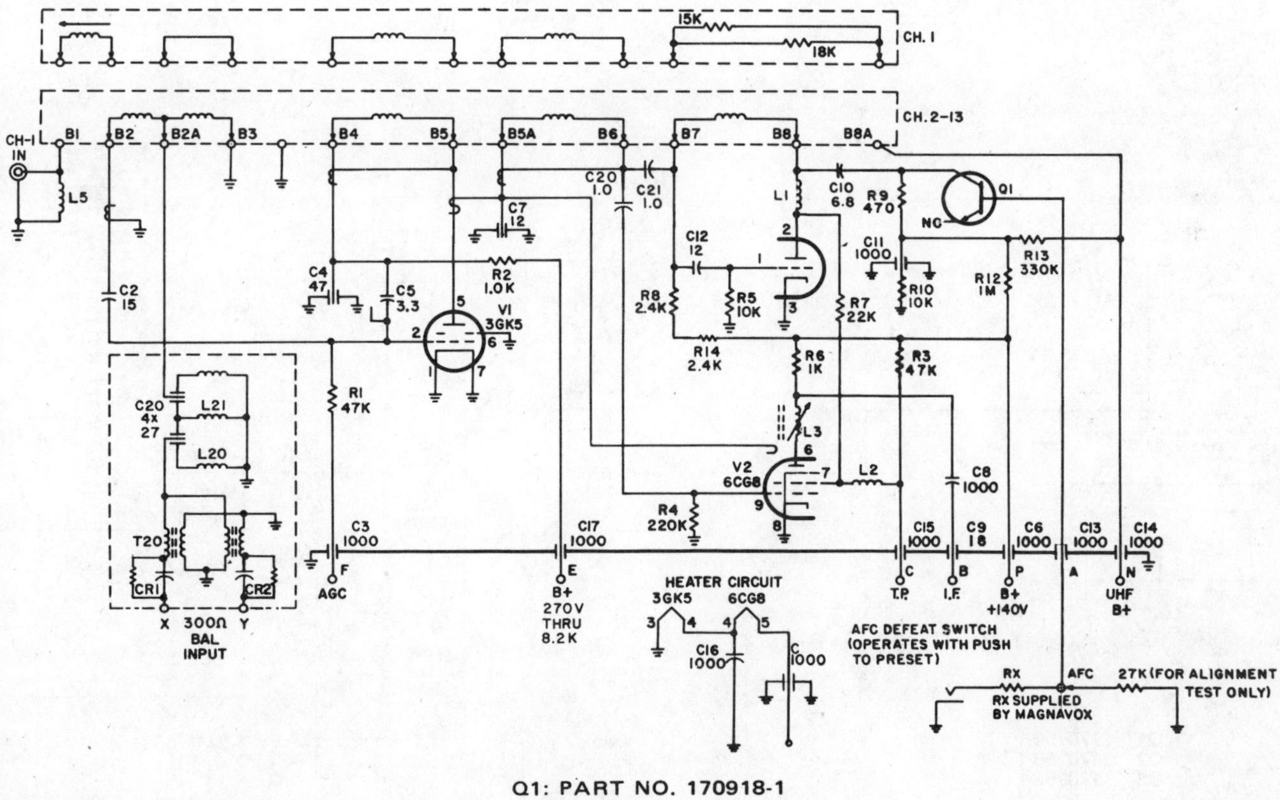

Fig. 6-10. VHF tuner schematic—340145.

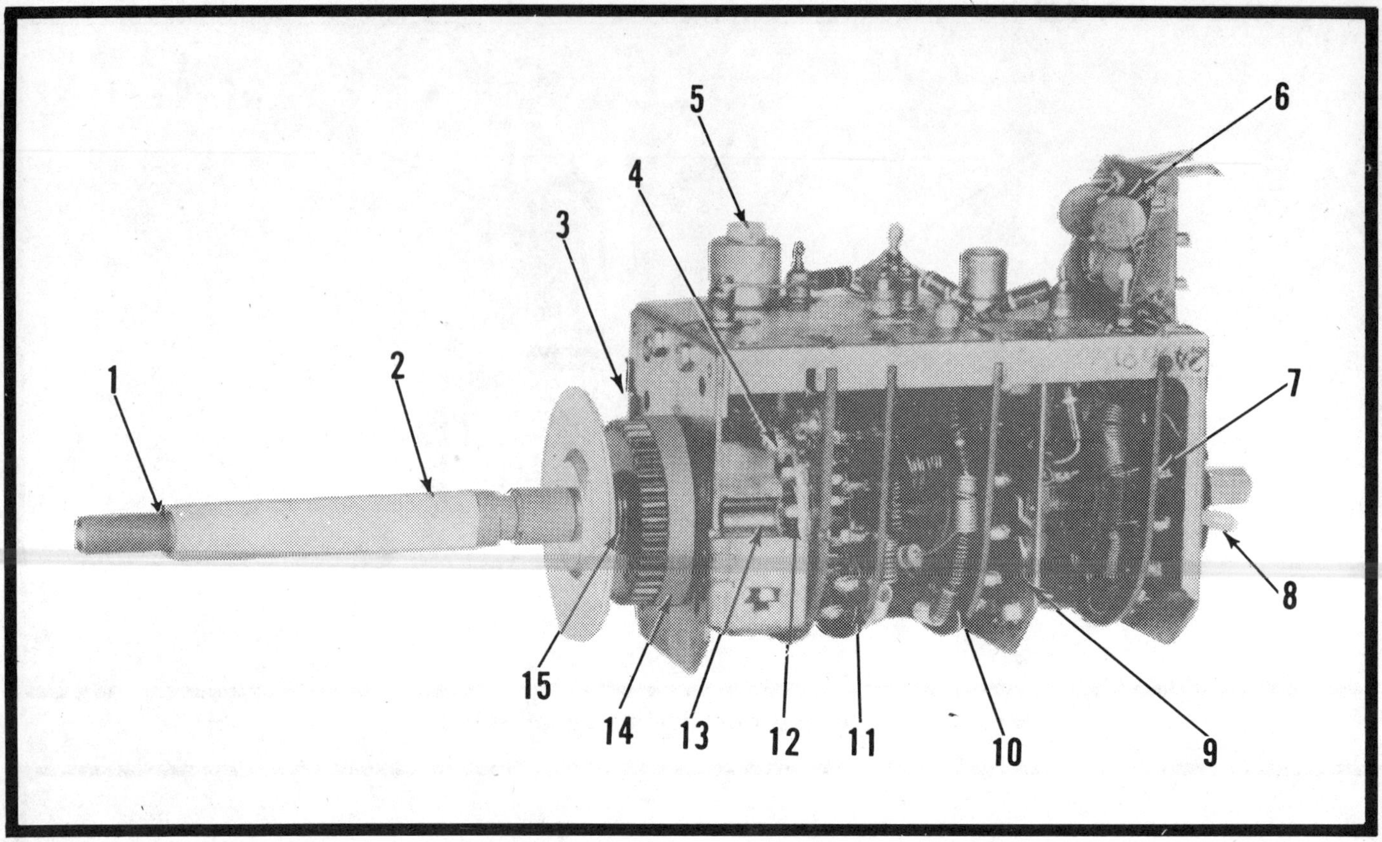

Fig. 6-11. Photo and part identification—340152.

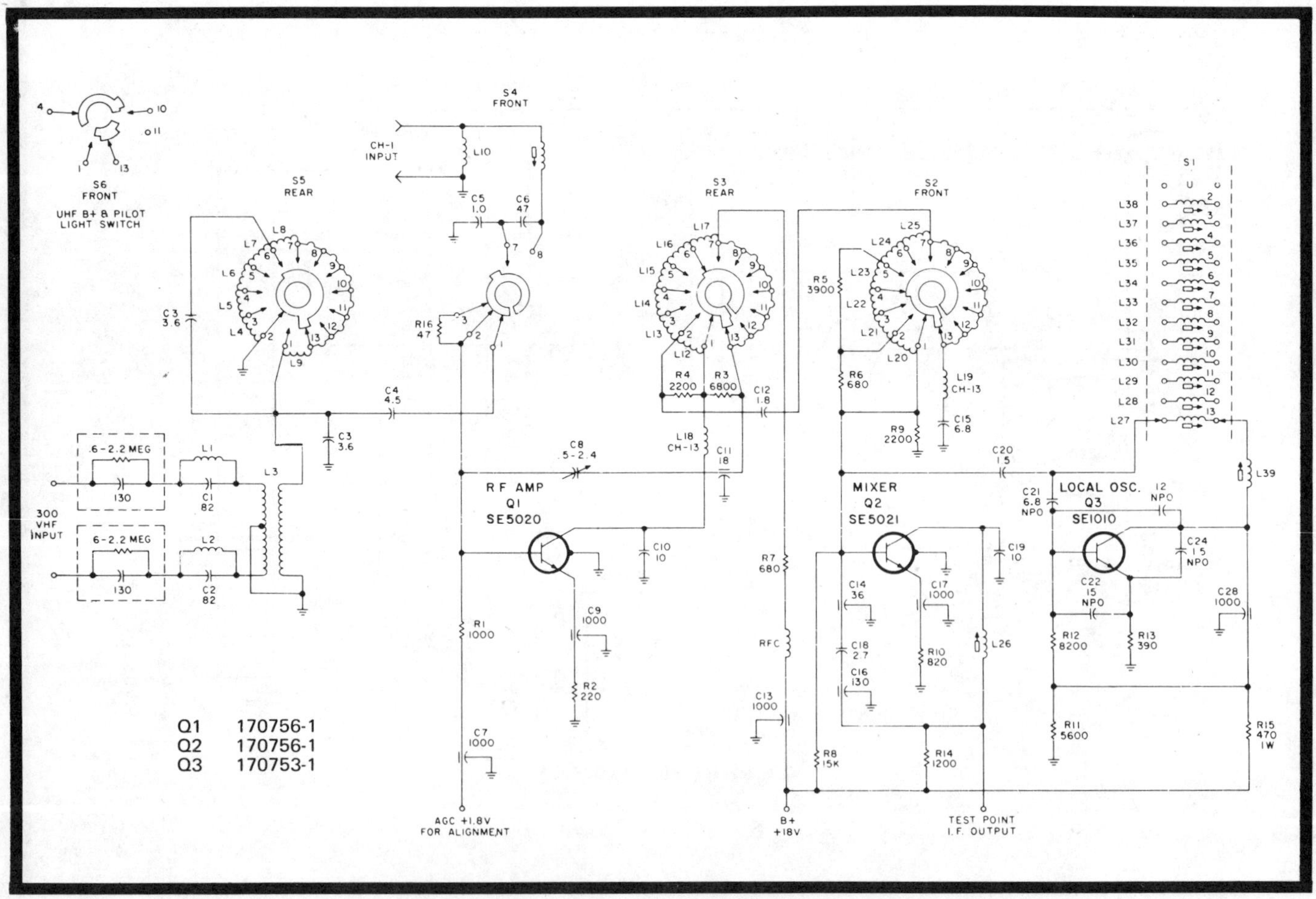

Fig. 6-12. VHF tuner schematic—340152.

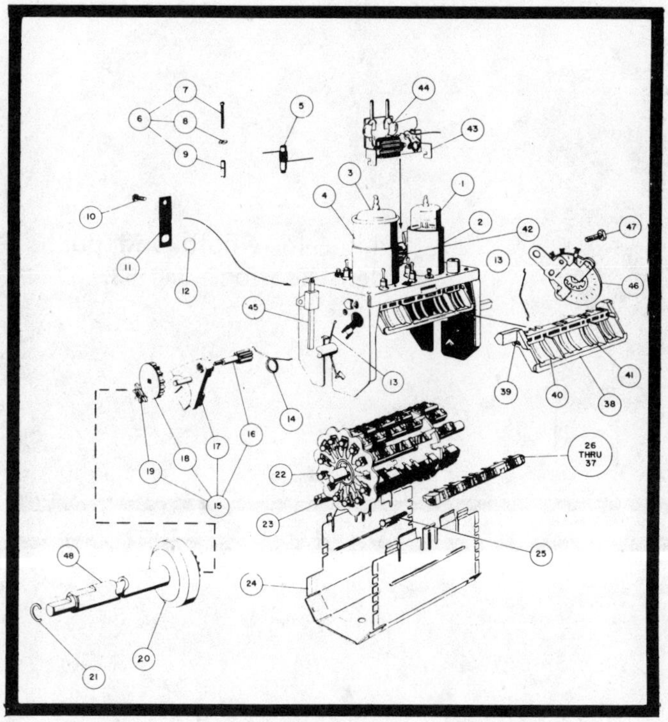

Fig. 6-13. Exploded view—340156.

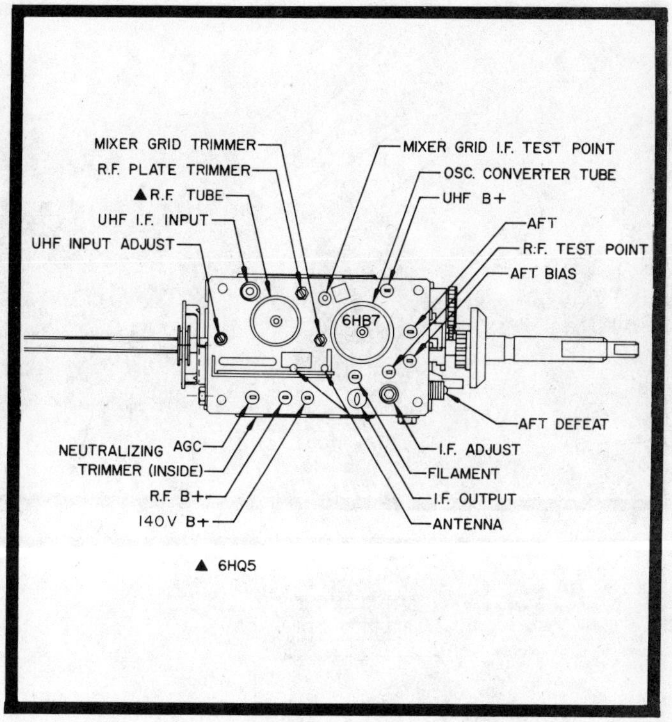

Fig. 6-14. VHF tuner layout—340156.

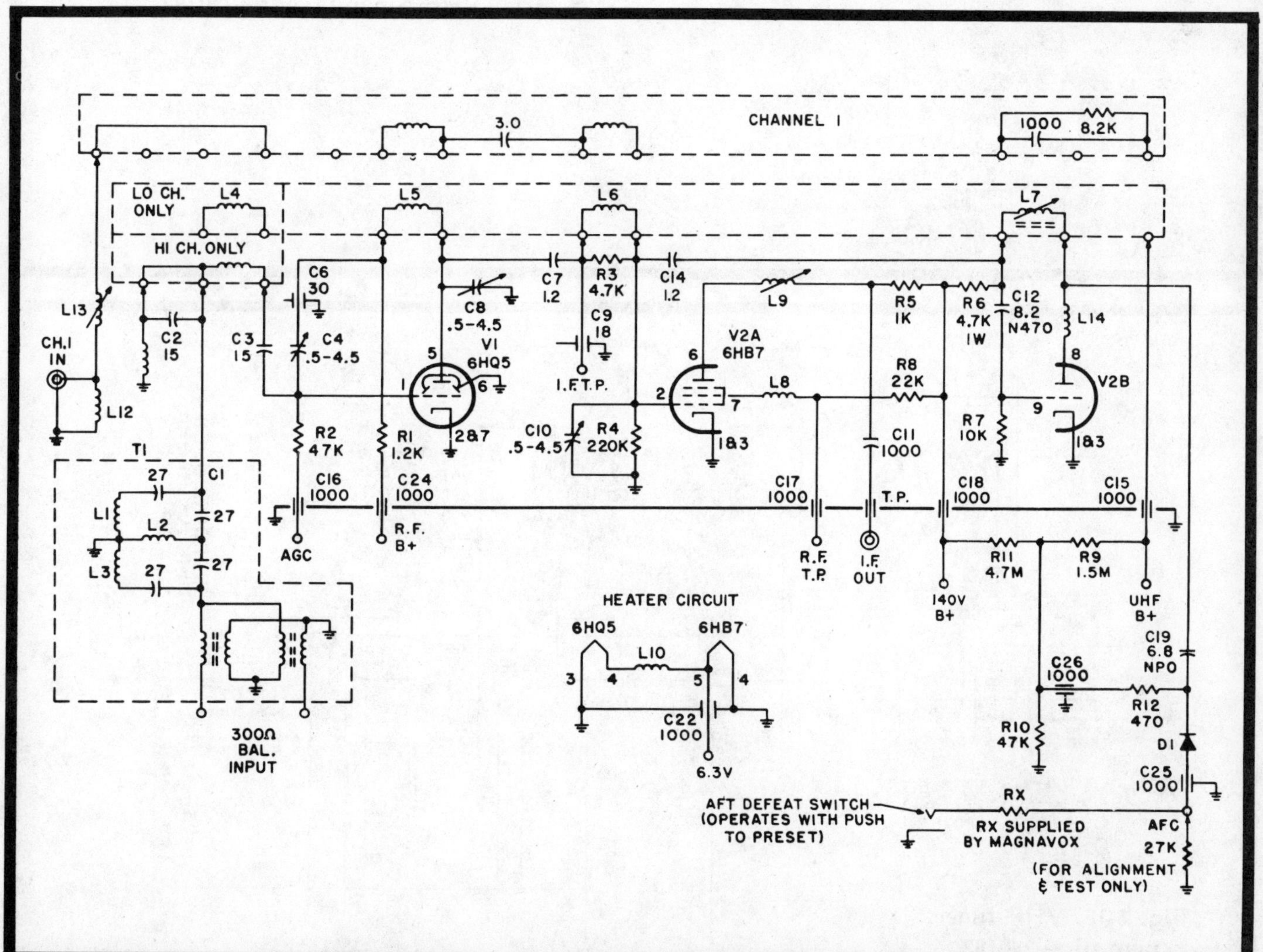

Fig. 6-15. VHF tuner schematic—340156.

Fig. 6-16. Photo and part identification—340183.

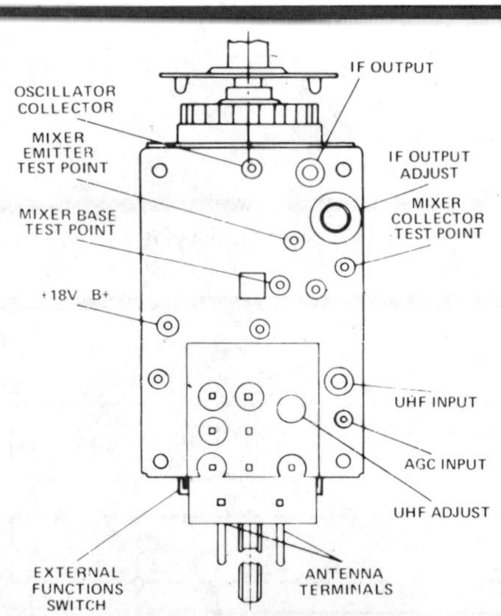

Fig. 6-17. VHF tuner layout—340183.

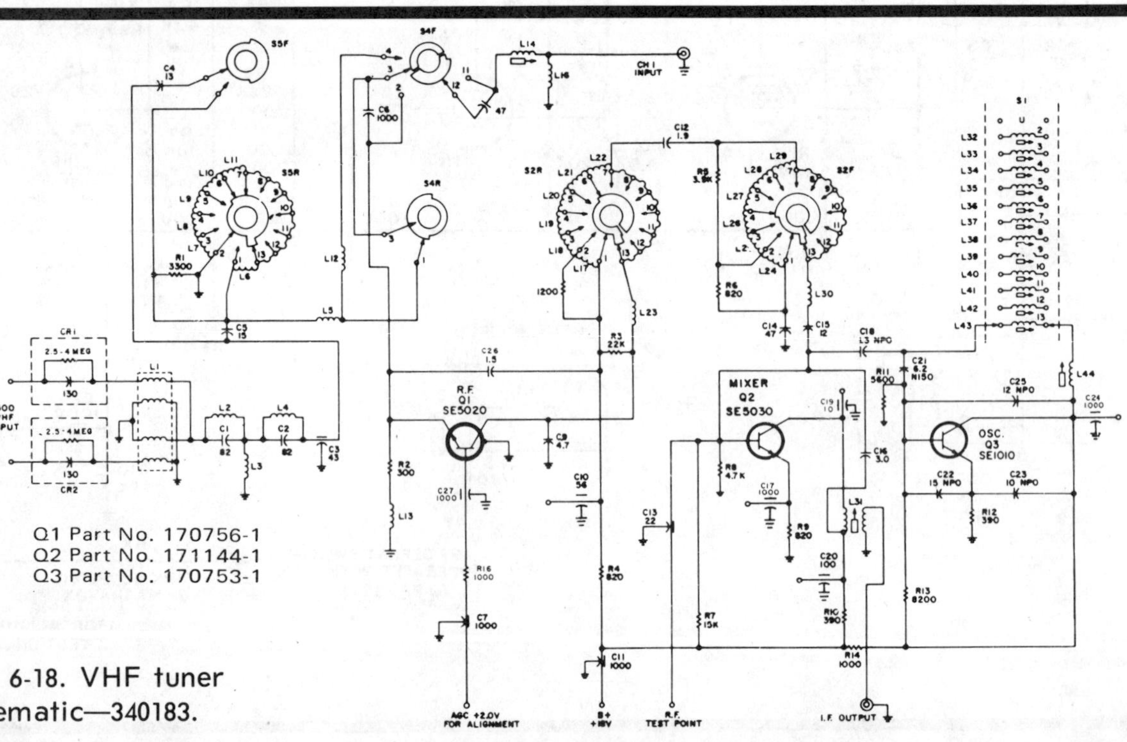

Q1 Part No. 170756-1
Q2 Part No. 171144-1
Q3 Part No. 170753-1

Fig. 6-18. VHF tuner schematic—340183.

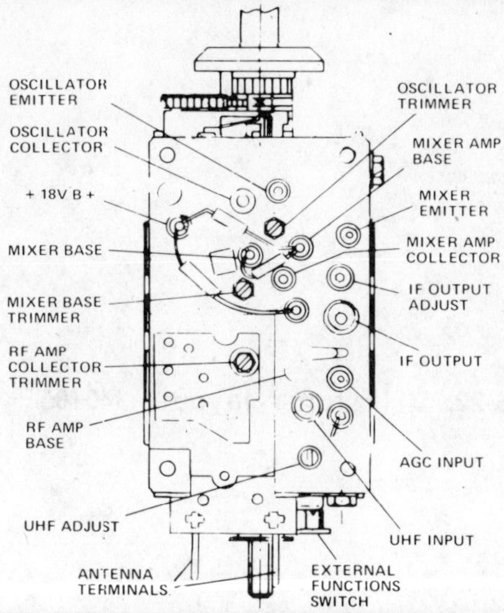

Fig. 6-19. VHF tuner layout—340184.

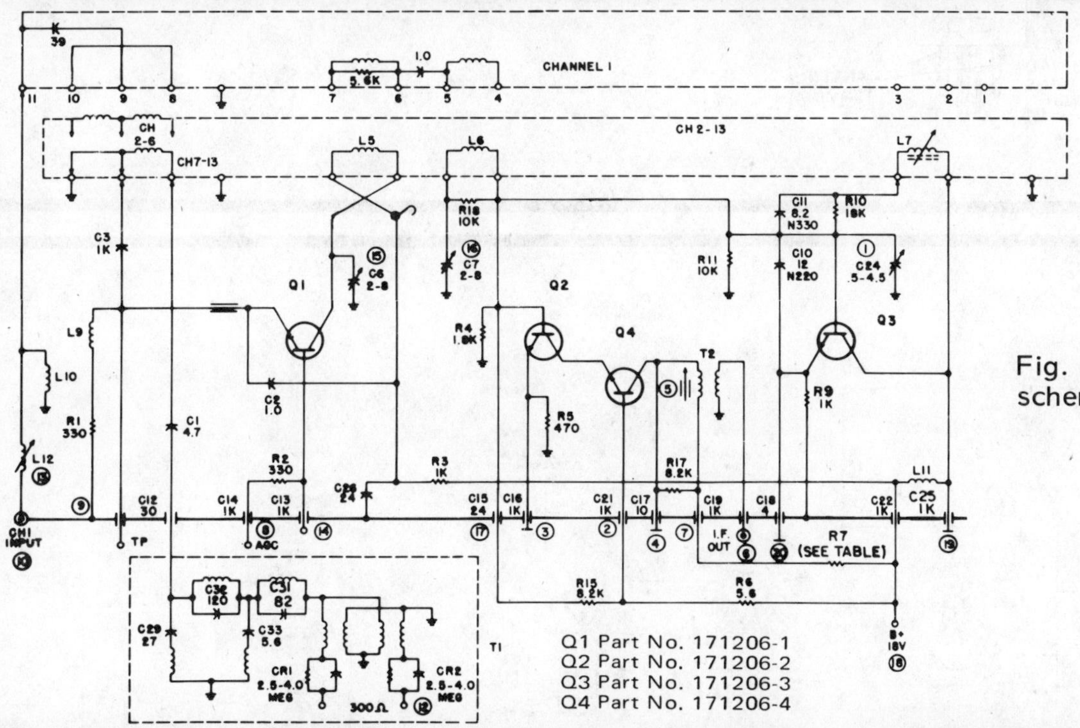

Fig. 6-20. VHF tuner schematic—340184.

Fig. 6-21. Photo and part identification—340185.

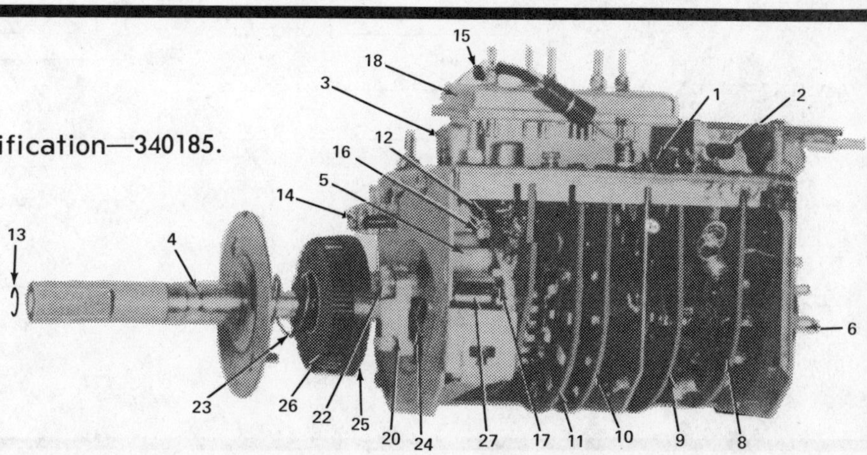

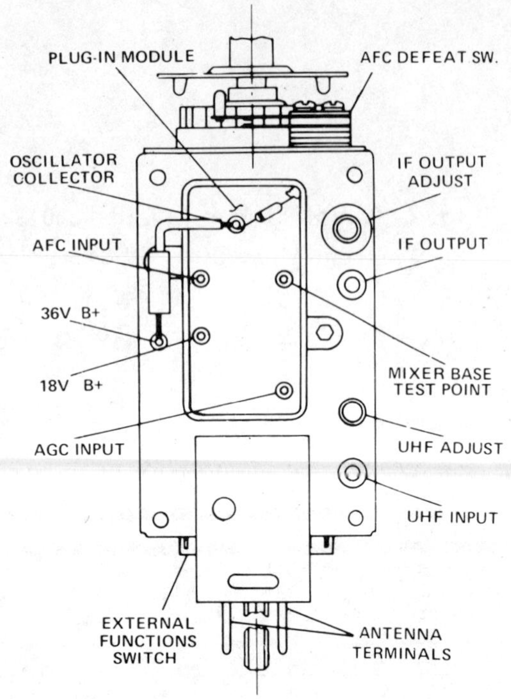

Fig. 6-22. VHF tuner layout—340185.

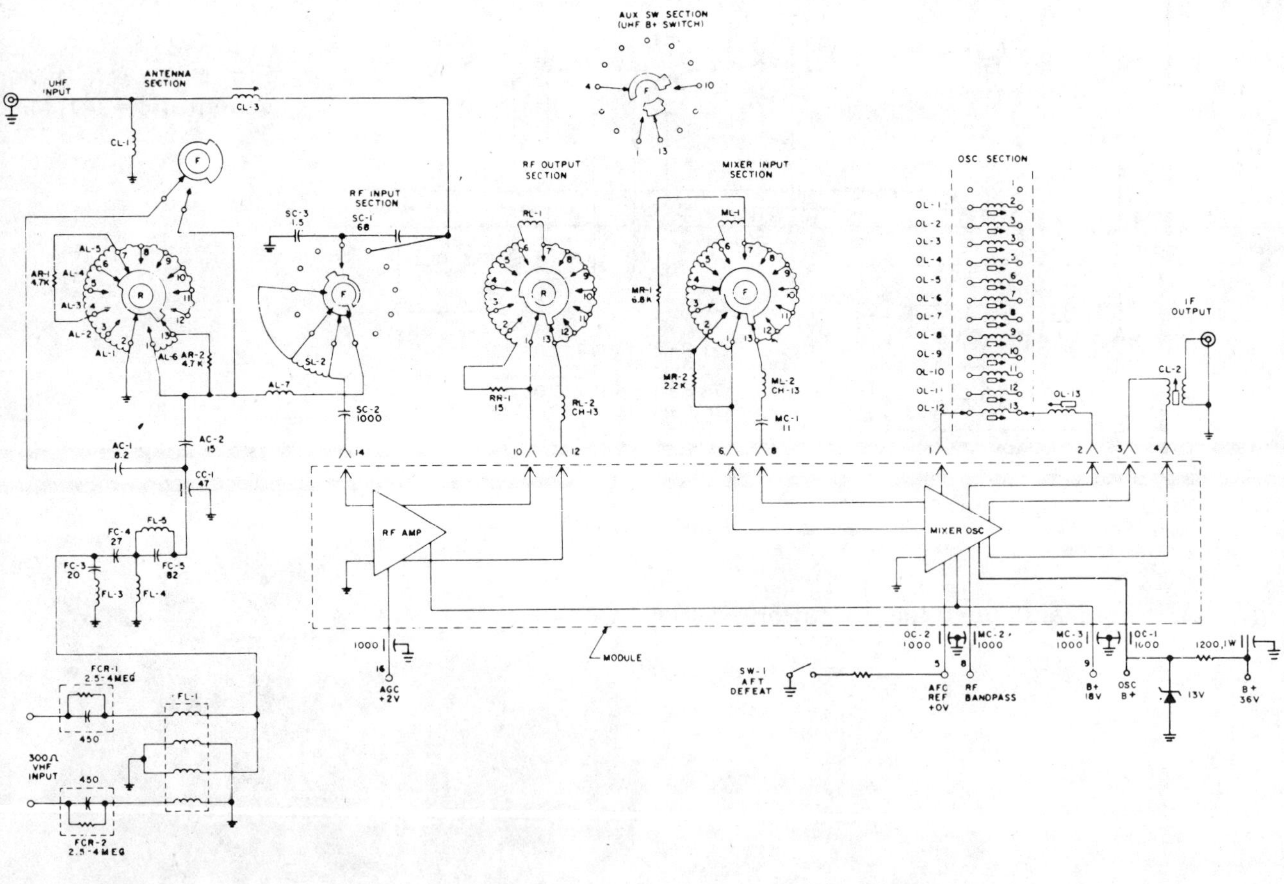

Fig. 6-23. VHF tuner schematic—340185.

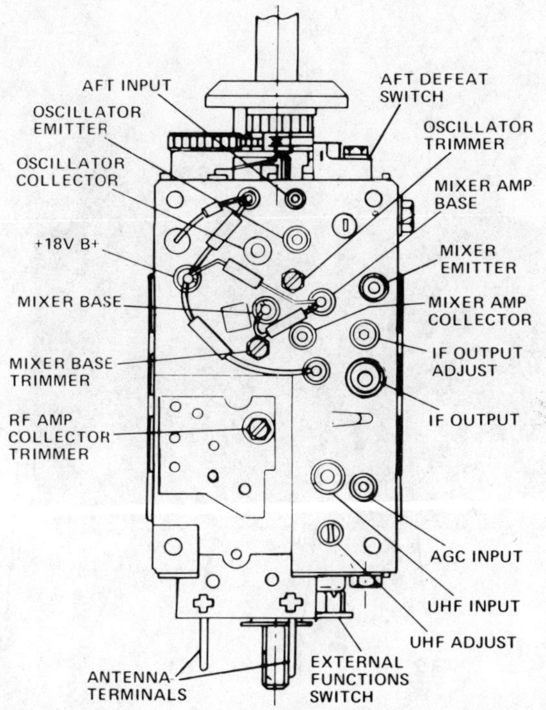

Fig. 6-24. VHF tuner layout—340186.

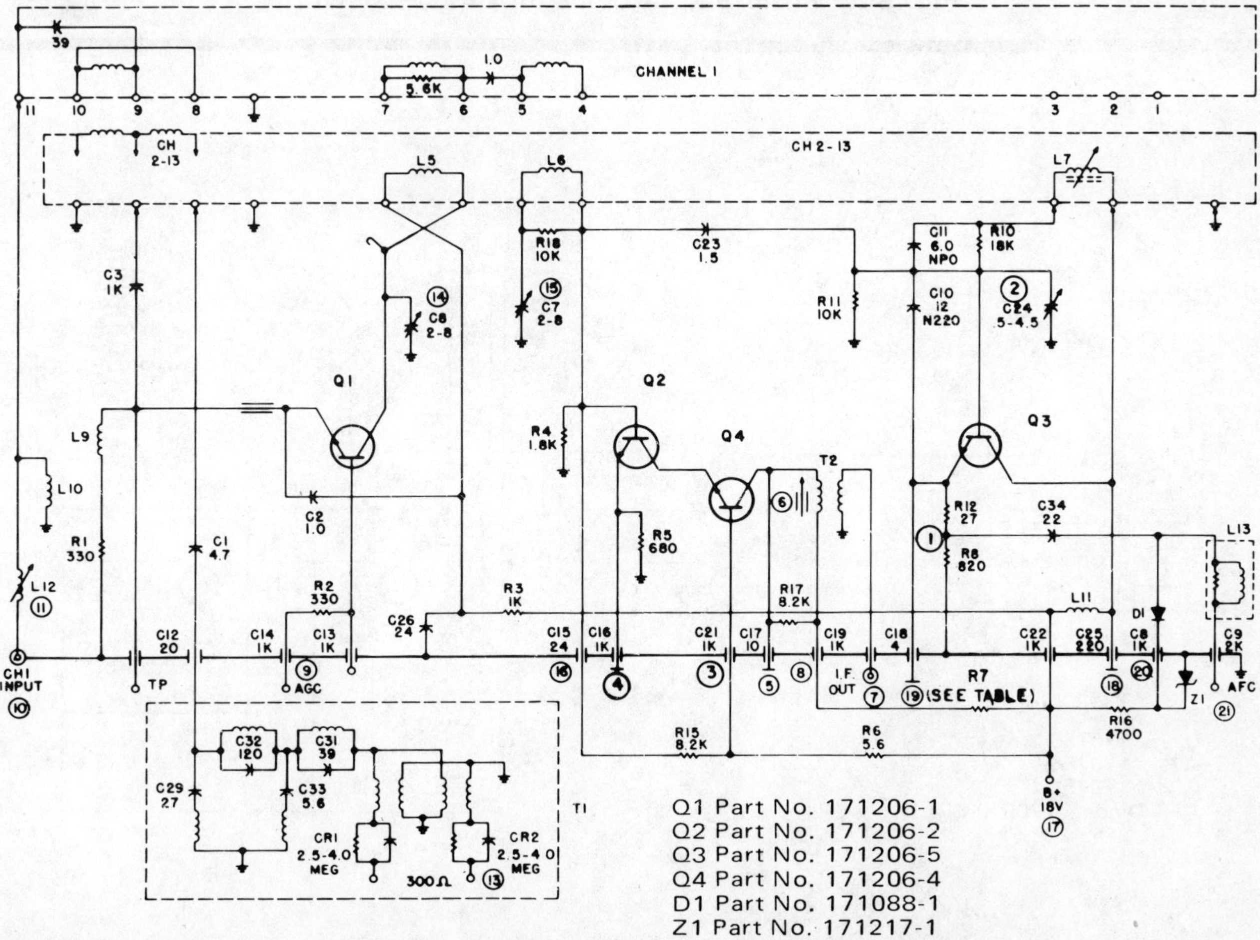

Fig. 6-25. VHF tuner schematic—340186.

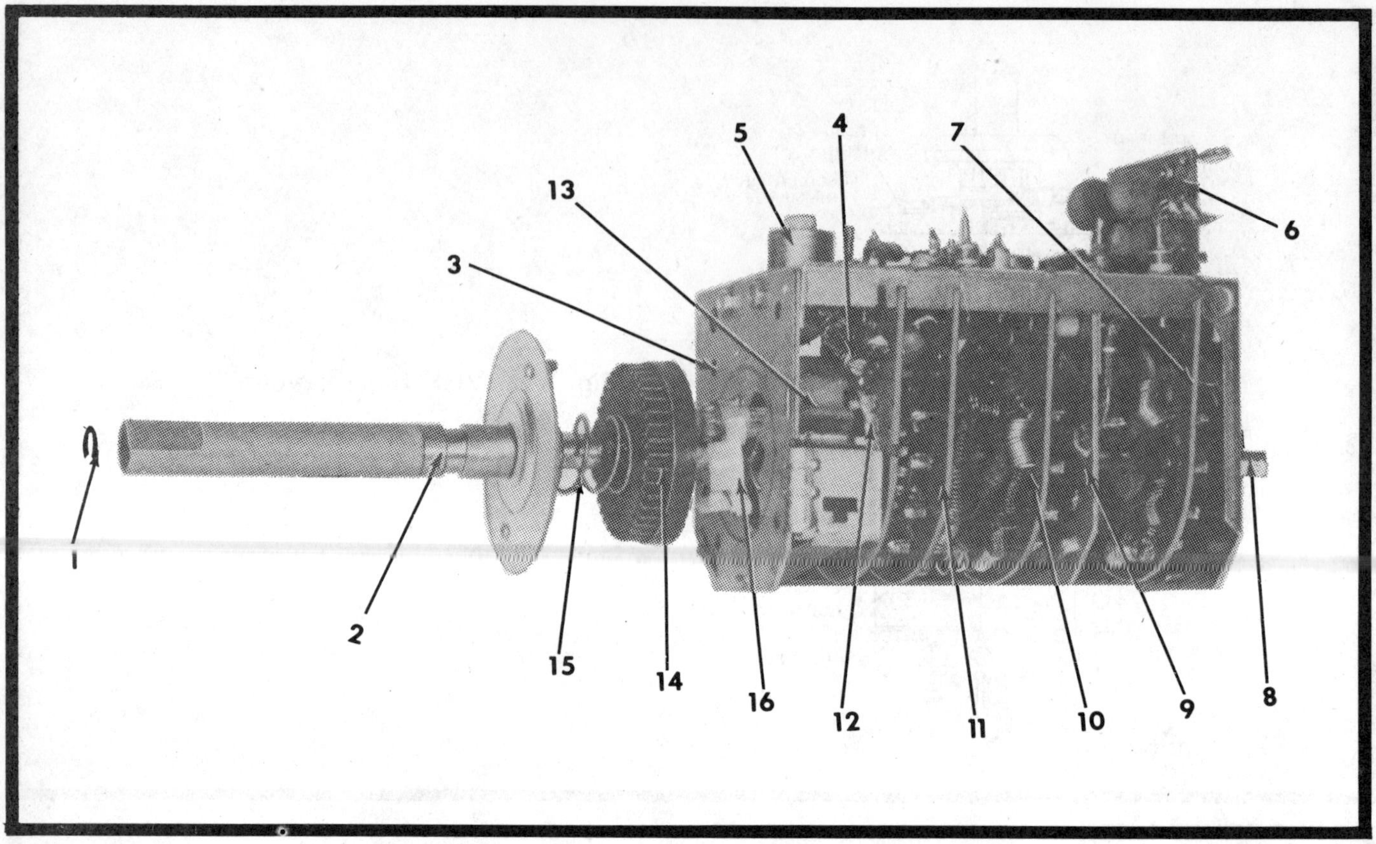

Fig. 6-26. Photo and part identification—340187.

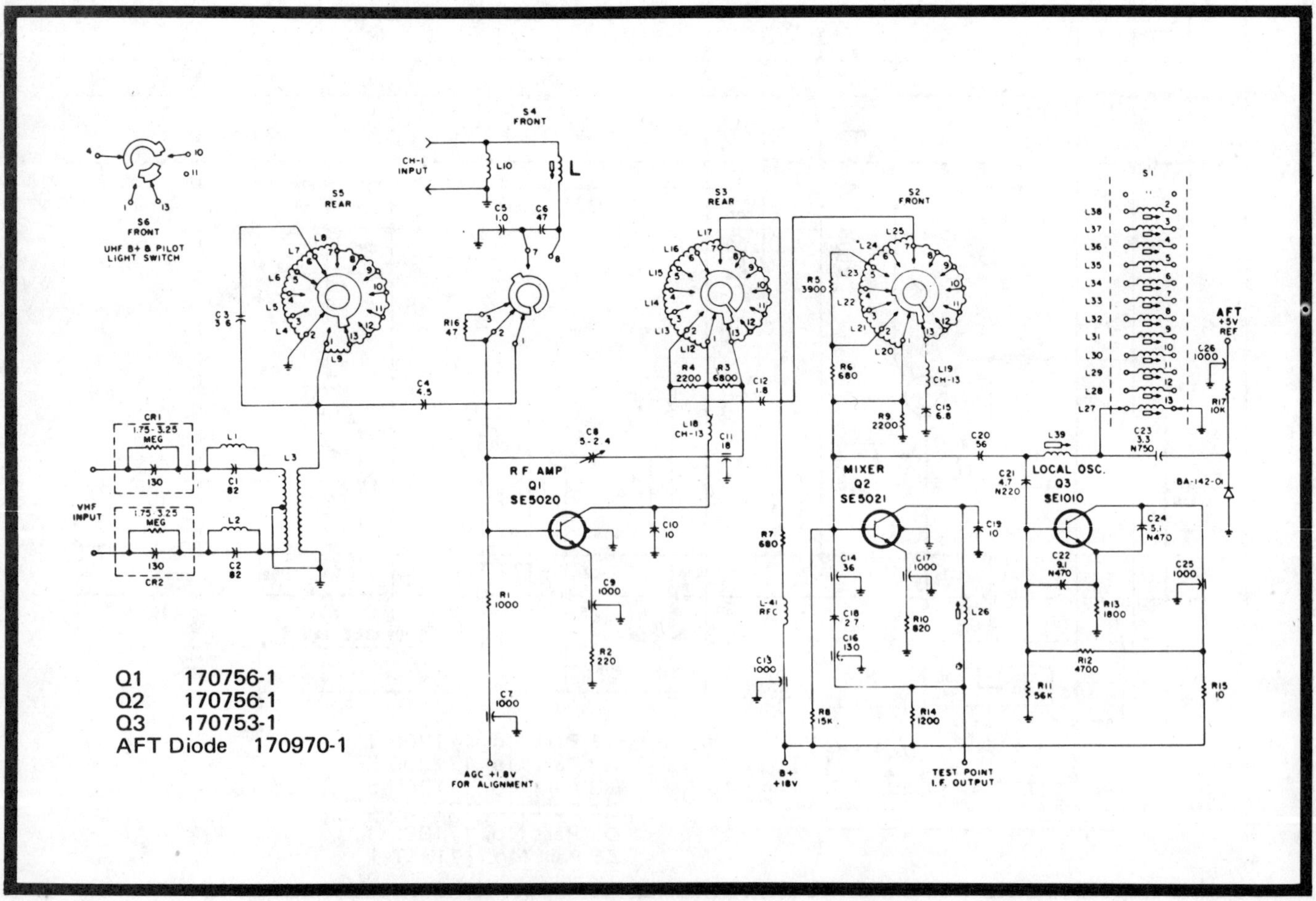

Fig. 6-27. VHF tuner schematic—340187.

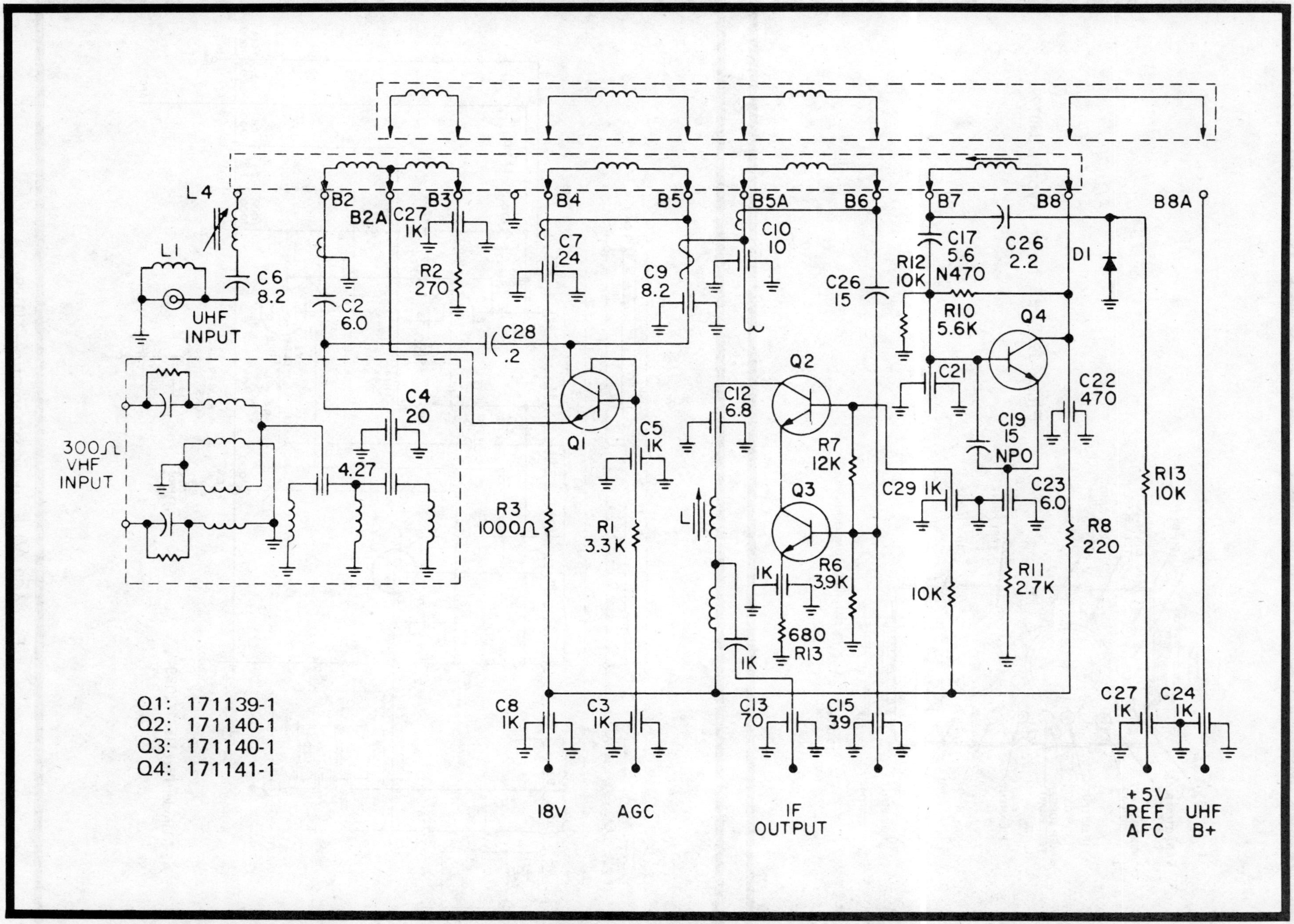

Fig. 6-28. VHF tuner schematic—340192.

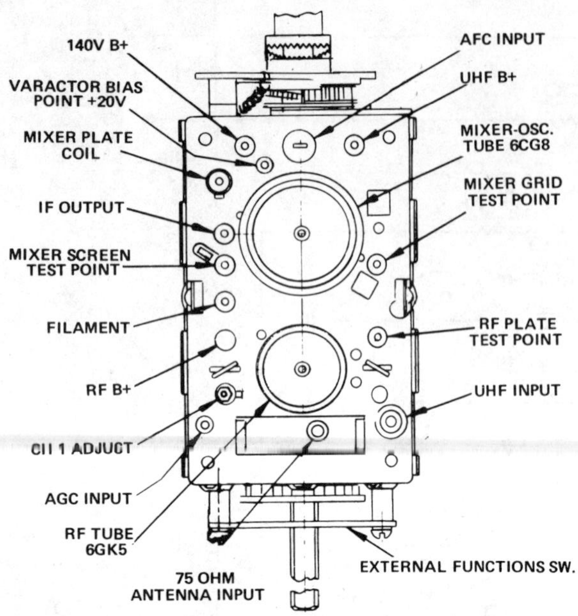

Fig. 6-29. VHF tuner layout—340199.

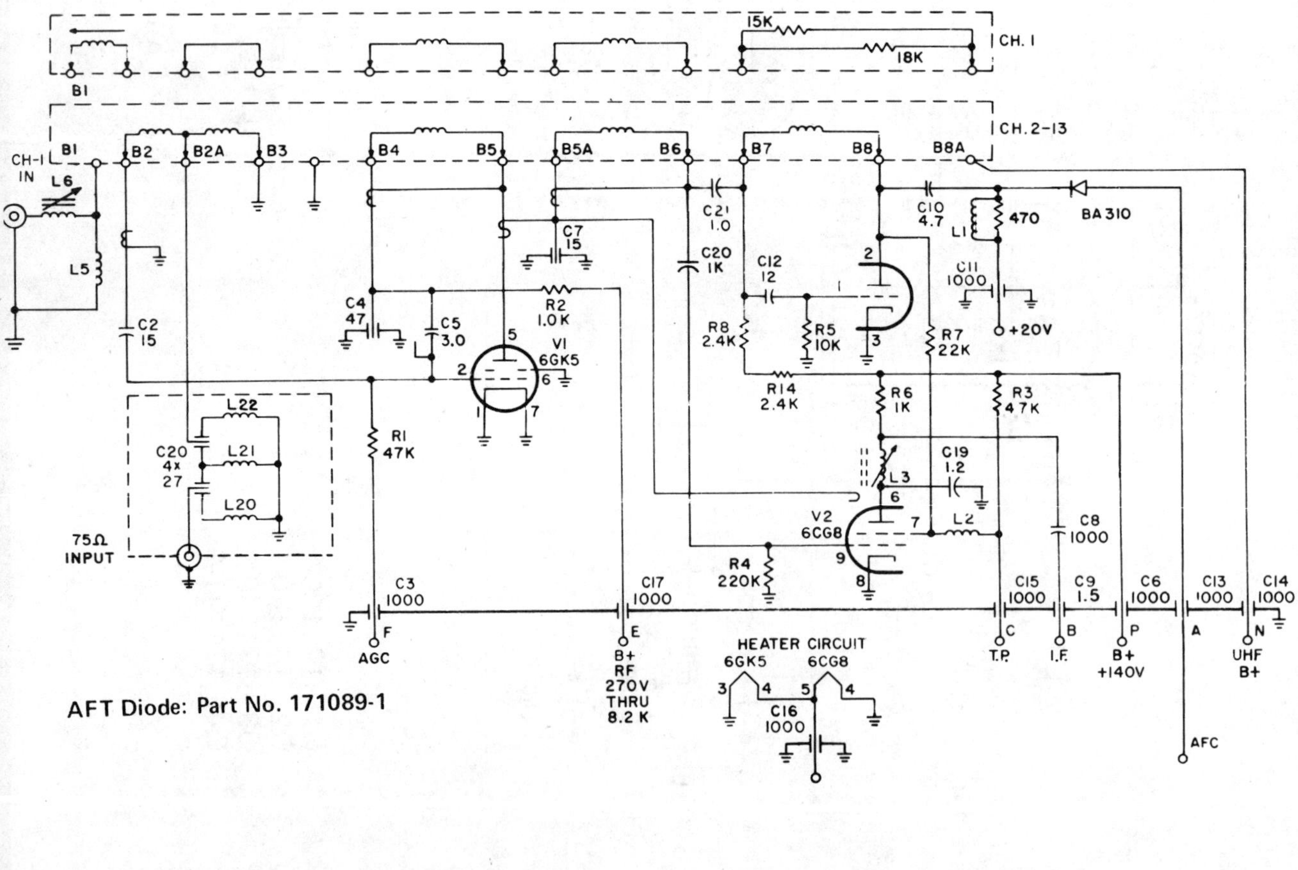

Fig. 6-30. VHF tuner schematic—340199.

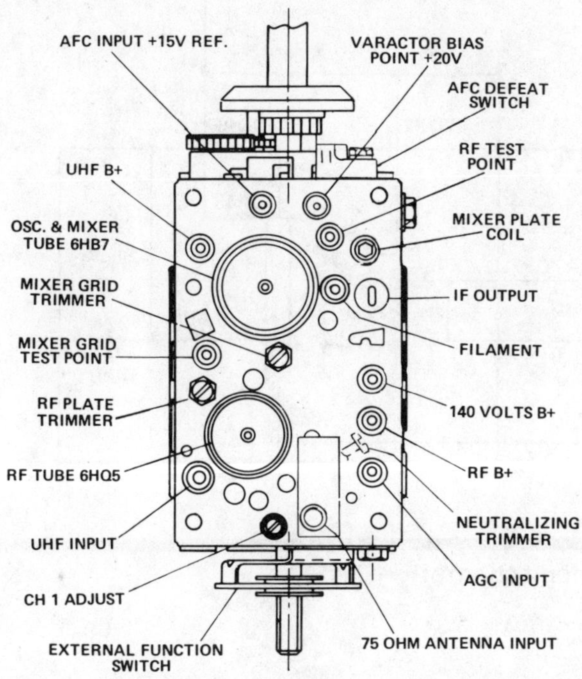

Fig. 6-31. Varactor hybrid tuner layout—340200.

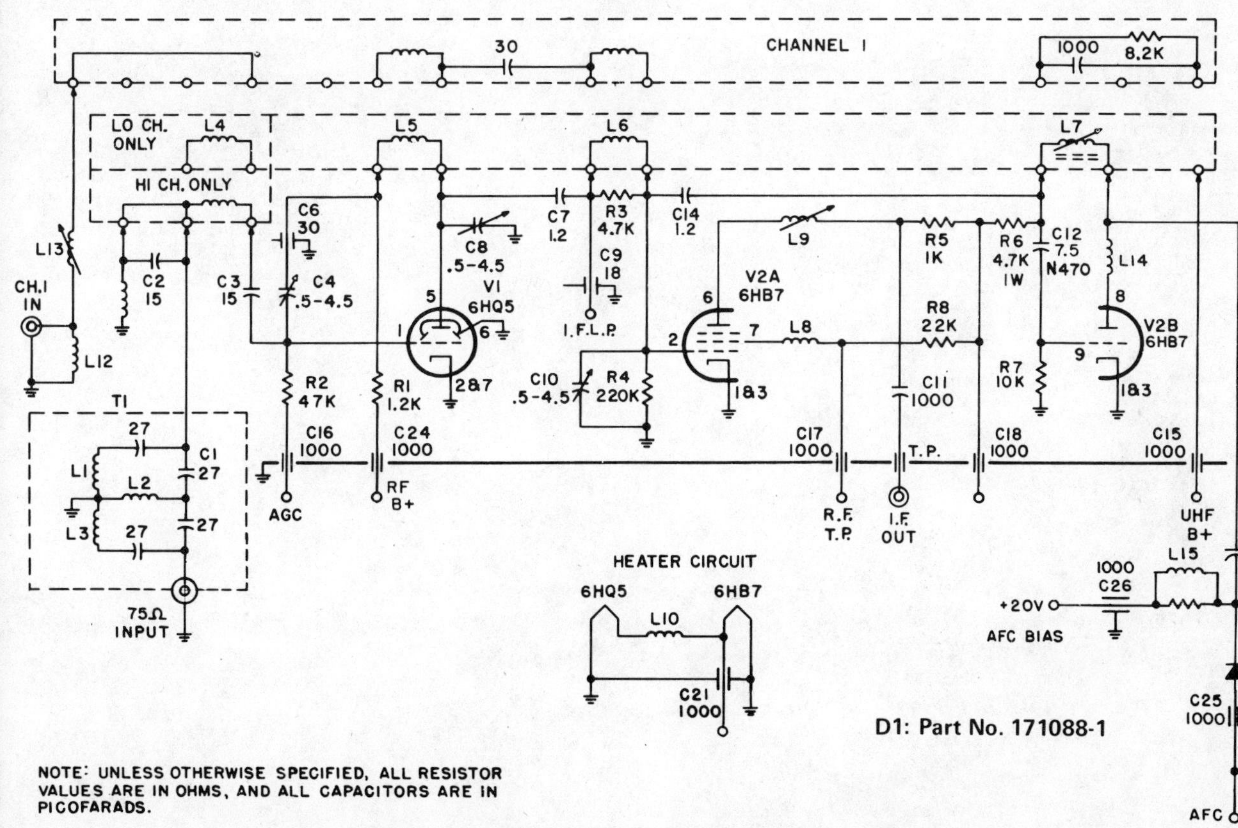

Fig. 6-32. Varactor hybrid tuner schematic—340200.

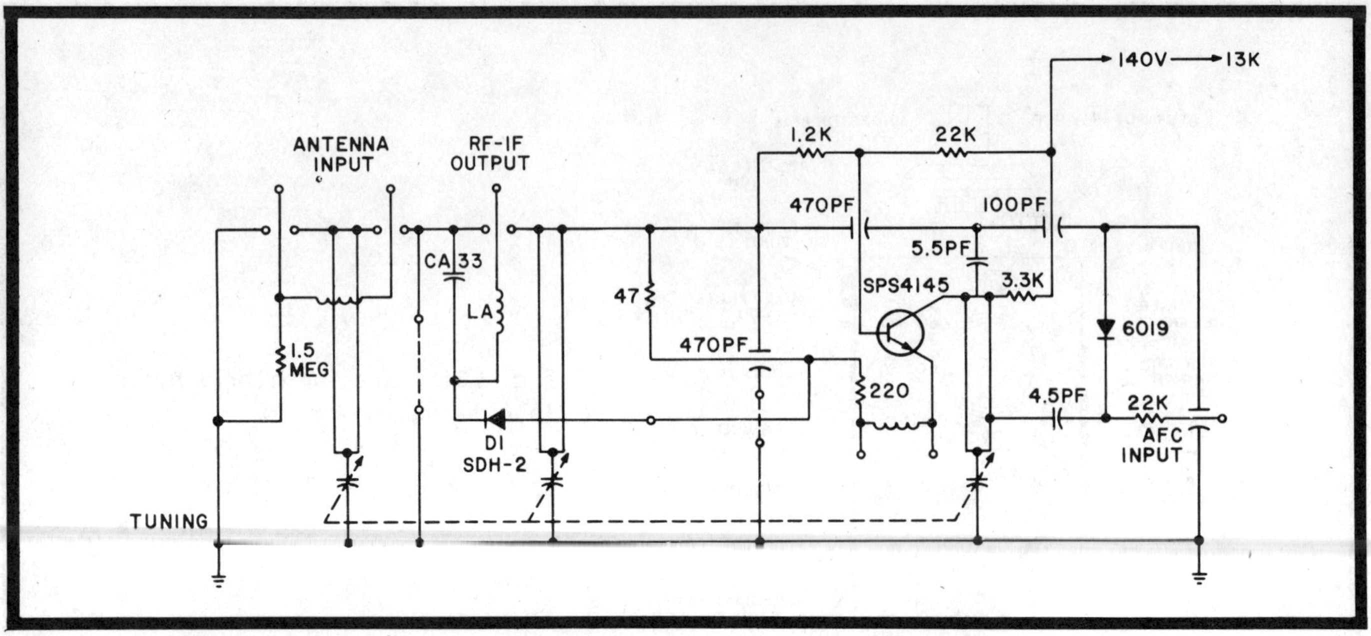

Fig. 6-33. UHF tuner schematic—340202.

Chapter 7

Motorola

In supplying the diagrams, charts, and alignment information for this chapter, Motorola service engineers have also generously provided some very useful service data that is generally applicable in the repair and maintenance of all tube-type Motorola UHF and VHF tuners. Except for light editing as required for general style and form, this material is presented herein as provided by Motorola.

VHF TUNER SERVICE INSTRUCTIONS

Due to the frequencies involved and the demand for exact lead and parts placement, the precaution should be taken to restore any circuit worked on to its original physical layout. Noting the placement of parts, the dressing of leads, and lead length is very important. A sketch noting layout and spacing is often helpful and will avoid trouble should you be called away before completing your job.

A check of the tuner alignment should be made after any repair work is done. Tuner shield should always be in place and secured when aligning or checking alignment of any tuner.

Servicing Wafers

The VHF tuner has been provided with removable wafers for the tuning sections. The major component parts comprising the actual tuning circuits are mounted to the appropriate wafer, and changing the wafer replaces most of the components that will affect the tuning.

To remove the tuner wafers, pull the channel selector shaft out the front of the tuner to the extent that the shaft clears the wafer to be replaced or removed for service.

Wafer Removal

1. Make note of the position of the wafers (in respect to tuner shaft) before removing tuner shaft. The small identifying notch in the shaft opening of each wafer and the 45-degree chamfer on the end of tuner shaft will aid in locating correct position.

2. Remove shaft tension spring located on the wafer shield.

3. Straighten the crimped end of the wafer retainer bar that protrudes through front plate assembly.

4. Remove the two hex screws that secure the front plate assembly to tuner chassis, and carefully slide tuner shaft (with front plate assembly) out of tuner.

5. Unsolder the wafer retainer bar from the shield and lift it off.

6. Unsolder all connections between the wafer being removed and the tuner base. Note dressing of leads and components **before lifting wafer out of tuner.**

Wafer Replacement

1. When replacing wafer, see that all wafer rotor contacts are on same channel. The small identifying notch in the shaft opening of each wafer will aid in locating positions.

2. Replace wafer in correct position; carefully make all necessary solder connections to the tuner base.

3. Determine correct rotational position of tuning shaft and insert shaft through wafers.

PREALIGNMENT INSTRUCTIONS

Alignment of a tuner is very seldom needed. Unless unmistakable evidence of tampering is found, it should be the last resort. If alignment is attempted on a tuner in which a faulty component exists, successful alignment will have to be repeated when the real cause of the trouble is finally corrected.

Prealignment tests of the tuner should include voltage and resistance measurements, routine checks for bad soldering connections and contacts, and visual inspection of circuits for overheated components and obvious wiring defects.

1. Check tubes. Use tube substitution method. Tubes may be changed in most cases without realignment, but if other than Motorola tubes are used, realignment may be necessary.

2. Check tuner voltages. Use tuner schematic (Fig. 7-1).

3. Check mechanical troubles.

4. Check response curve (Fig. 7-2). Response curves are the quality indicators of tuner performance and must be checked to insure acceptable performance.

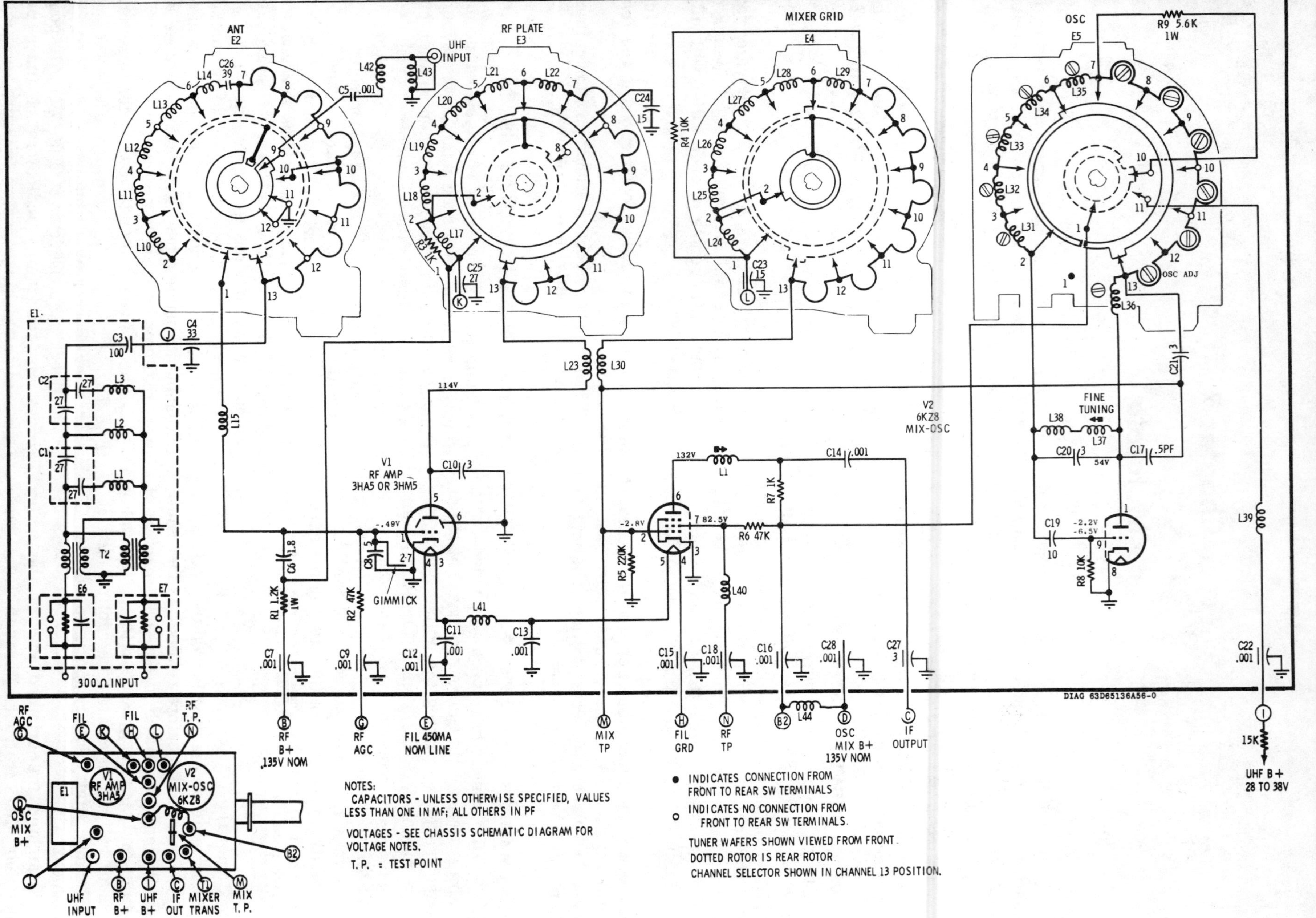

Fig. 7-1. VHF tuner TT-393Y (Motorola).

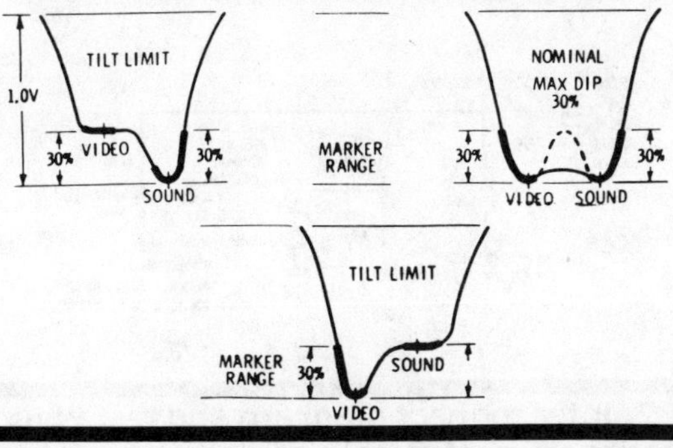

Fig. 7-2. Alignment response curves (Motorola).

VHF TUNER CMTT-393Y ALIGNMENT

Before alignment is attempted, make certain oscillator-to-mixer-grid injection voltage is adequate. Connect VTVM to mixer test point (M). See the tuner's respective schematic diagram for correct grid injection voltage. If the mixer injection voltage is excessively high or low, correct by replacing mixer-oscillator tube, if necessary. Check tuner B+, oscillator plate and grid voltage, and the components associated with the oscillator portion of the mixer-oscillator tube. Replace questionable components.

If necessary, remove tuner from bracket and reposition for better accessibility. Secure in this position. Make certain a good ground is provided, then connect a variable bias supply to tuner agc line of feedthrough capacitor on top of tuner, point (G). Set bias to —2V.

CAUTION

Do not remove tuner cover. Make coil adjustment through holes provided in tuner shield.

Connect sweep generator, properly terminated to antenna input terminals. See Fig. 7-3 for sweep generator termination network.

Maintain 0.5V peak-to-peak output on scope.

Refer to Fig. 7-2 for marker placement on response curve. For coil location, see the tuner's respective component location detail. See schematic diagram for detail of adjustment and test point (TP) locations.

If excessive hum is present on scope, distorting the response curve being viewed, connect an adequate filter capacitor from oscillator and rf B+ (point (B)) to chassis ground.

Disable horizontal output by unplugging the two yellow yoke leads to eliminate rf radiation interference. Connect a 47K voltage normalizing resistor.

1. Set tuner to channel 13. Refer to channel chart of Fig. 7-4 for sweep and marker generator settings for each channel. (The abbreviation "Mc" on chart refers to frequency, and is same as "MHz" or megahertz.)

2. Connect generator to antenna terminal. Set generator for 10 MHz, sweep width at center of channel 13 (213 MHz). Set marker generator as indicated in last column of Fig. 7-4.

NOTE: BE SURE TO TERMINATE GENERATOR AS SHOWN IN FIG. 7-3.

VHF/UHF TUNERS

Chan.	Ant. Coil	RF Plate Coil	Mixer Coil
6	L14	L22	L29
5	L13	L21	L28
4	L12	L20	L27
3	L11	L19	L26
2	L10	L18	L25

Spread or compress the antenna, RF plate and mixer coils of the respective channels for proper marker placement of video and sound marker on response curve as shown in Figure #3. See channel chart, Figure #2, for correct video and sound marker frequencies of the respective channels.

Fig. 7-4. Motorola low-band coil identification.

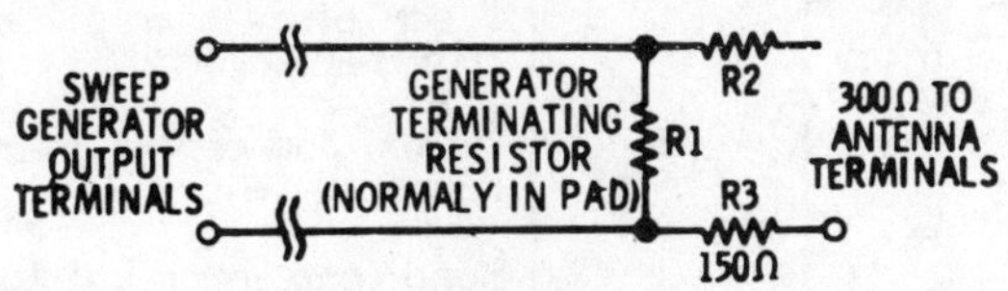

GEN. OUTPUT	R1	R2
50	50	125
75	75	115
150	150	75
300	Connect to ant. ter.	

Use the combination of resistors shown in table to match generator output impedance. Use nearest RMA values.

Fig. 7-3. Generator matching network for Motorola alignment.

Channel	Sweep Gen	MARKER GENERATOR	
		Sound Mc	Video Mc
2 (54-60 Mc)	57 Mc	59.75	55.25
3 (60-66 Mc)	63 Mc	65.75	61.25
4 (66-72 Mc)	69 Mc	71.75	67.25
5 (76-82 Mc)	79 Mc	81.75	77.25
6 (82-88 Mc)	85 Mc	87.75	83.25
7 (174-180 Mc)	177 Mc	179.75	175.25
8 (180-186 Mc)	183 Mc	185.75	181.25
9 (186-192 Mc)	189 Mc	191.75	187.25
10 (192-198 Mc)	195 Mc	197.75	193.25
11 (198-204 Mc)	201 Mc	203.75	199.25
12 (204-210 Mc)	207 Mc	209.75	205.25
13 (210-216 Mc)	213 Mc	215.75	211.25

Fig. 7-5. Channel chart (Motorola).

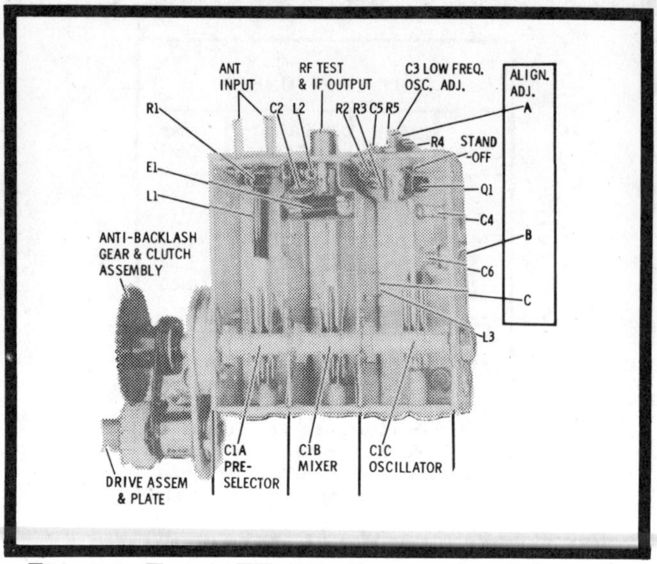

Fig. 7-6. Tuner TT-615 alignment points and part locations.

3. Connect scope through 47K resistor to rf test point (N) on top of tuner.

4. Adjust L23 and L30 to place video marker (211.25 MHz) on or near peak of response curve (Fig. 7-2); then place audio marker (215.75 MHz) on or near peak of response curve. If bandwidth of response is too narrow, adjust by decreasing the spacing of L23 and L30; this increases the coupling between the two inductances.

5. Set tuner to channel 7 and set generator to center frequency for this channel (177 MHz).

6. Set marker generator as required (see last column of Fig. 7-4).

7. Make sure scope is connected as in step 3, and touch up L23 and L30 as necessary to obtain proper marker placement (Fig. 7-2). The video frequency is 175.25 MHz and the audio frequency is 179.75 MHz.

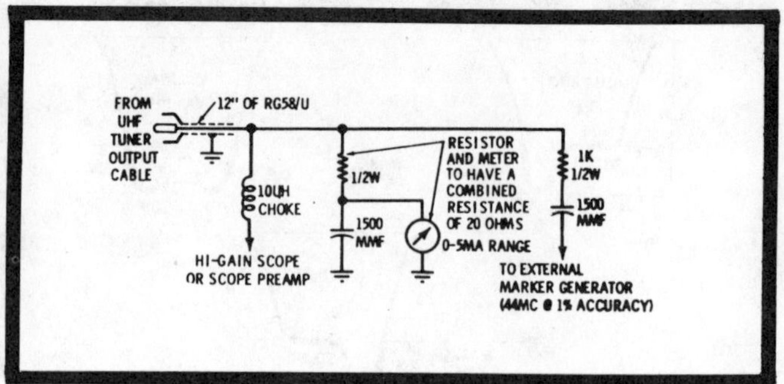

Fig. 7-7. Interconnect diagram and test equipment connections for Motorola alignment.

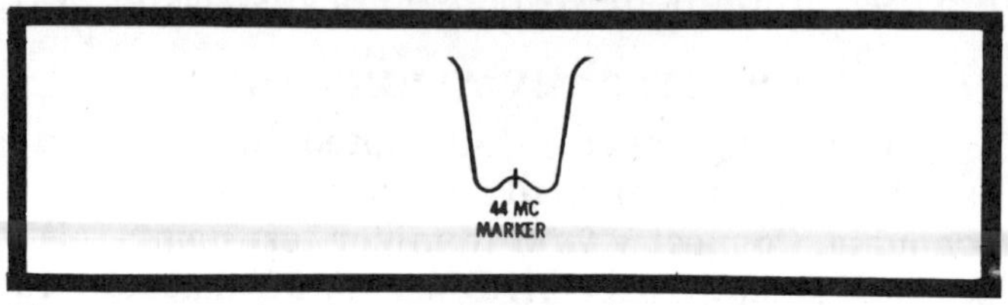

Fig. 7-9. Ideal bandpass response curve (Motorola).

NOTE: REPEAT STEPS 1 THROUGH 7 AS OFTEN AS NECESSARY TO GET CHANNELS 7 AND 13 WITHIN THE LIMITS SHOWN. CHANNELS 8 THROUGH 12 WILL AUTOMATICALLY FALL WITHIN PRESCRIBED LIMITS.

Channels 2 through 6 may be aligned in the same manner as the high-band channels. Use the procedure described in the above steps, with these exceptions:

a. Set generator in each case to the center frequency of the channel being aligned. (Refer to channel chart of Fig. 7-4.)

b. Rather than L23 and L30, the adjustments should be the antenna coil, the rf plate coil, and the mixer coil. The identification of these adjustments is shown in the chart of Fig. 7-5.

STEP	TUNER KNOB DRIVE SHAFT	SWEEP GENERATOR	INDICATOR	ADJUST	ADJUST FOR AND/OR REMARKS
1.	Fully clockwise (tuning gang fully closed)	Sweep at 468 MHz and external marker set at 44 MHz	Scope at tuner output (Fig. 7-7)	A	Set marker to center of curve by adjusting low frequency oscillator trimmer screw.
2.	Fully counterclockwise	Sweep at 900 MHz		B	Adjust for center marker response (Fig. 7-9).
				C	Bend loop in or out for a reading of 0.75 to 2.25 mA as indicated on 0-5 mA meter (Fig. 7-7).
3.	Turn slowly clock wise across entire band to stop	Track with tuner	0-5 mA meter	C	Bend loop until crystal current is within limits of 0.75 to 3.5 mA across band.

IMPORTANT NOTE: It is NOT recommended that any additional adjustments be made other than those given below, otherwise, proper alignment may be impossible. Remove tuner cover but not oscillator cover; adjustments B & C are accessible through openings in oscillator cover.

Fig. 7-8. UHF tuner alignment procedure. (Be sure to read preliminary instructions in text before performing alignment on Motorola tuners.)

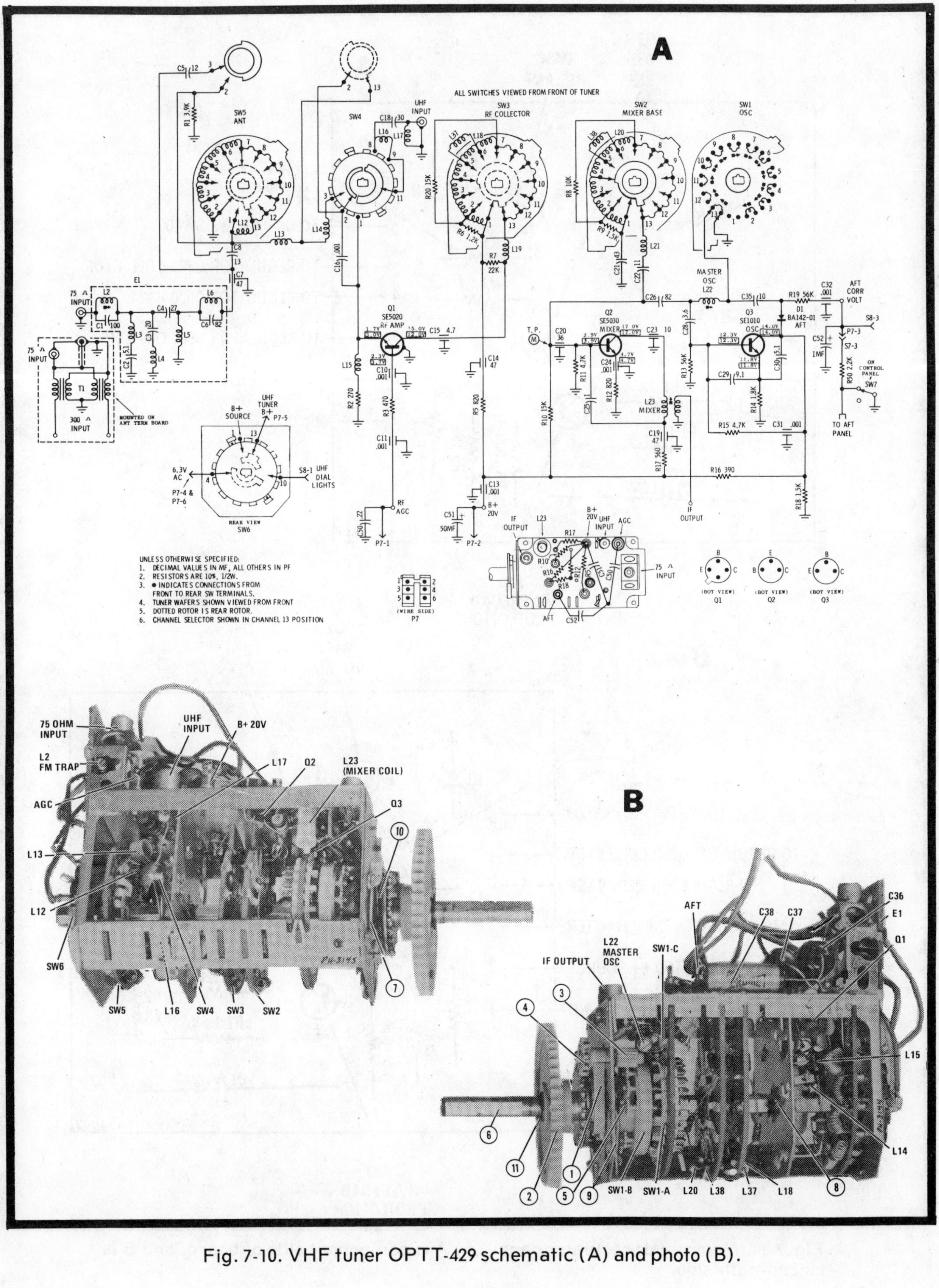

Fig. 7-10. VHF tuner OPTT-429 schematic (A) and photo (B).

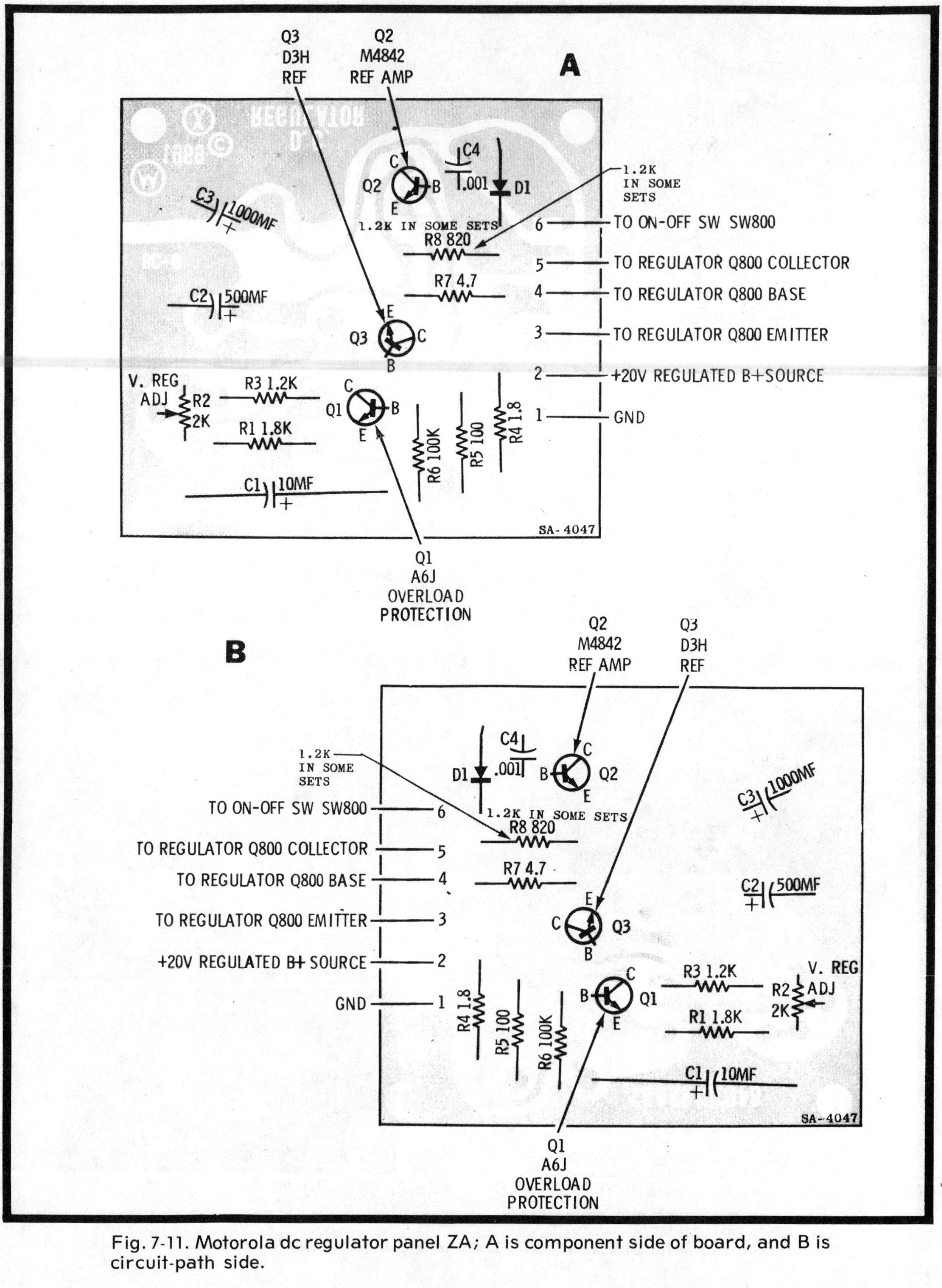

Fig. 7-11. Motorola dc regulator panel ZA; A is component side of board, and B is circuit-path side.

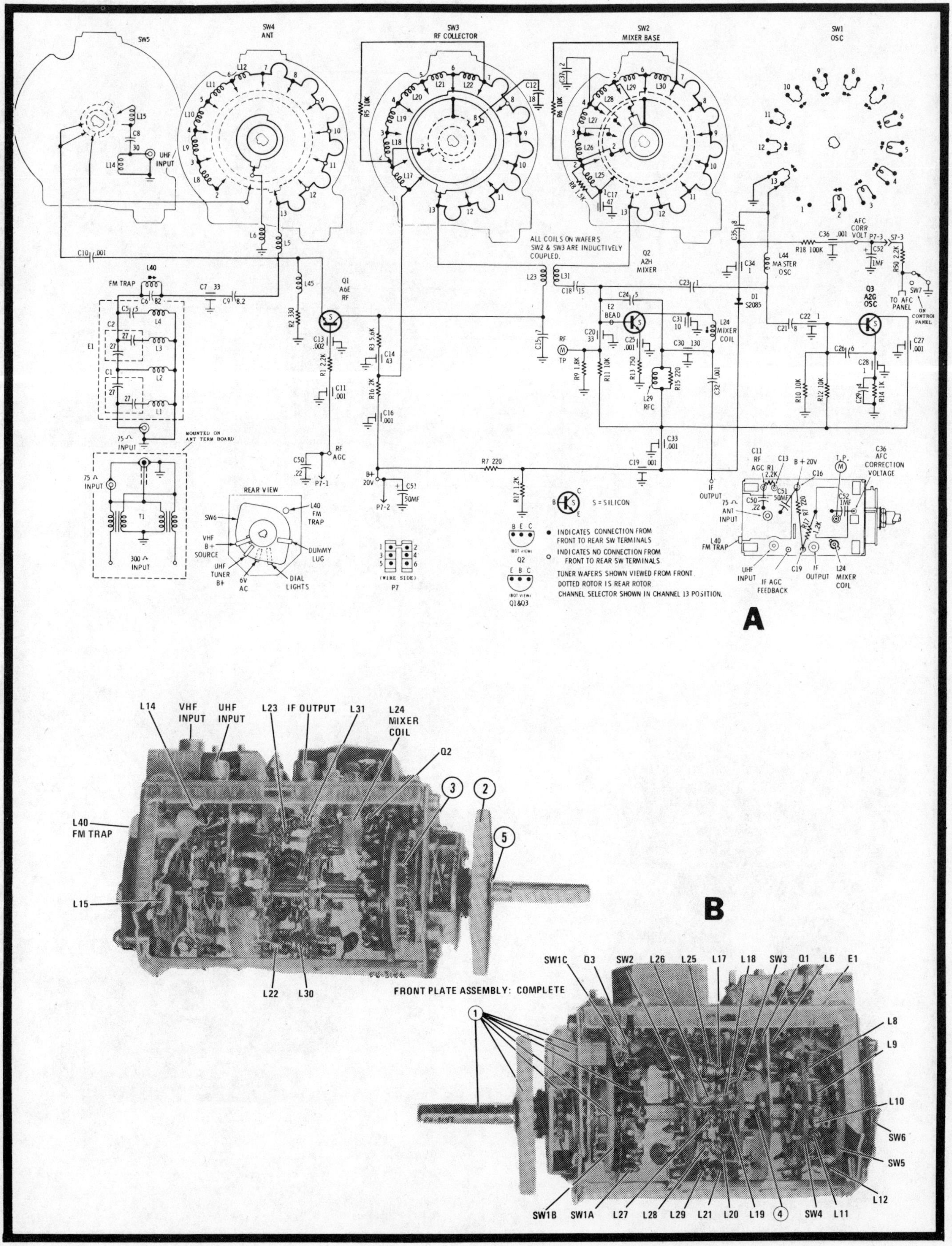

Fig. 7-12. VHF tuner OPTT-430 schematic (A) and photos (B).

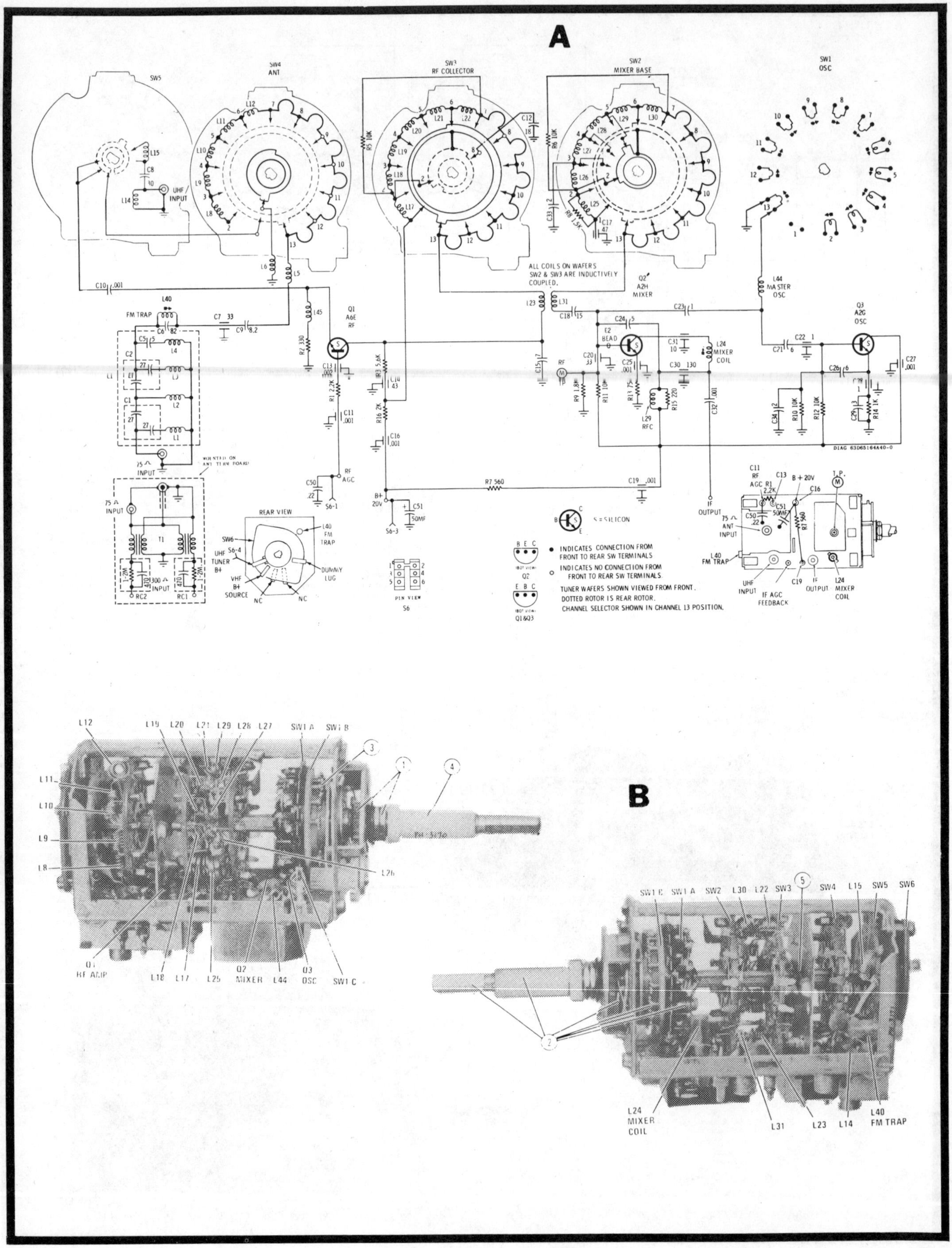

Fig. 7-13. VHF tuner CPTT-424 schematic (A) and photos (B).

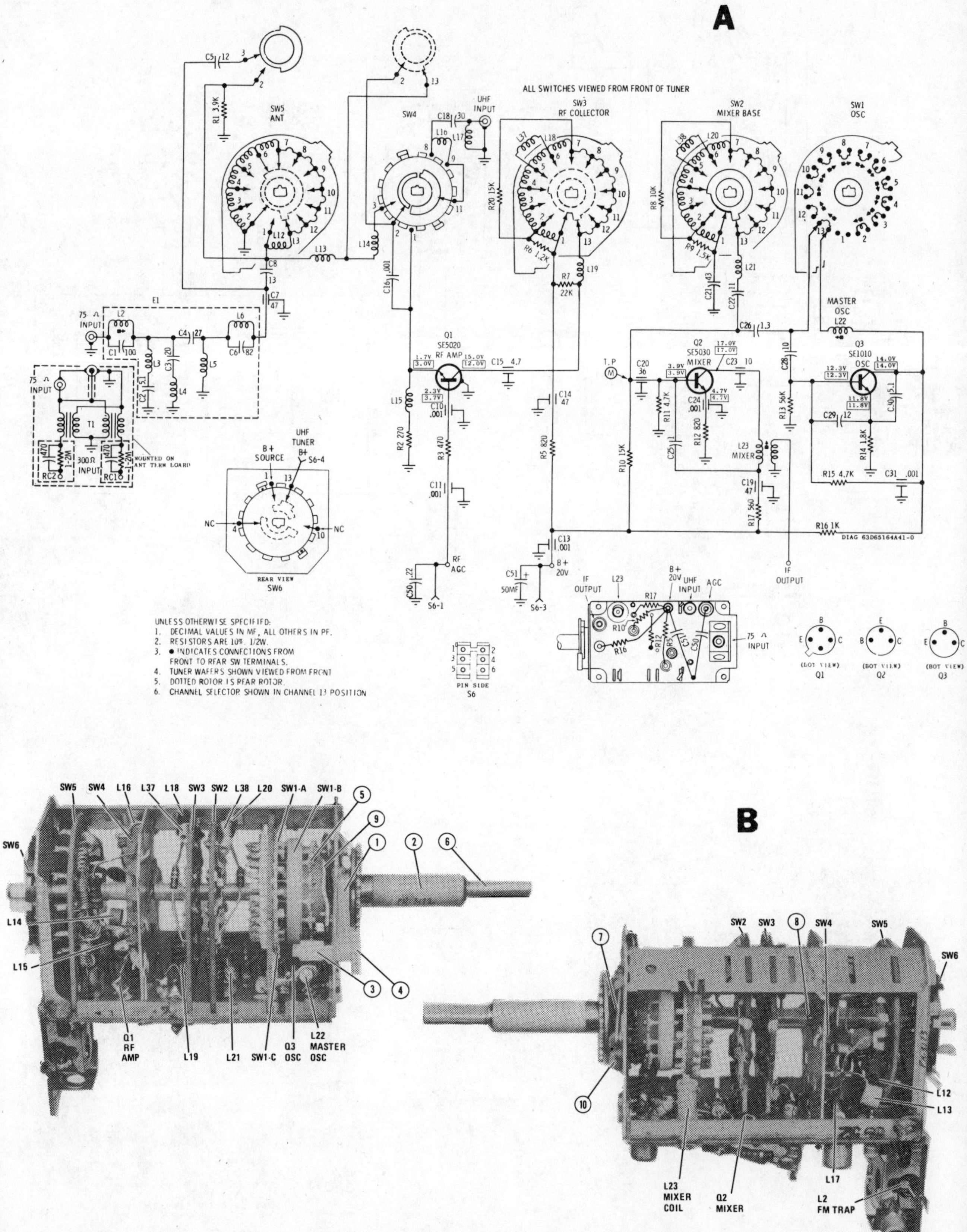

Fig. 7-14. VHF tuner CPTT-425 schematic (A) and photos (B).

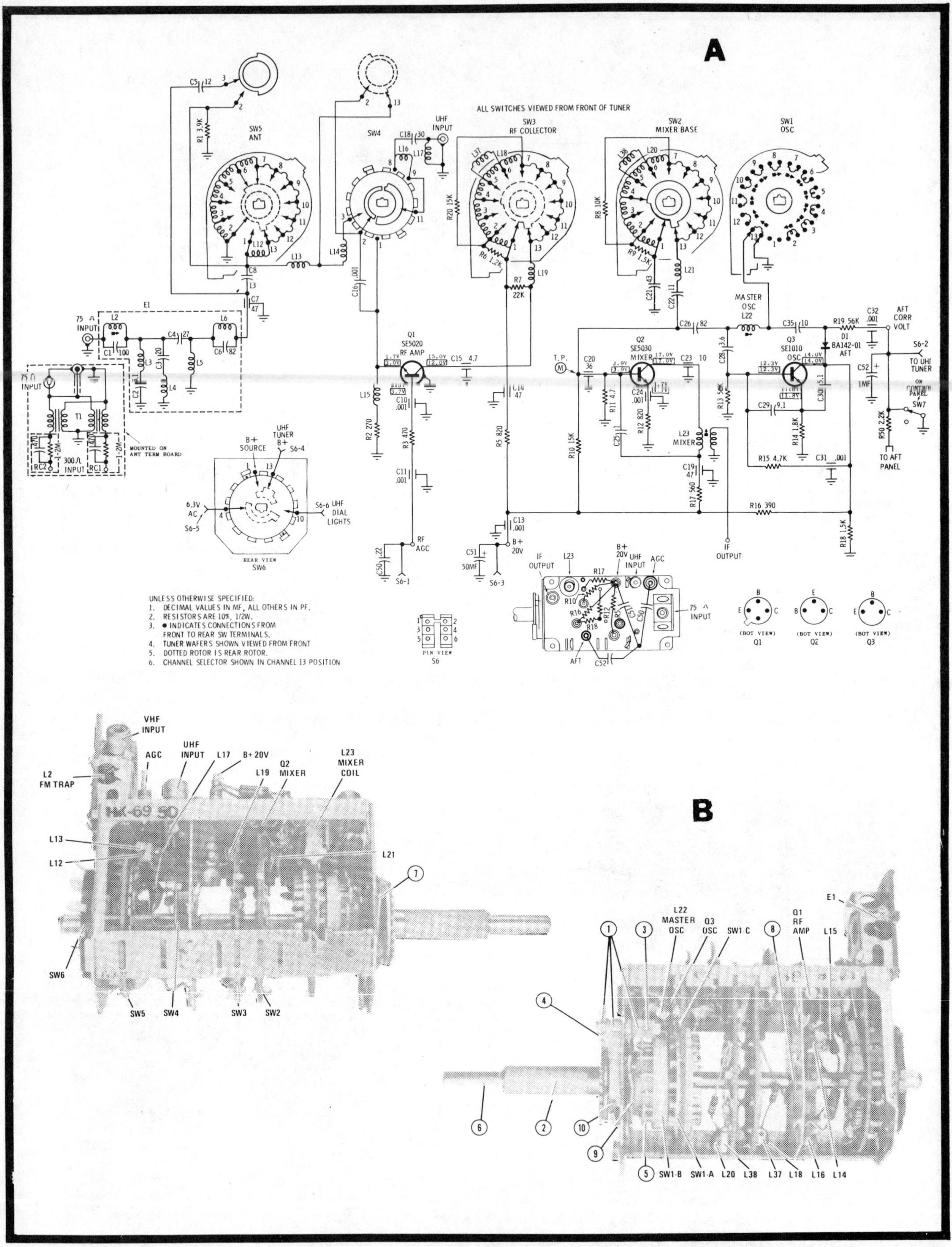

Fig. 7-15. VHF tuner CPTT-429 schematic (A) and photos (B).

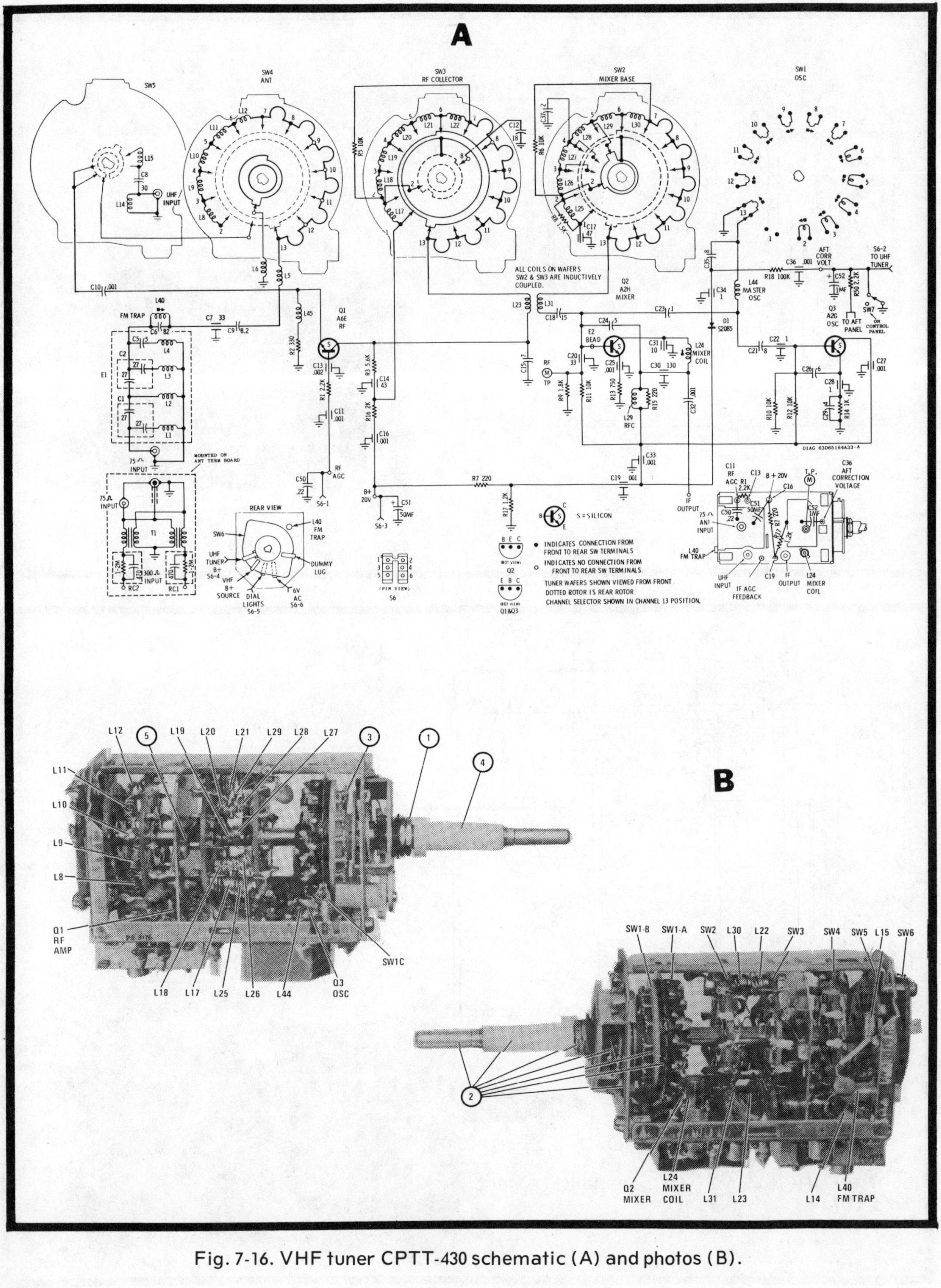

Fig. 7-16. VHF tuner CPTT-430 schematic (A) and photos (B).

A

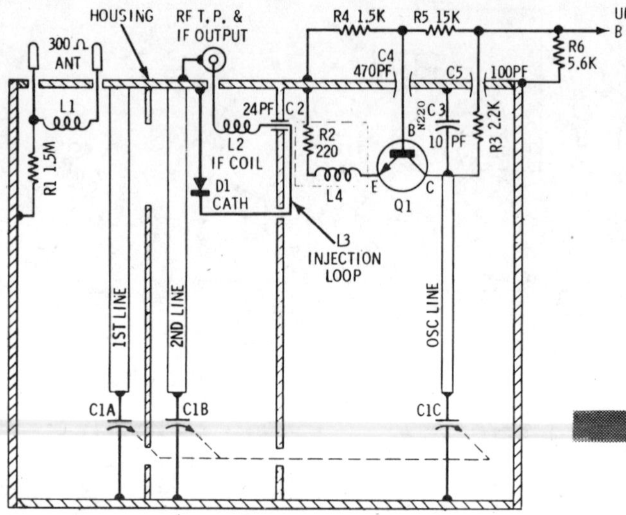

Fig. 7-17. UHF tuner TT-640 schematic (A) and photo(B).

B

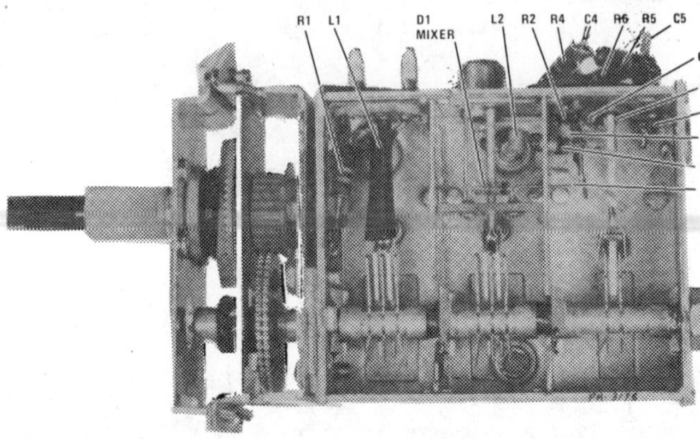

A

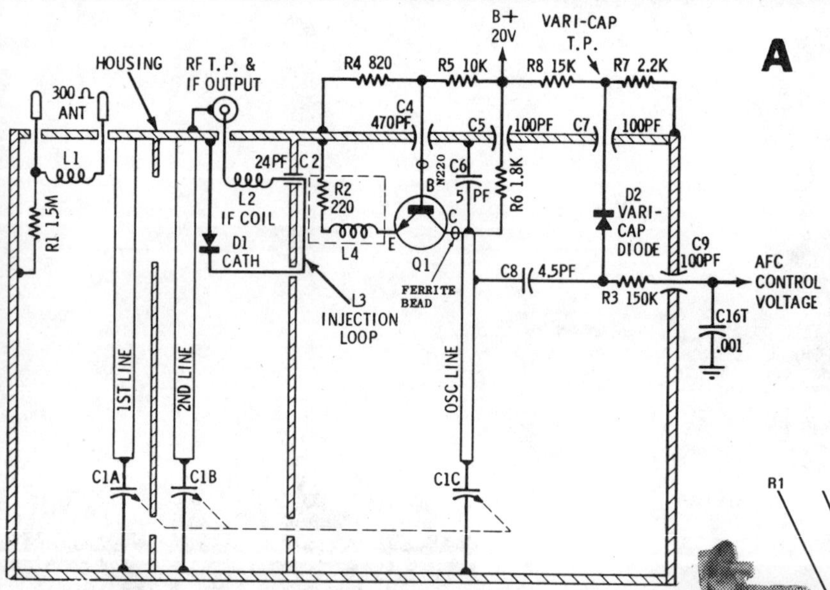

B

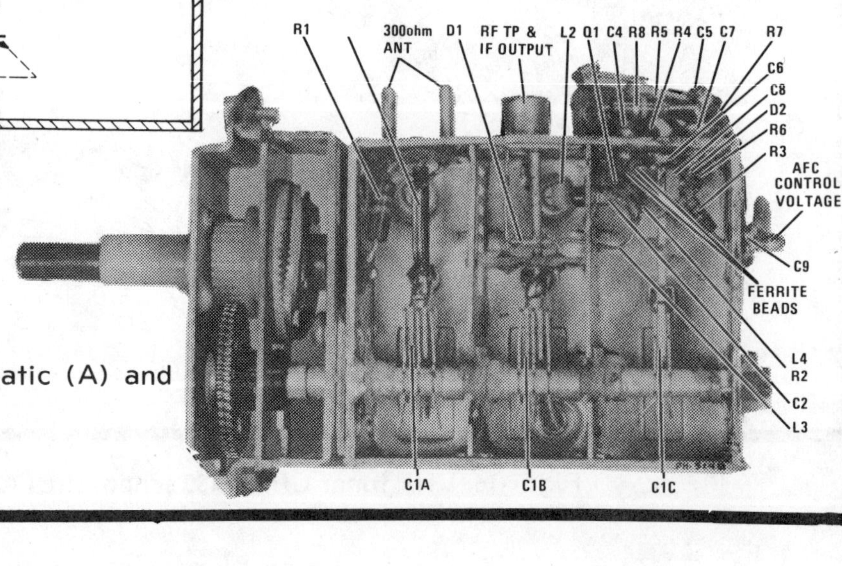

Fig. 7-18. UHF tuner TT-645 schematic (A) and photo (B).

A

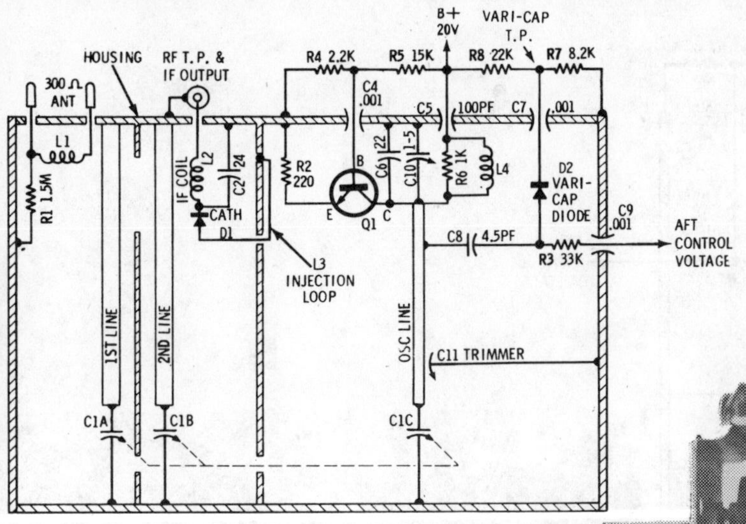

B

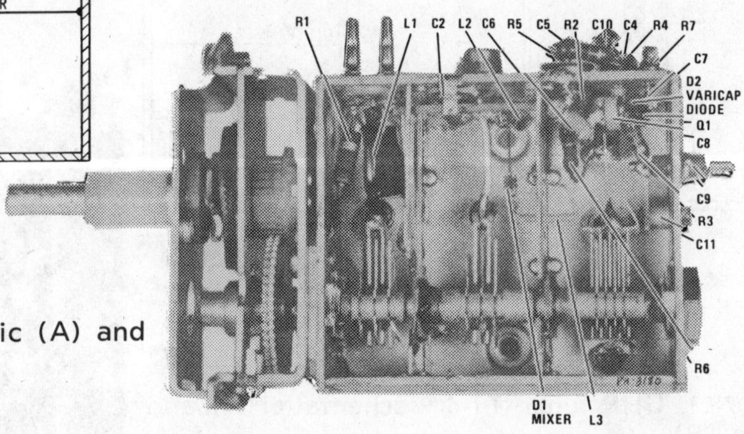

Fig. 7-19. UHF tuner TT-646 schematic (A) and photo (B).

A

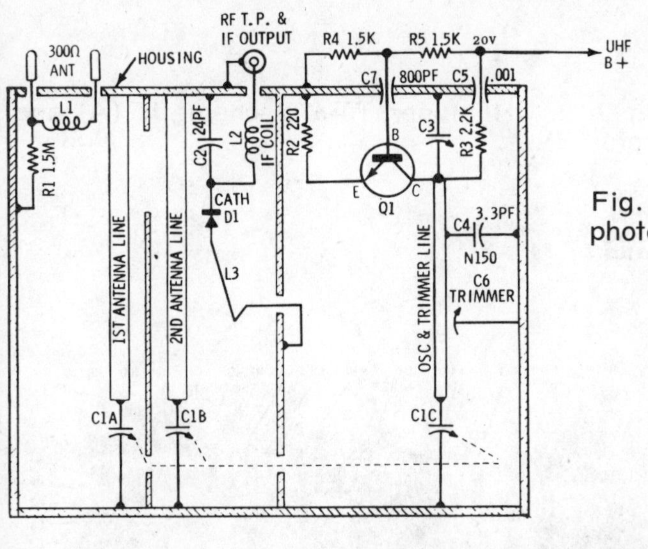

Fig. 7-20. UHF tuner TT-647 schematic (A) and photo (B).

B

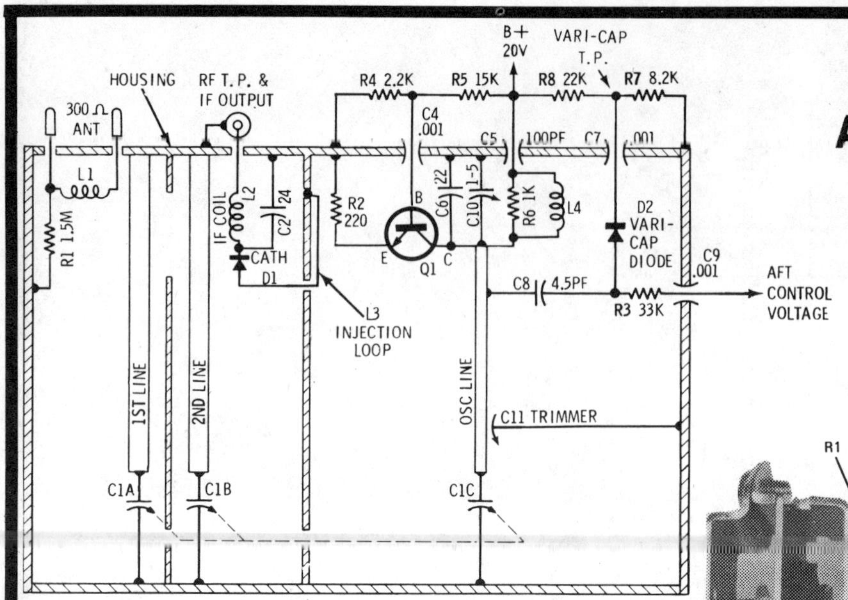

Fig. 7-21. UHF tuner TT-649 schematic (A) and photo (B).

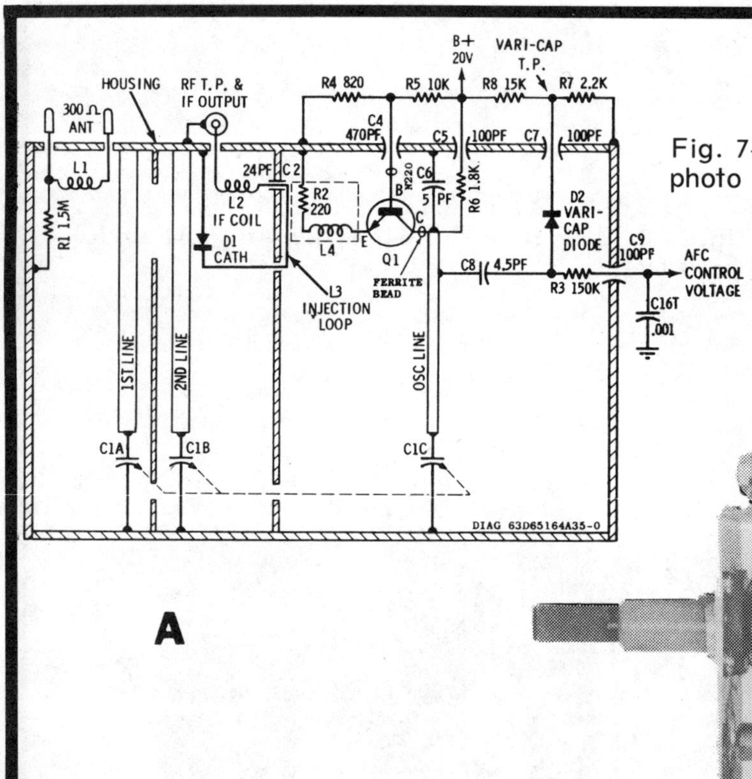

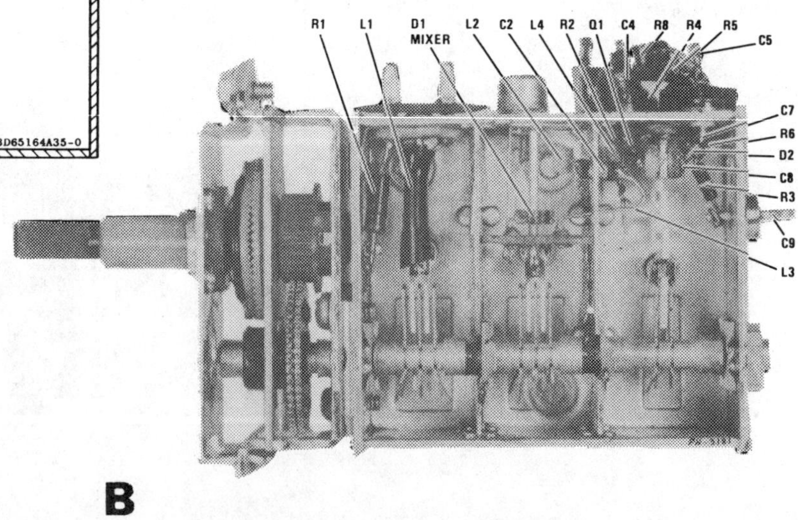

Fig. 7-22. UHF tuner TT-650 schematic (A) and photo (B).

Chapter 8

Philco-Ford

The service and alignment procedures presented herein are generally applicable to all Philco-Ford sets, even though the tuners covered in this chapter specifically are:

TT151D	TT182AD
TT152C	TT183AH
TT152CD	TT192
TT244	TT700
TT180C	

Complete alignment data for the Model 76-14099-5 tuner (TT180C) is provided. While this information may not be applicable to all other tuners, the basic procedure may serve as a guide for aligning all comparable units.

The information in this chapter has been provided through the courtesy of Philco-Ford.

While it is simpler to include diagrams, schematics, photos, and other figures as they are referenced in the text, such an approach doesn't always make it easier to find a particular diagram you might be interested in. The following table will help you quickly find the schematics or part layout drawings you need in servicing Philco sets. Find the specific tuner of interest in the left columns, then read horizontally to locate the figure numbers in this chapter that contain the information you need for servicing. In the first column, a designator (U or V) has been added to identify the tuner as being either UHF or VHF.

VHF BANDPASS ALIGNMENT (TT180C)

It is recommended that the tuner be removed from the cabinet for proper alignment; this is to provide greater accessibility to certain of the rf padding adjustments. The tuner shield must be on during final alignment check. The larger portion of shield cover may be removed during alignment procedure; but to facilitate alignment, it is advisable to use an "adjustment" cover with additional holes to allow access to the adjustments indicated below.

1. **Rf sweep marker gen**: Connect to antenna input through an antenna-input matching network (maximum required output of generator 20 mV p-p).

2. **Oscilloscope**: Connect high-gain scope to C16T (mixer base test point) through rf detector (Fig. 8-1). Keep body of detector away from feedthrough and ground towards right front corner of tuner. A scope with vertical deflection sensitivity of approximately 2.5 mV per inch is needed to view the response curve. This may require a preamplifier ahead of the scope.

Table 8-1.

TYPE	TUNER	SCHEMATIC DIAGRAM	EXPLODED VIEW	PARTS LAYOUT	WIRING DIAGRAM	COMPANION TO
U	TT151D	Fig. 8-19		Fig. 8-18	Fig. 8-21	TT182AD
U	TT152C TT152CD	Fig. 8-15A Fig. 8-15B	Fig. 8-25	Fig. 8-20 Fig. 8-14		TT182AD TT183AH
U	TT244	Fig. 8-17B	Fig. 8-17A	Fig. 8-16	Fig. 8-16	TT192
V	TT180C	Fig. 8-8	Fig. 8-7	Fig. 8-2		76-14099-5U
V	TT182AD	Fig. 8-26	Fig. 8-21	Fig. 8-21	Fig. 8-21	TT151D
V	TT183AH	Fig. 8-24	Fig. 8-25	Fig. 8-25		TT152C(D)
V	TT192	Fig. 8-22	Fig. 8-23	Fig. 8-16	Fig. 8-16	TT244
VU	TT700	Fig. 8-27	Fig. 8-28	Fig. 8-29	Fig. 8-30 Fig. 8-31	

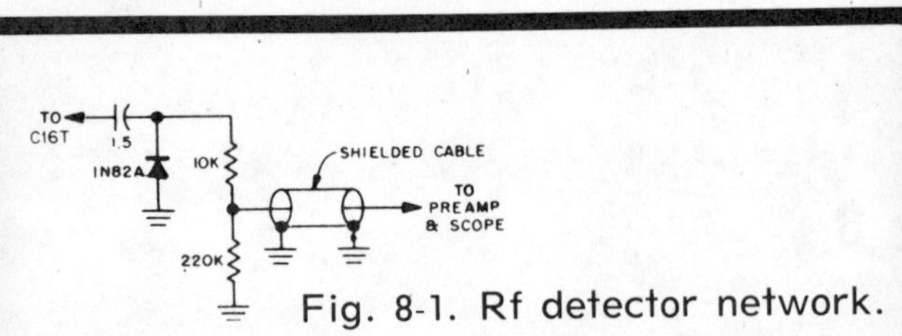

Fig. 8-1. Rf detector network.

3. Apply 1.4 to 1.5V dc bias to C10T (agc) for alignment and +20V dc to C9T for B+. Apply 6.5V dc to AFC feedthrough "A" and "B" (C29T and C35T). Batteries may be used to supply these voltages, the total current drain being only 14 mA.

To keep hum level at a minimum, use a large sheet of metal on which to connect all equipment grounds. The i-f coil (T2T) should not be touched during alignment. If moved, it will be necessary to readjust the overall i-f padding. To minimize distortion of the viewed rf curve due to i-f kickback, plug into the i-f output jack connector (J1T) jig shown in Fig. 8-2. The actual alignment steps are presented in tabular form in Fig. 8-3.

AUTOSHIFT TUNER DIAGRAMS

The exploded view of Fig. 8-7 shows both UHF (—5U) and VHF (—5V) tuners as they are integrated into the overall tuner assembly. The circled part numbers shown in the diagram may be referenced against like numbers for this assembly shown in the parts lists at the end of this book. The schematic applicable to the VHF tuner portion appears in Fig. 8-8. Figure 8-9 shows the UHF tuner.

COLOR AFPC ALIGNMENT (TT180C)

Set tint control to center of its range and turn killer threshold control fully counterclockwise. Be certain the horizontal deflection circuits are operating and the agc control is set properly.

Connect color bar generator to receiver antenna terminals. Adjust receiver for normal color reception.

Connect oscilloscope as prescribed.

Connect vacuum-tube (or other high-impedance) voltmeter in series with 470K ohm resistor to lug M40.

Perform alignment steps of Fig. 8-10.

OPERATIONAL CHECK AND ADJUSTMENTS (UHF)

Because of the simplicity of the electrical design of the UHF tuner (Fig. 8-9) the operational check and adjustments provided in troubleshooting are limited to the following.

If the picture is weak, snowy, or completely absent and the UHF tuner is suspected, the first check that may be made is to see if the diode mixer (1N82AG) is being adequately driven by the local oscillator. To do this, disconnect the i-f cable which joins the VHF and UHF units. Connect a milliammeter (0-5 or 0-10 mA scale) between jack J1U on the UHF tuner and ground. If the oscillator is supplying energy to the diode, the diode current reading should between 0.5 and 4 mA. Too little or too large a reading will result in picture degradation, especially if the current falls to very low values. If the current is not correct, the probability of trouble lies with the diode or transistor oscillator stage.

UHF Tuner 76-14099-5U

The oscillator supply voltage as measured at C1U (Fig. 8-13) should be approximately 12V with respect to chassis ground. The transistor voltages may be checked using a VTVM having an isolation resistor

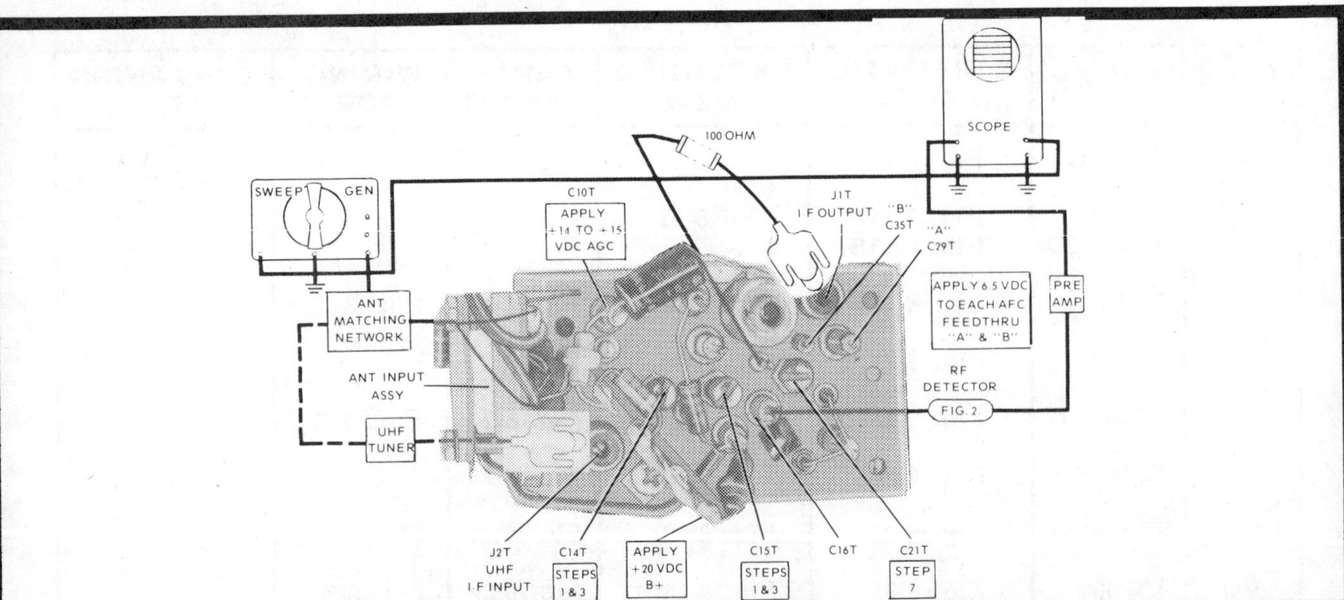

Fig. 8-2. Test equipment connections and top view of tuner.

STEP	SWEEP/MARKER GENERATOR: SWEEP DIAL SETTING	SWEEP/MARKER GENERATOR: MARKER DIAL SETTING	TUNER SET ON	PROCEDURE	REMARKS
1	Ch. 13 (213MHz with 10MHz sweep width)	Set first to 211.15 MHz and note position of marker on curve. Then set to 215.75MHz and note position of marker on curve	Ch. 13	Adjust C14T and C15T for a symmetrical bandpass (Fig. 8-4A).	Figures 8-5 and 8-6 should be referred to as necessary during alignment.
2	Ch. 7 (177MHz with 10MHz sweep width)	177MHz	Ch. 7	Reduce R-F stage gain to a minimum by reverse biasing to zero volts (C10T connected to ground). Apply maximum available signal from sweep generator to tuner and use maximum amount of scope gain consistent with hum level. Adjust L12T for a null point centered about 177MHz or for minimum output. (Refer to Fig. 8-5.)	Tuner R-F stage is neutralized by steps 1 and 2.
3	Ch. 13	211.25MHz and 215.75MHz settings required	Ch. 13	Apply 1.4 to 1.5V dc bias. Recheck and re-adjust Ch. 13, if necessary, to conform with Fig. 8-4A.	
4	Ch. 12, 11, 10, 9, 8 & 7	Ch. 12 - 205.25MHz - 209.75MHz Ch. 11 - 199.25MHz - 203.75MHz Ch. 10 - 193.25MHz - 197.75MHz Ch. 9 - 187.25MHz - 191.75MHz Ch. 8 - 181.25MHz - 185.75MHz Ch. 7 - 175.25MHz - 179.75MHz	Ch. 12 Ch. 11 Ch. 10 Ch. 9 Ch. 8 Ch. 7	View response curves on other high channels and see if they conform with limits set out by Fig. 8-4B. If Channel 7 thru 13 passbands are satisfactory, proceed with step 5.	If Channel 12 thru Ch. 7 still do not fall within limits of Fig 8-4B the coils on the strips L13T, L14T & L15T can be spiked or tweaked. (L13T for maximum amplitude and L14T & L15T for centering and tilt).
5	Ch. 6	83.25MHz and 87.75 MHz settings required	Ch. 6	Adjust Channel 6 strip coils L14T & L15T for a symmetrical bandpass with carriers set as shown in Fig. 8-4C.	This special alignment applies only to Ch. 6.
6	Ch. 5, 4, 3 and 2	Ch. 5 - 77.25MHz - 81.75MHz Ch. 4 - 67.25MHz - 71.75MHz Ch. 3 - 61.25MHz - 65.75MHz Ch. 2 - 55.25MHz - 59.75MHz	Ch. 5 Ch. 4 Ch. 3 Ch. 2	Channels 5, 4, 3 and 2 should have the shape described by Fig. 8-4B. If they do not conform to limits indicated, adjustment may be made by "tweaking" or "spiking" the appropriate coils on the strips (L13T, L14T & L15T).	L13T for maximum amplitude L14T & L15T for centering and tilt.
7	Ch. 13 (213MHz)	41.25MHz - 47.75 MHz	Ch. 13	Turn fine tuning CW 5 full turns, then CCW 2 full turns (to center fine tuning slug on Ch. 13 strip). Adjust C21T to correct frequency.	
8	Ch. 48 (677MHz)	41.25MHz and 45.75MHz	Ch. 1 (UHF I-F Ch.)	Ch. 48 Adjust T1T for symmetrical response as in Fig. 8-4D. (See Fig. 8-5 for location of adjustment.)	Adjustment of Channel 1 coils is ordinarily unnecessary. Valley and/or tilt should not exceed 35%. Either carrier should not drop down on the curve skirts more than 10% from peaks.

NOTE: Spiking or tweaking means increasing or decreasing the space between turns of the coil.

Fig. 8-3. VHF tuner bandpass alignment procedure.

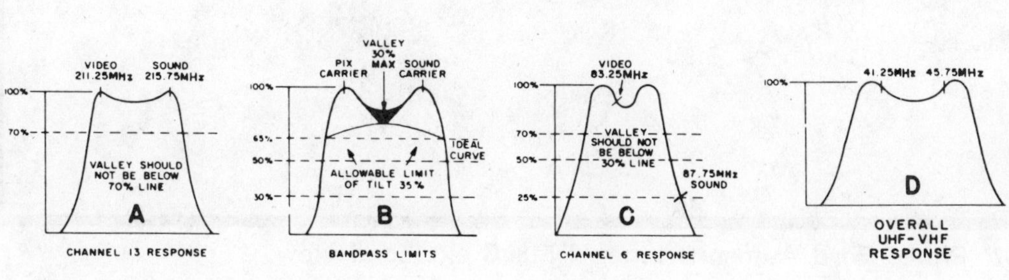

Fig. 8-4. Response curves for bandpass alignment procedure.

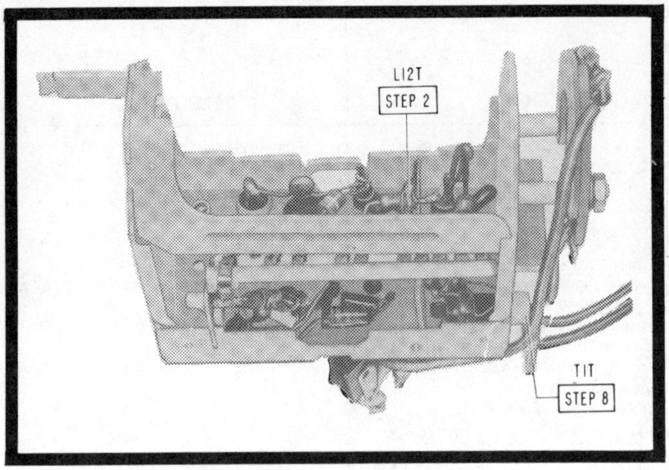

Fig. 8-5. Locations of adjustments to be made in steps 2 and 8 of alignment procedure.

(10K to 50K) attached directly to the probe end. The emitter-to-ground voltage should be 2.0 to 2.5V. Collector-to-ground voltage will be 12V dc with the probe connected to the oscillator tuning gang stator. If these voltages deviate widely from the foregoing values, the probable cause is a defective transistor. This may be replaced, but care should be taken to wire it in as closely as possible to its original position, lead dress, and lead length. If the transistor

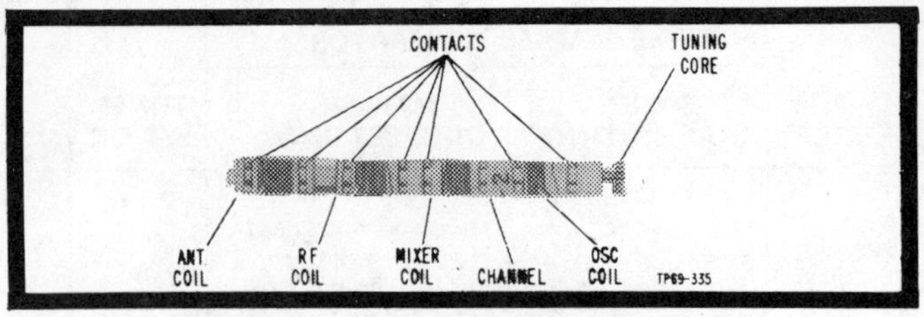

Fig. 8-6. Channel strip layout.

operating voltages seem normal, a defective diode is indicated. In changing the diode, the same wiring and lead dress precautions should be observed as for transistor replacement.

Replacing the transistor oscillator may upset the tracking of the tuner. If the desired channel is located near the low end of the board, between 14 and 45, the oscillator frequency can be readjusted by means of trim tab C9U. In this case, the tuner should be set to the desired channel and trim tab C9U (Fig. 8-13) adjusted accordingly. Channels at the high end of the band must be set by adjusting the rotor plate segments of C8U tuning capacitor.

UHF Tuner TT152CD

The oscillator supply voltage as measured at C1U (Fig. 8-14) should be approximately 12V with respect to chassis ground. The transistor voltages may be

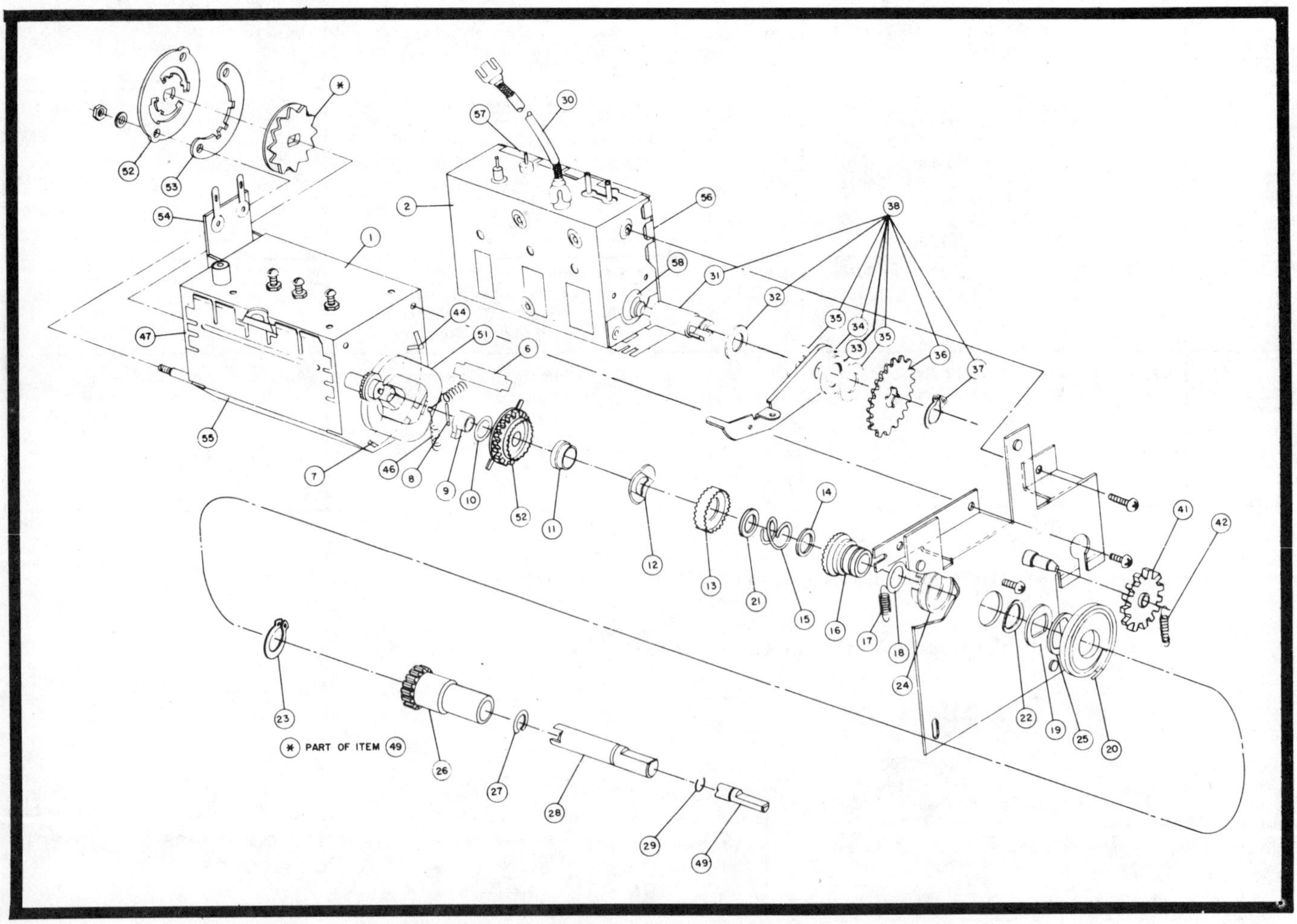

Fig. 8-7. Philco-Ford Autoshift tuner TT180C exploded view.

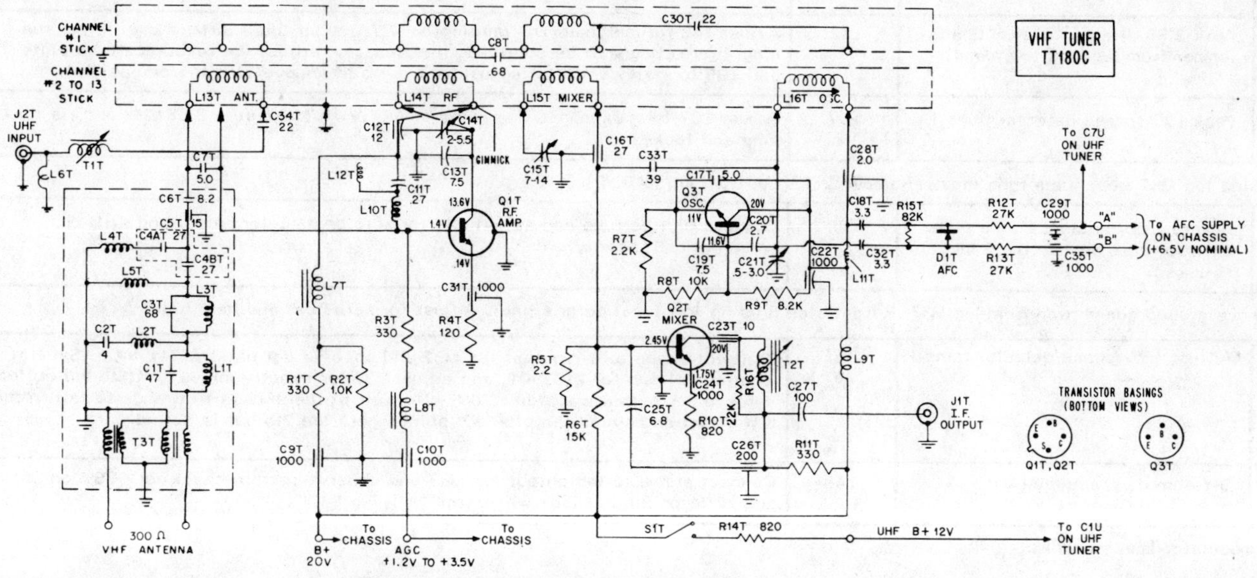

NOTES:
CHASSIS: 20QT75
LINE VOLTAGE: 120V.A.C.
ACTIVE FILTER: 20V.D.C.
TUNER POSITION: CH. # 1 (UHF)
VOLTAGE READINGS: B & K 175 V.T.V.M.
RESISTANCE READINGS: B & K 120 V.O.M.
(TAKEN IN CIRCUIT WITH POWER OFF)

* TRANSISTOR RESISTANCE CHART (USE X100 SCALE)

TRAN.	FUNCT.	E To GND. + −	B To GND. + −	C To GND. + −	C To E + − (− +)	C To B + − (− +)	B To E + − (− +)
Q1T	RF	120Ω	1700Ω	700Ω	800Ω (800Ω)	3.3K (1600Ω)	1700Ω (3K)
Q2T	MIX	820Ω	1650Ω	650Ω	1500Ω (1500Ω)	2.8K (1600Ω)	1700Ω (3K)
Q3T	OSC.	2.4K	2.2K	300Ω	1800Ω (2.7K)	5K (1850Ω)	1750Ω (7.5K)

* DETERMINE POLARITY OF VOLTAGE ACROSS TEST LEADS OF OHMMETER BEFORE TAKING RESISTANCE READINGS.

Fig. 8-8. Philco-Ford VHF tuner 76-14099-5V (TT180C) schematic.

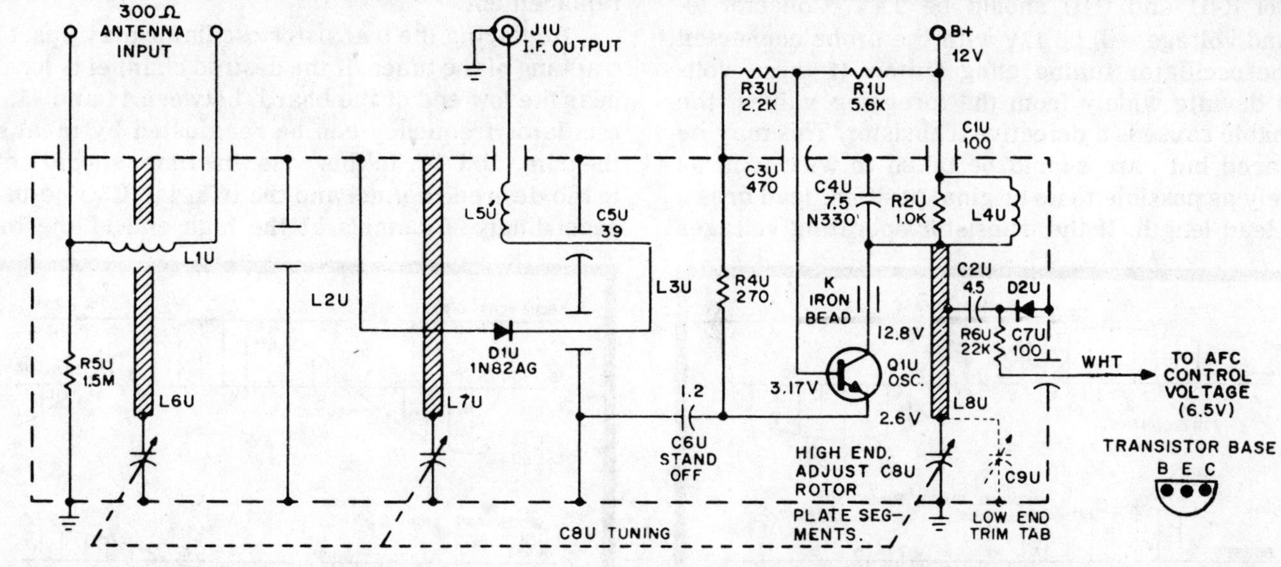

NOTES:
CHASSIS: 20QT75
LINE VOLTAGE: 120 V.A.C.
ACTIVE FILTER: 20 V.D.C.
TUNER POSITION: CH. #1 (UHF)
VOLTAGE READINGS: B&K 175 V.T.V.M.
RESISTANCE READINGS: B&K 120 V.O.M. (TAKEN IN CIRCUIT WITH POWER OFF)

*TRANSISTOR RESISTANCE CHART (USE X100 SCALE)

TRAN.	FUNCT.	E TO GND + −	B TO GND + −	C TO GND + −	C TO E + − (− +)	C TO B + − (− +)	B TO E + − (− +)
Q1U	OSC.	270Ω	1600Ω	950Ω	1250Ω (1250Ω)	2.2K (1600Ω)	1600Ω (2.1K)

*DETERMINE POLARITY OF VOLTAGE ACROSS TEST LEADS OF OHMMETER BEFORE TAKING RESISTANCE READINGS.

Fig. 8-9. Philco-Ford UHF tuner 76-14099-5U schematic.

	STEP	ADJUST	REMARKS
1	Peak 3.58 MHz osc. trans. (Place jumper from lug M44 to ground.)	L42	Adjust L42 for maximum DC reading on V.T.V.M. If the 3.58 MHz osc. is not running, no reading will be obtained. If necessary, adjust the reactance tube plate coil L39 to start osc. after adjustment is made remove jumper from lug M44.
2	Peak L37 (phase detector trans.)	L37	Adjust L37 for maximum DC reading on V.T.V.M. Make sure 3.58 MHz osc. is running and locked in.
Ground lug M37 (reactance tube input). Remove V.T.V.M. from lug M40.			
3	Adjust L39 (reactance tube plate coil) for zero beat of the 3.58 MHz osc.	L39	Observe picture tube and adjust L39 for zero beat (color bars stand still on screen or drift slowly).
Remove ground connection from lug M37. Repeat step 3 using low level color signal. Adjust for zero beat and best hold.			
4	Adjust L37 (phase detector trans.)	L37	Connect scope to R-Y output lug M52 and observe bar pattern Fig. 8-11. Set tint control fully up for 21ST90T, and adjust L37 for exactly minus 30' (5th bar nulled); refer to R-Y output waveform "B", allow all of the excess tint range to fall in the CW direction beyond the plus 30' point where the 7th bar is nulled.
5	B-Y demod, phase check.	L43	Connect scope to B-Y output lug M57 and observe pattern-check for –25% on 3rd bar and +25% on 9th bar. See waveform "A", Fig. 8-12.
Demodulator phase checks: Check the three color outputs at the CRT grids for proper matrixing. The waveforms and amplitudes should conform to Fig. 8-12. A. Check R-Y waveform at lug M52. B. Check G-Y waveform at lug M53. C. Check B-Y waveform at lug M57 (phase should be correct within ± ½ bar).			
6	Adjust VR102A (killer threshold)	VR102A	Adjust with a snowy raster. Adjust VR102A so that color snow just disappears. Check with color signal to assure setting is not killing on color.

Fig. 8-10. Color AFPC alignment procedure.

checked using a VTVM having an isolation resistor (10K to 50K) attached directly to the probe end. The emitter-to-ground voltage as measured at the junction of R4U and C4U should be 2.4V. Collector-to-ground voltage will be 12V with the probe connected to the oscillator tuning gang stator. If these voltages deviate widely from the foregoing values, the probable cause is a defective transistor. This may be replaced but care should be taken to wire it in as closely as possible to its original position, lead dress, and lead length. If the transistor operating voltages seem normal, a defective diode is indicated. In changing the diode the same wiring and lead dress precautions should be observed as for transistor replacement.

Replacing the transistor oscillator may upset the tracking of the tuner. If the desired channel is located near the low end of the board, between 14 and 45, the oscillator frequency can be readjusted by means of the trim tab C7U. In this case, the tuner should be set to the desired channel and the trim tab C5U adjusted accordingly. Channels at the high end of the band

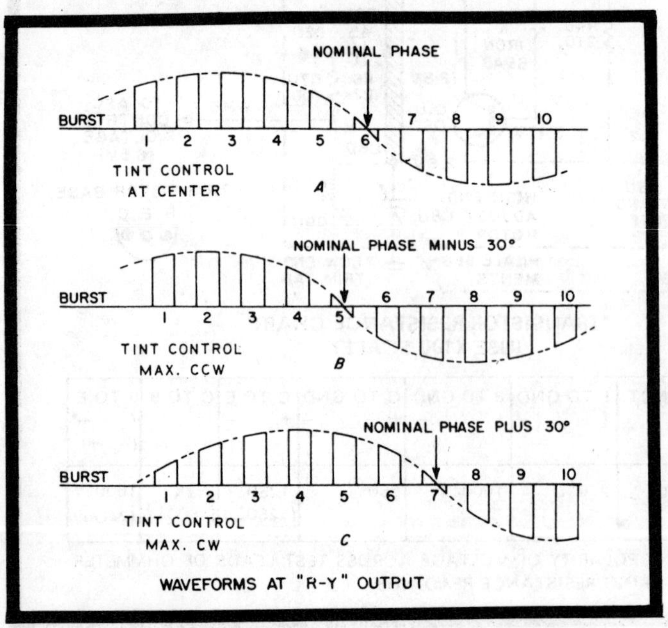

Fig. 8-11. Waveforms at R-Y output.

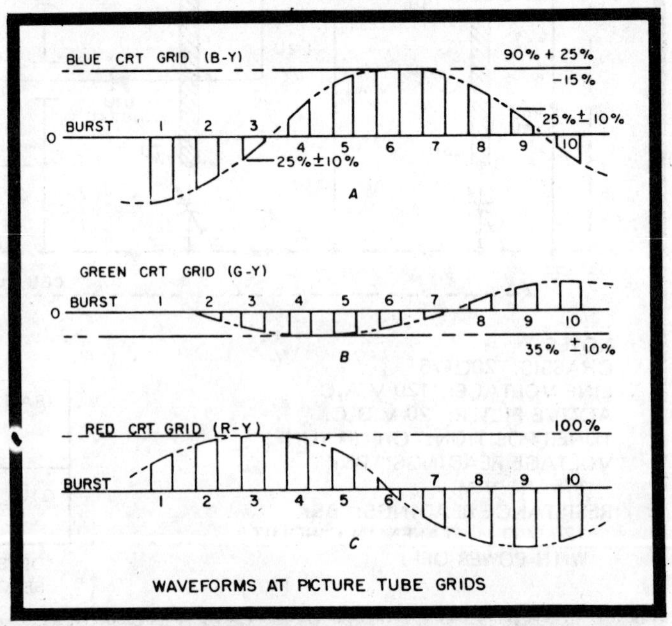

Fig. 8-12. Waveforms at picture-tube grids.

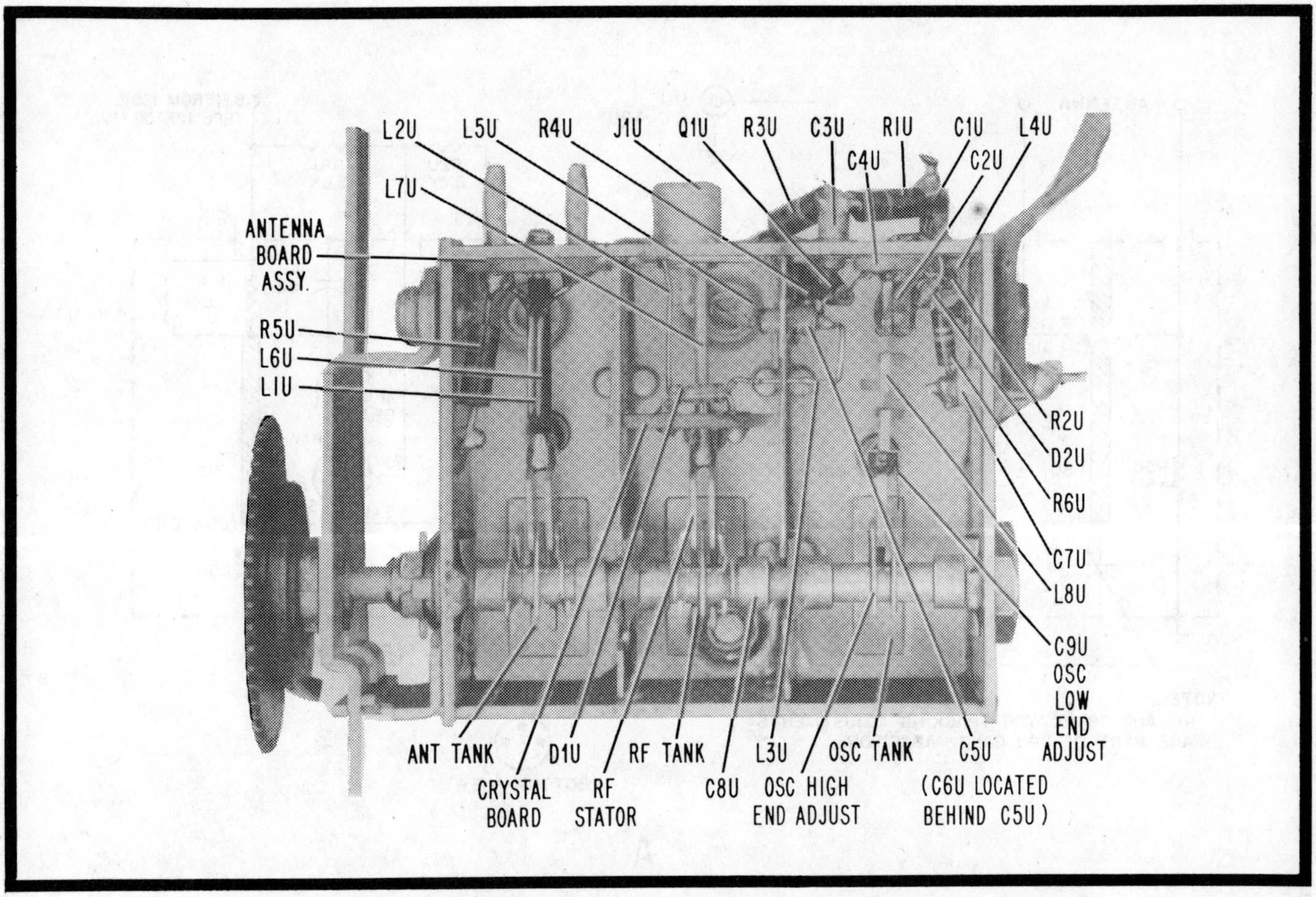

Fig. 8-13. Component layout, Philco-Ford UHF tuner 76-14099-5U.

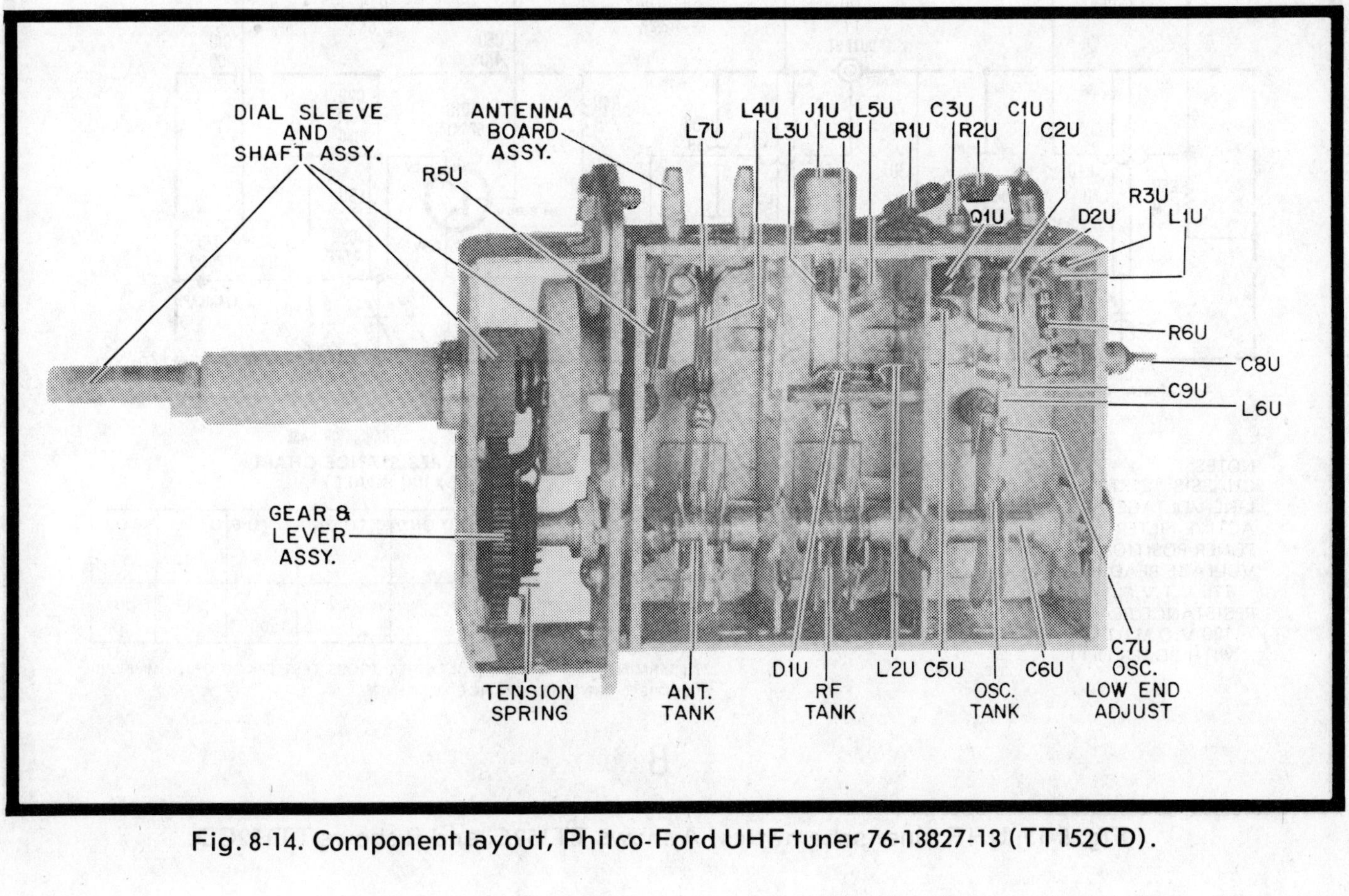

Fig. 8-14. Component layout, Philco-Ford UHF tuner 76-13827-13 (TT152CD).

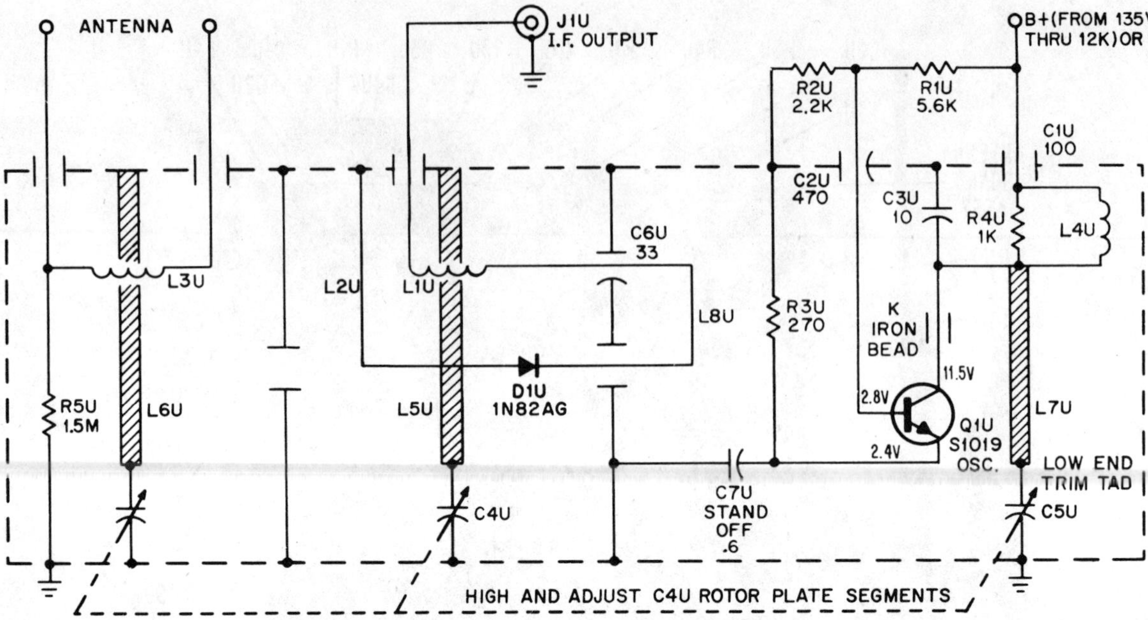

NOTE:
RF AND OSCILLATOR TRACKING ADJUSTMENTS ARE PART OF C4U GANG CAPACITOR

TRANSISTOR BASING

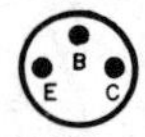

BOTTOM VIEW

A

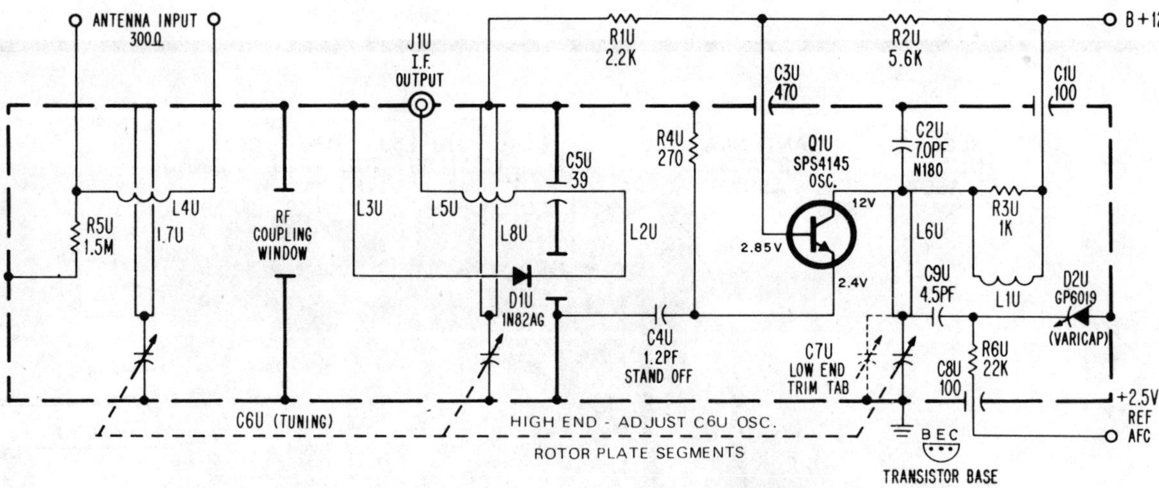

NOTES:
CHASSIS: 21KT41
LINE VOLTAGE: 120 V.A.C.
ACTIVE FILTER: 20 V.D.C.
TUNER POSITION: CH. #1 (UHF)
VOLTAGE READINGS: B&K 175 V.T.V.M.
RESISTANCE READINGS: B&K 120 V.O.M. (TAKEN IN CIRCUIT WITH POWER OFF)

*TRANSISTOR RESISTANCE CHART (USE X100 SCALE)

TRAN.	FUNCT.	E TO GND + −	B TO GND + −	C TO GND + −	C TO E + − (− +)	C TO B + − (− +)	B TO E + − (− +)
Q1U	OSC.	275Ω	1700Ω	1000Ω	1200Ω (1200Ω)	2.2K (1700Ω)	1600Ω (2.3K)

*DETERMINE POLARITY OF VOLTAGE ACROSS TEST LEADS OF OHMMETER BEFORE TAKING RESISTANCE READINGS.

B

Fig. 8-15. UHF tuner schematics; A shows TT152C, and B shows TT152CD.

must be set by adjusting the rotor plate segments of C6U tuning capacitor. (See schematics, Fig. 8-15A and B.)

The oscillator supply voltage as measured at C1U (Fig. 8-16) should be approximately 20V with respect to chassis ground. The transistor voltages may be checked using a VTVM having an isolation resistor (10K to 50K) attached directly to the probe end. If the voltages deviate widely from the values shown on the schematic, the probable cause is a defective transistor. This may be replaced but care should be taken to wire it in as closely as possible to its original position, lead dress, and lead length. If the transistor operating voltages seem normal, a defective diode is indicated. In changing the diode, the same wiring and lead dress precautions should be observed as for transistor replacement.

Replacing the transistor may upset the tracking of the tuner. The frequency can be readjusted by setting the tuner to each desired channel (starting at the highest and working to the lowest) and adjusting the outer rotor plate segments in the oscillator tank circuit.

Avoid closing the gap between the rotor and stator plates less than 0.010 inch to prevent shorting out of the plates.

Mechanical Parts Replacement. Replacement of mechanical parts for the UHF detent tuner is limited to replacement of the channel film readout assembly, channel drum drive gear, outer fine tune shaft, drive gear spacer, compression ring, and roll pin. Any other malfunction in the drive mechanism will require replacement of the complete UHF detent tuner. (Refer to Fig. 8-17 for exploded view.) To change readout assembly:

1. Remove volume, on-off, and tuner knobs.
2. Remove cabinet back and antenna connections.
3. Disconnect three cable connectors to tuner assembly and i-f output cable on VHF tuner; then remove complete tuner assembly (4 mtg. screws).
4. Set VHF channel indicator dial to UHF position (clear window on VHF channel indicator dial will be at 12 o'clock); then remove VHF dial and bracket assembly (3 screws).
5. Slide channel indicator dial light and socket assembly up from access hole in UHF channel indicator drum.
6. Unsolder B+ and AFC leads on UHF tuner.
7. Remove UHF tuner (3 screws).
8. Remove nut and lockwasher which secure channel film readout assembly and drum spacer block to tuner mounting bracket. Retain drum spacer block for mounting new readout assembly.
9. Set UHF tuner to channel 14 position (rotor and stator plates fully meshed). Rotate drum and readout assembly and center channel 14 indicator below the notch in the top of the drum; then reassemble readout assembly to the mounting bracket using the spacer block, lockwasher, and nut previously removed in step 8.
10. Remount UHF tuner (3 screws).
11. Resolder B+ and AFC leads to UHF tuner.
12. Remount channel indicator dial light and socket assembly in UHF channel indicator drum.
13. Remount VHF channel indicator dial and bracket assembly (3 screws).

When mounting the VHF dial and bracket, be sure that the VHF tuner is in the UHF position and the indicator mark in the clear window lines up with the embossed mark on the tuner mounting panel.

14. Remount complete tuner assembly (4 screws) in cabinet. Reconnect the three cable connectors on the tuner assembly to the chassis and the secondary controls; then connect the i-f output cable to the VHF tuner.
15. Reconnect antenna connections and replace cabinet back.
16. Replace control knobs.

UHF Tuner 76-13873-4 (TT151D)

The oscillator supply voltage as measured at C1U (Fig. 8-18) should be approximately 12V with respect to chassis ground. The transistor voltages may be checked using a VTVM having an isolation resistor (10K to 50K) attached directly to the probe end. The emitter-to-ground voltage should be 2.0 to 2.5V. Collector-to-ground voltage will be 11-12V with the probe connected to the oscillator tuning gang stator. If these voltages deviate widely from the foregoing values, the probable cause is a defective transistor. This may be replaced but care should be taken to wire it in as closely as possible to its original position, lead dress, and lead length. If the transistor operating voltages seem normal, a defective diode is indicated. In changing the diode the same wiring and lead dress precautions should be observed as for transistor replacement. (See schematic, Fig. 8-19.)

Replacing the transistor oscillator may upset the tracking of the tuner. If the desired channel is located near the low end of the board, between 14 and 45, the oscillator frequency can be readjusted by means of trim tab C5U (Fig. 8-20). In this case, the tuner should be set to the desired channel and the trim tab C5U adjusted accordingly. Channels at the high end of the band must be set by adjusting the rotor plate segments of C4U tuning capacitor.

Figure 8-21 shows an overall tuner assembly exploded view for UHF tuner TT151D and VHF tuner TT182AD.

OTHER PHILCO-FORD DATA

Figures 8-22 through 8-31 are schematics, wiring diagrams, layout drawings, and other useful graphic material for Philco-Ford models. The following captions describe the material presented as the balance of this chapter.

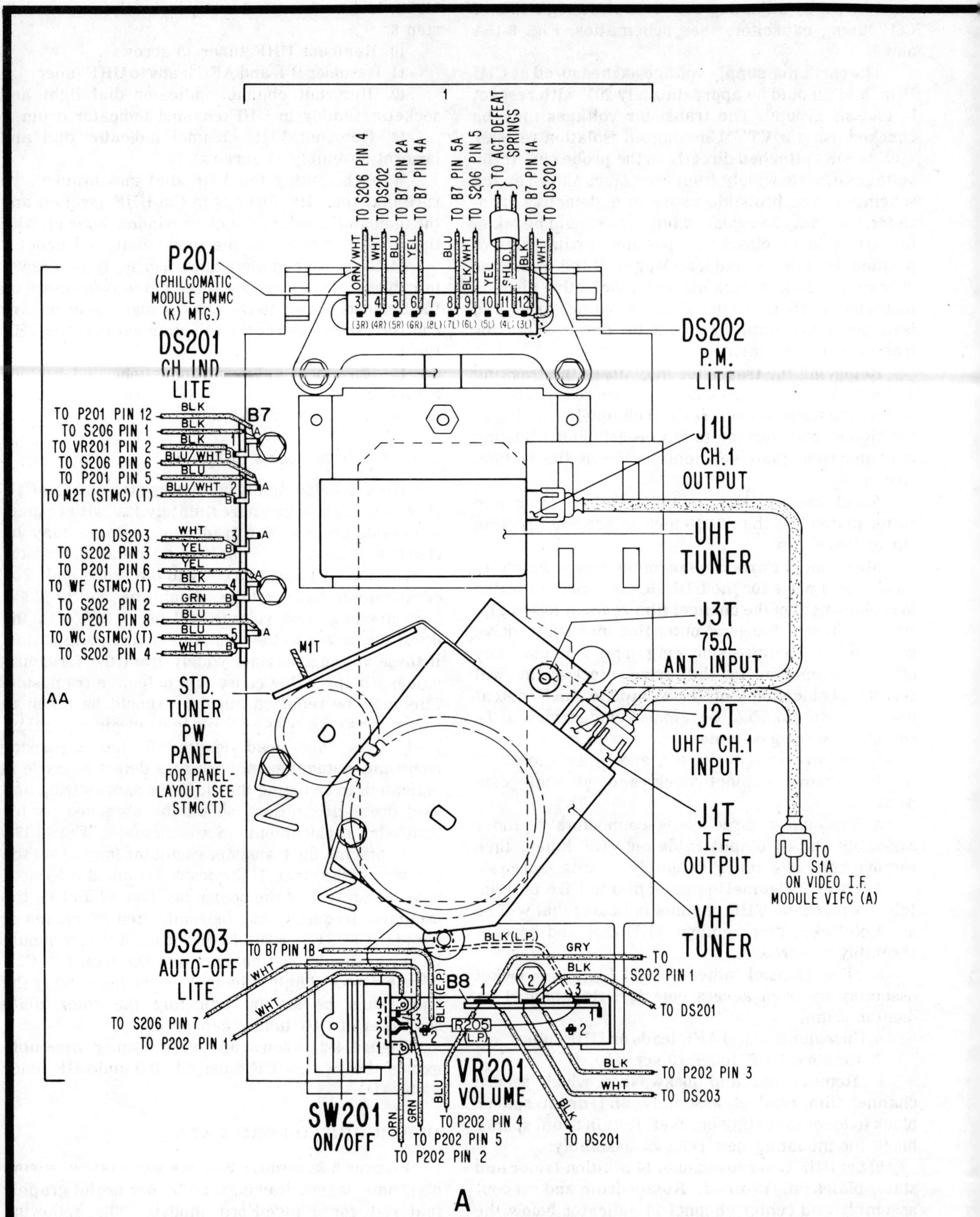

Fig. 8-16. Wiring diagram (A) and parts layout (B) for UHF tuner

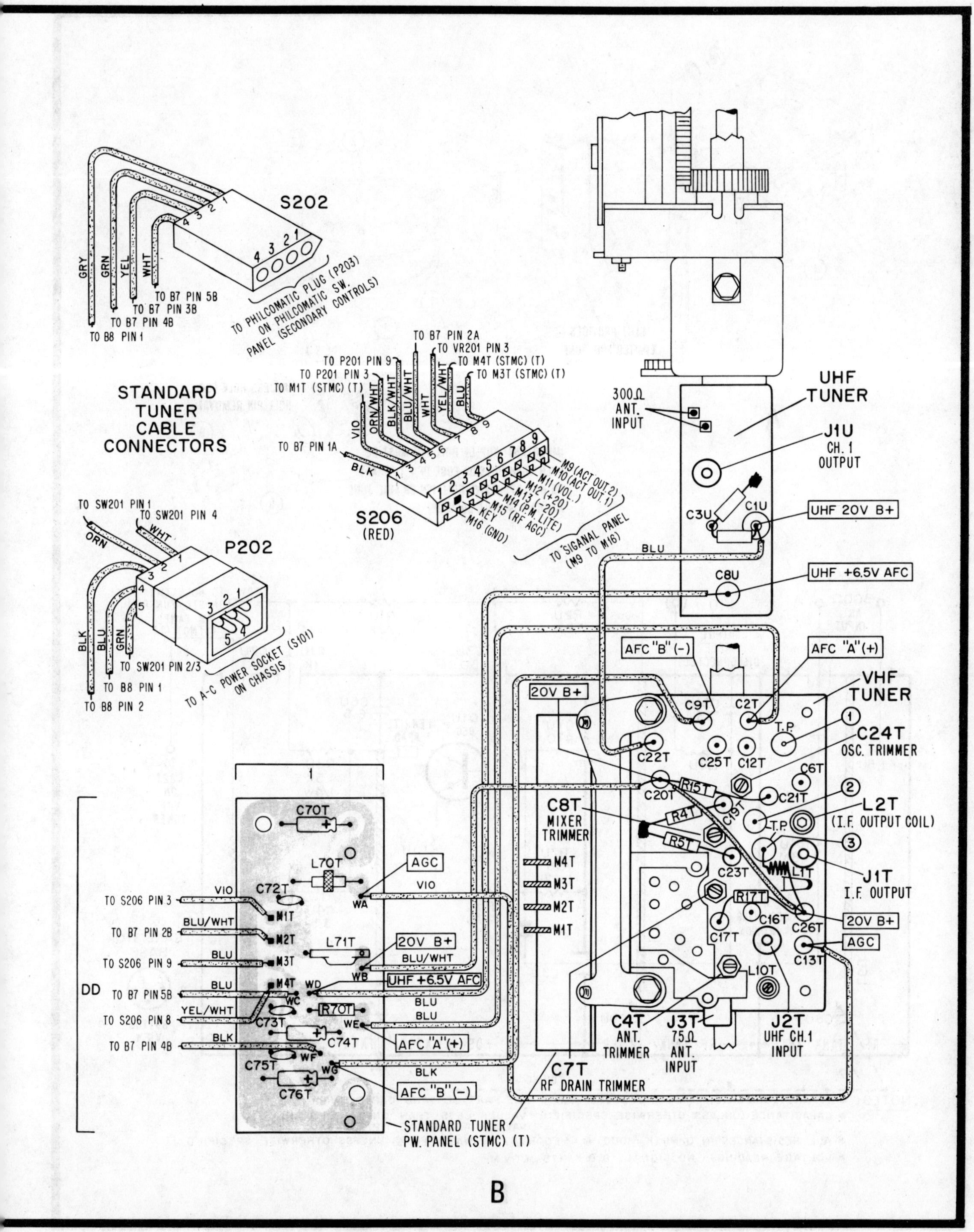

TT244 and VHF tuner TT192, Philco-Ford 3CS90 & 3CY90 chassis.

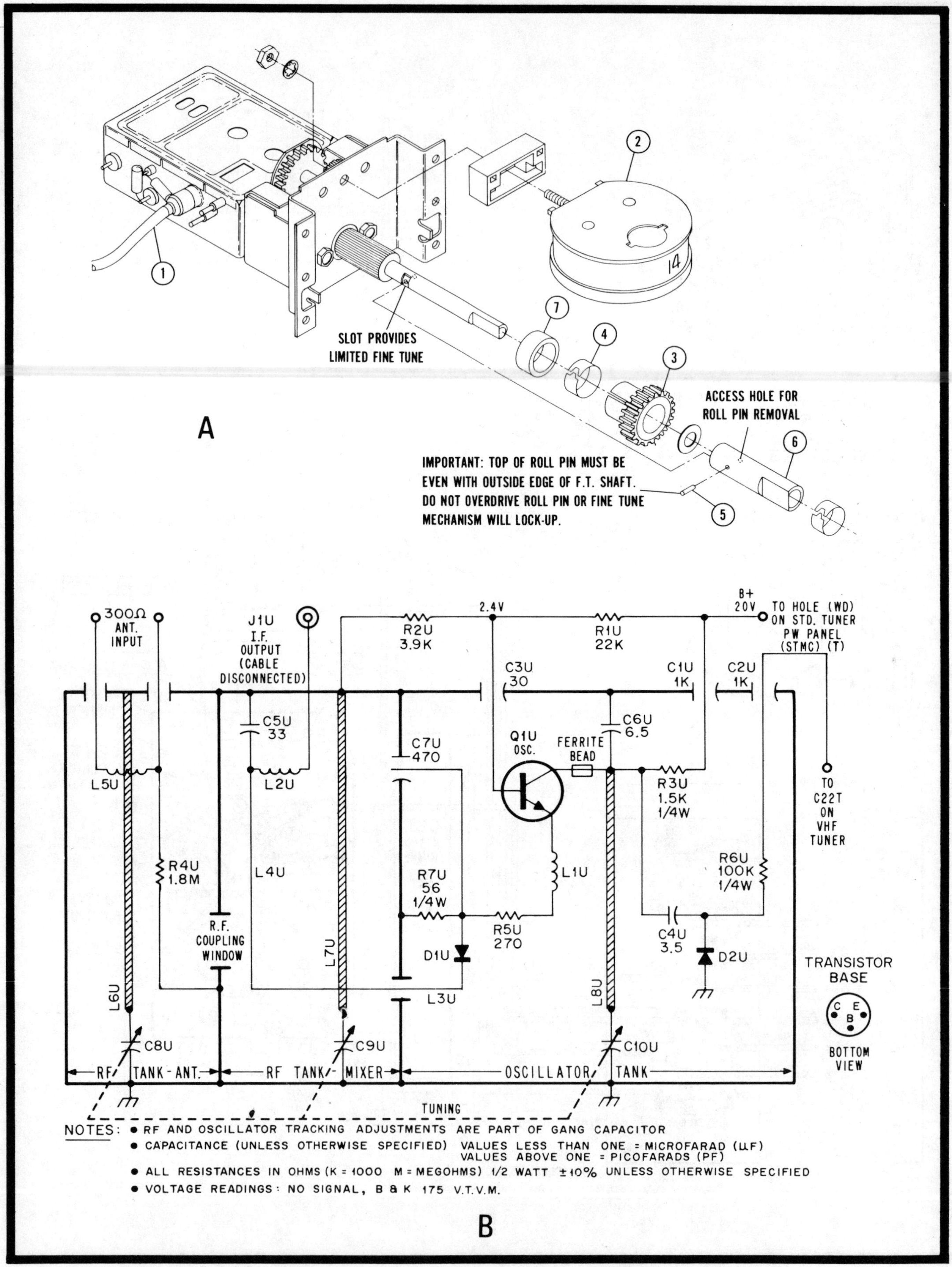

Fig. 8-17. Philco-Ford UHF tuner TT244 exploded view (A) and schematic (B).

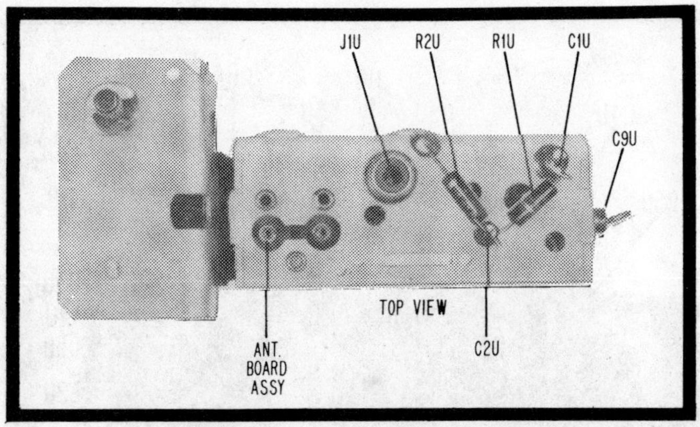

Fig. 8-18. Component layout, Philco-Ford UHF tuner TT151D.

Fig. 8-21. Parts layout, Philco-Ford VHF TT182AD and UHF tuner TT151D.

Fig. 8-22. Philco-Ford VHF tuner TT192 schematic.

Fig. 8-23. Philco-Ford VHF tuner TT192 exploded view.

Fig. 8-24. Philco-Ford VHF tuner TT183AH schematic.

Fig. 8-25. Philco-Ford tuners TT183AH (VHF) and TT152C(D) (UHF) exploded view of complete tuner assembly.

Fig. 8-26. Philco-Ford VHF tuner TT182AD schematic.

Fig. 8-27. Philco-Ford VVC tuner TT700 schematic.

Fig. 8-28. Philco-Ford VVC tuner TT700 primary control wiring and instrumentation assembly.

Fig. 8-29. Philco-Ford VVC tuner TT700 voltage checkpoints and parts layout.

Fig. 8-30. Top of tuner layout, cable connections, and interconnect wiring for Philco-Ford VVC tuner TT700 (3CS91 and 3CY91 chassis).

Fig. 8-31. VVC tuning unit and primary control wiring (TT700), Philco-Ford 3CS91 and 3CY91 chassis.

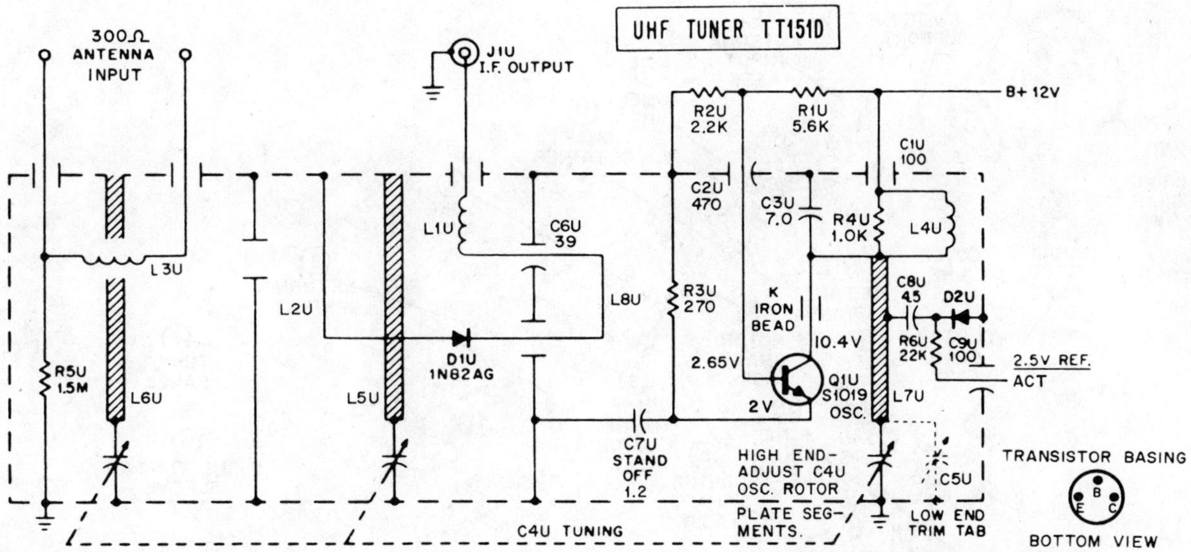

NOTES:
CHASSIS: 21ST90T
LINE VOLTAGE: 120 V.A.C.
VHF TUNER POSITION: CH. #1 (UHF)
VOLTAGE READINGS: B & K 175 V.T.V.M.
RESISTANCE READINGS: B & K 120 V.O.M.
(TAKEN IN CIRCUIT WITH POWER OFF)
ACT: OFF
NO SIGNAL

NOTE:
RF AND OSCILLATOR TRACKING ADJUSTMENTS ARE PART OF GANG CAPACITOR.

***TRANSISTOR RESISTANCE CHART (USE X100 SCALE)**

TRAN.	FUNCT.	E TO GND. + −	B TO GND. + −	C TO GND. + −	C TO E + − (− +)	C TO B + − (− +)	B TO E + − (− +)
Q1U	OSC.	250Ω	1600Ω	775Ω	1050Ω (1050Ω)	2.1K (1.5K)	1600Ω (2K)

*DETERMINE POLARITY OF VOLTAGE ACROSS TEST LEADS OF OHMMETER BEFORE TAKING RESISTANCE READINGS.

Fig. 8-19. Philco-Ford UHF tuner TT151D schematic.

Fig. 8-20. Component layout, Philco-Ford UHF tuner TT152C.

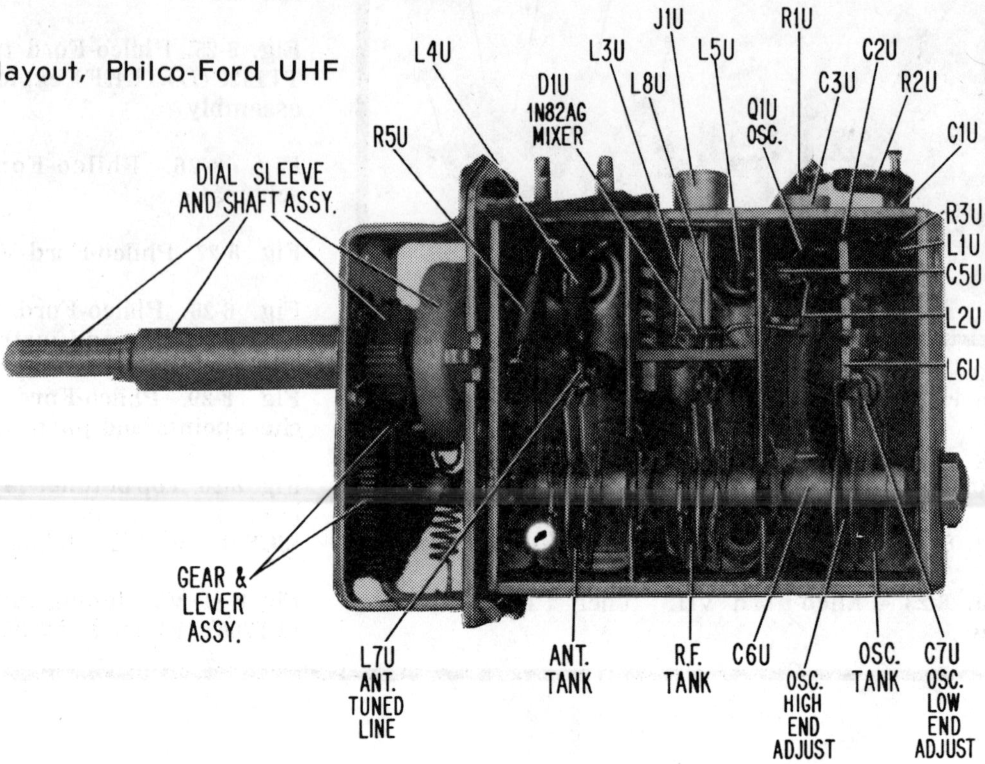

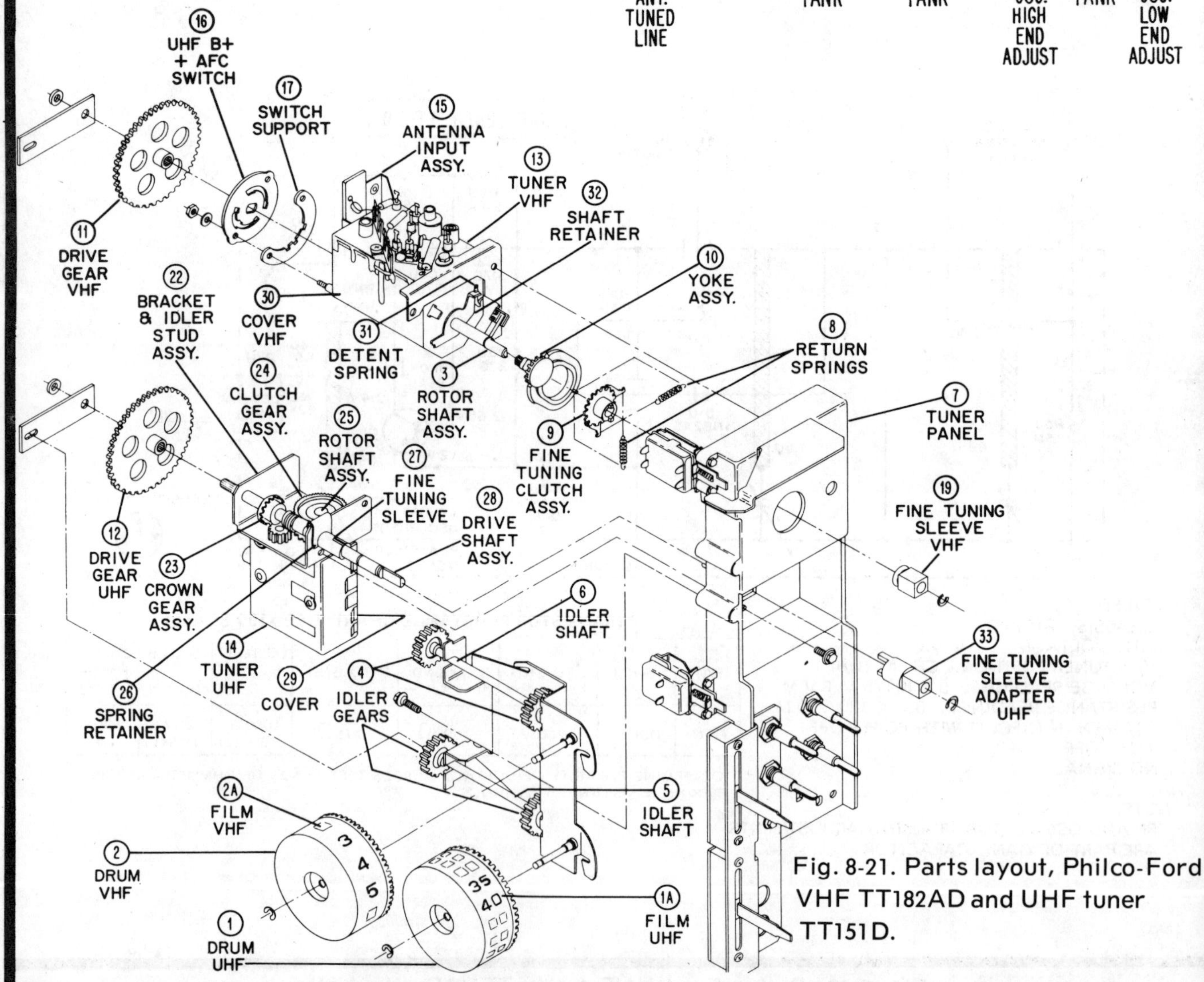

Fig. 8-21. Parts layout, Philco-Ford VHF TT182AD and UHF tuner TT151D.

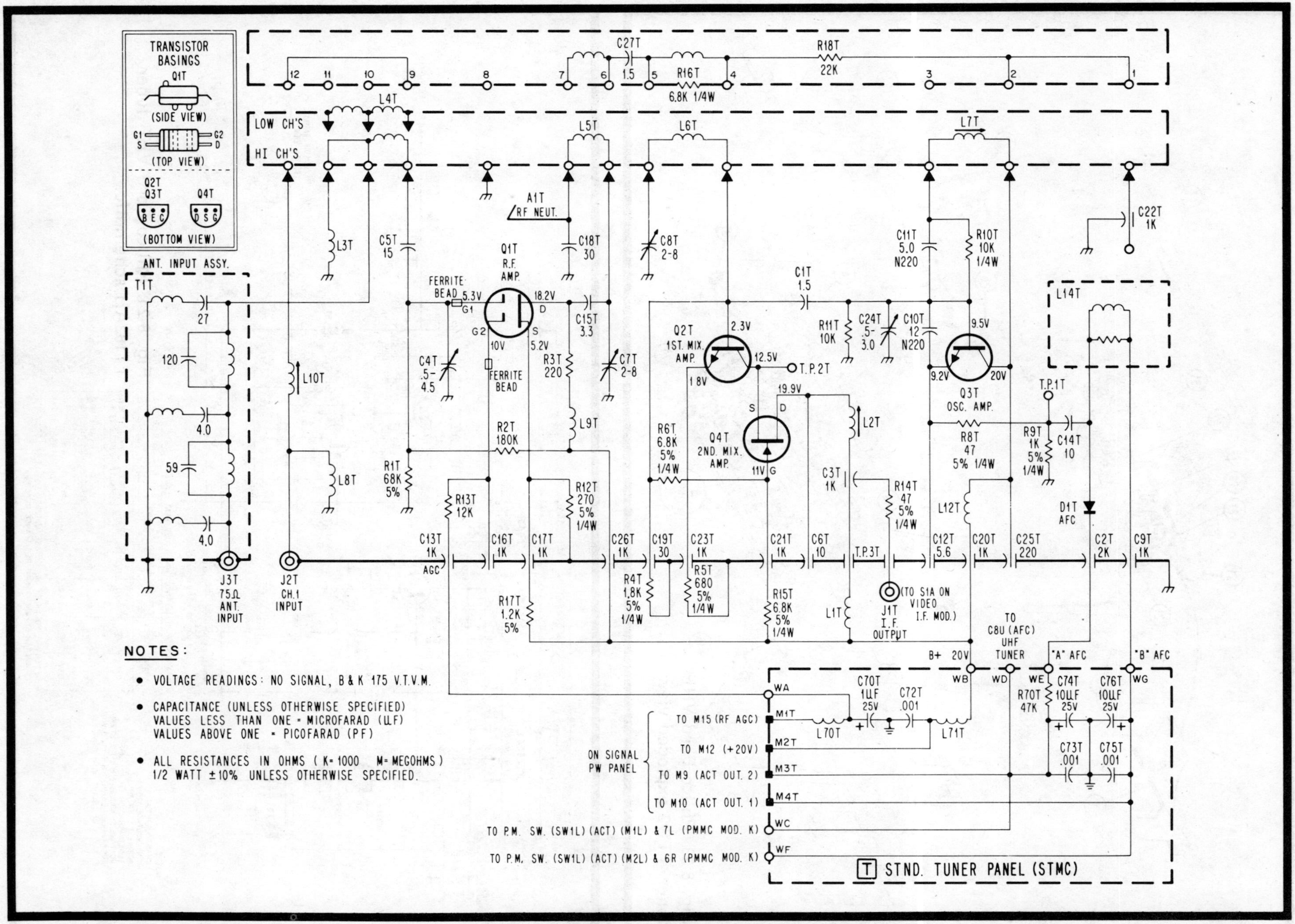

Fig. 8-22. Philco-Ford VHF tuner TT192 schematic.

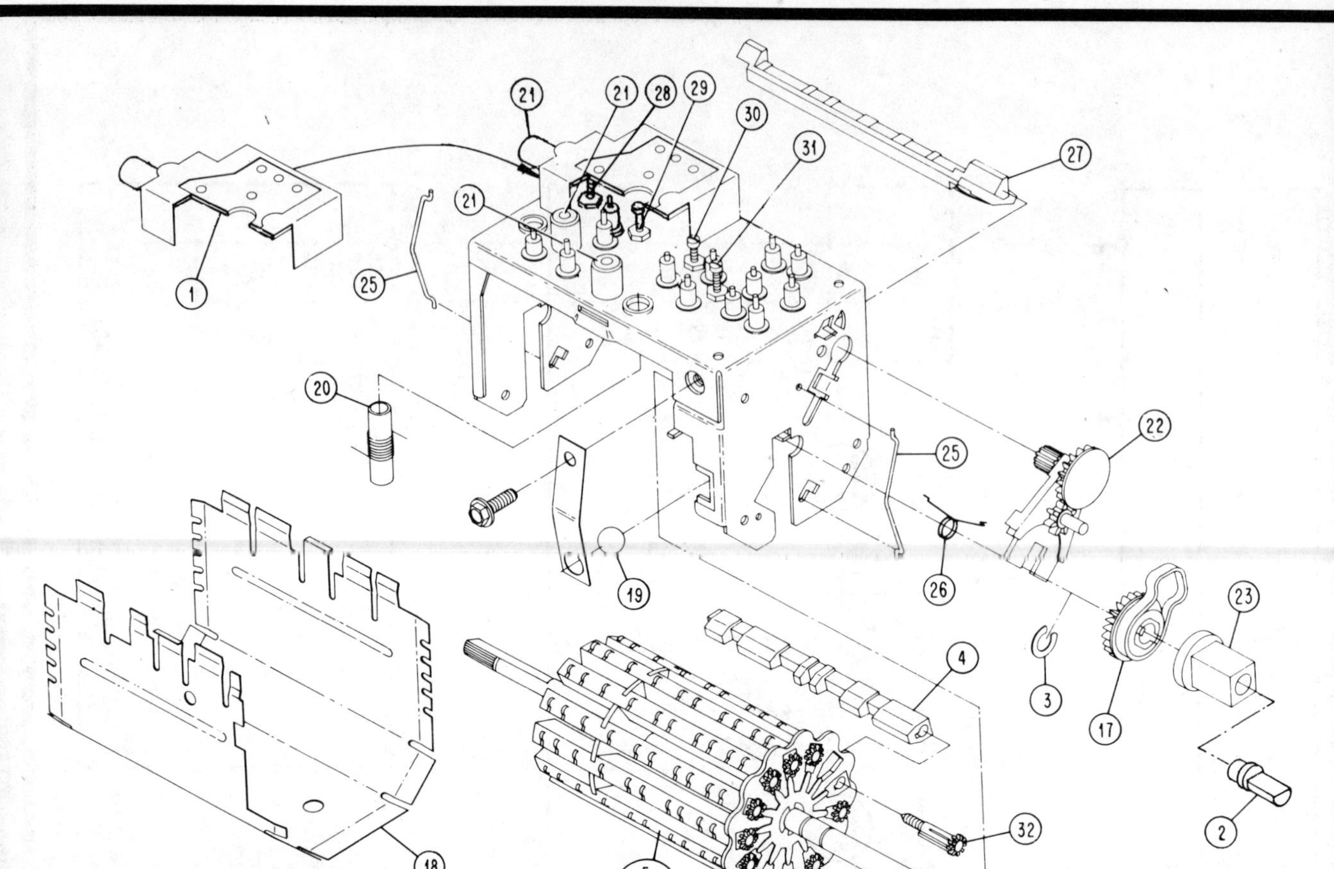

Fig. 8-23. Philco-Ford VHF tuner TT192 exploded view.

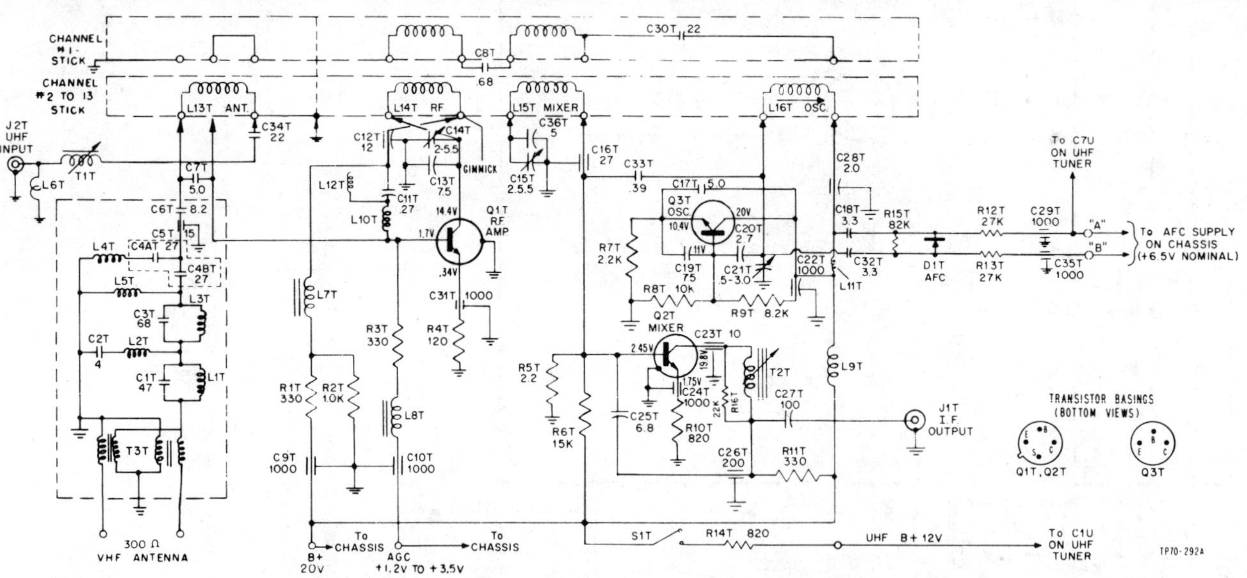

NOTES:
CHASSIS: 21KT41
LINE VOLTAGE: 120V.A.C.
ACTIVE FILTER: 20V.D.C.
TUNER POSITION: CH. # 1 (UHF)
VOLTAGE READINGS: B & K 175 V.T.V.M.
RESISTANCE READINGS: B & K 120 V.O.M.
(TAKEN IN CIRCUIT WITH POWER OFF)

* TRANSISTOR RESISTANCE CHART (USE X100 SCALE)

TRAN.	FUNCT.	E To GND + −	B To GND + −	C To GND + −	C To E + − (− +)	C To B + − (− +)	B To E + − (− +)
Q1T	RF	120Ω	1650Ω	400Ω	500Ω (500Ω)	3.4K (1650Ω	1700Ω (3.1K)
Q2T	MIX	850Ω	1550Ω	600Ω	1500Ω (1500Ω)	2.6K (1500Ω)	1750Ω (2.9K)
Q3T	OSC.	2.4K	2.1K	250Ω	1800Ω (2.7K)	4.8K (1.8K)	1.9K (7.0K)

* DETERMINE POLARITY OF VOLTAGE ACROSS TEST LEADS OF OHMMETER BEFORE TAKING RESISTANCE READINGS.

Fig. 8-24. Philco-Ford VHF tuner TT183AH schematic.

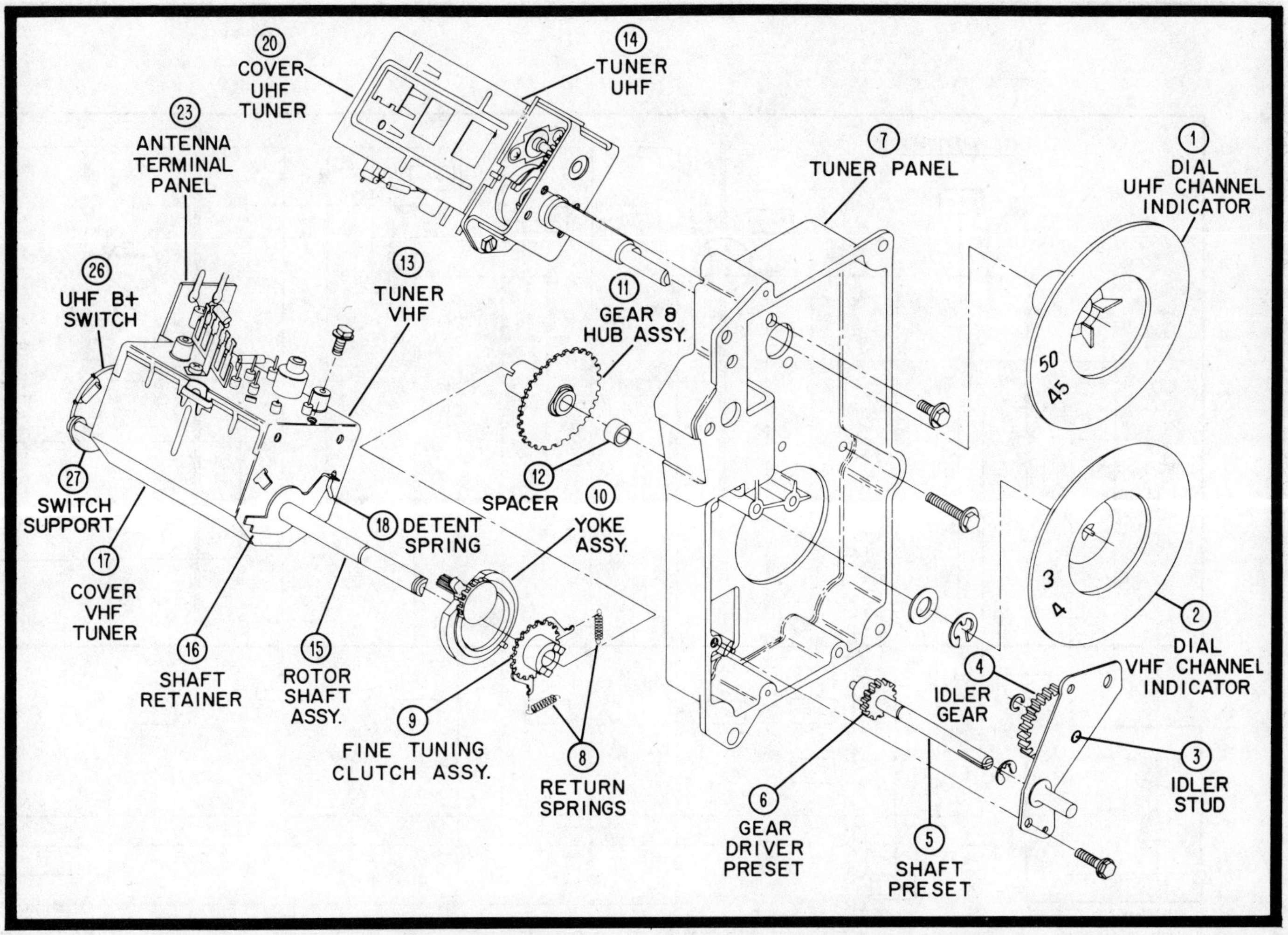

Fig. 8-25. Philco-Ford tuners TT183AH (VHF) and TT152C(D) (UHF) exploded view of complete tuner assembly.

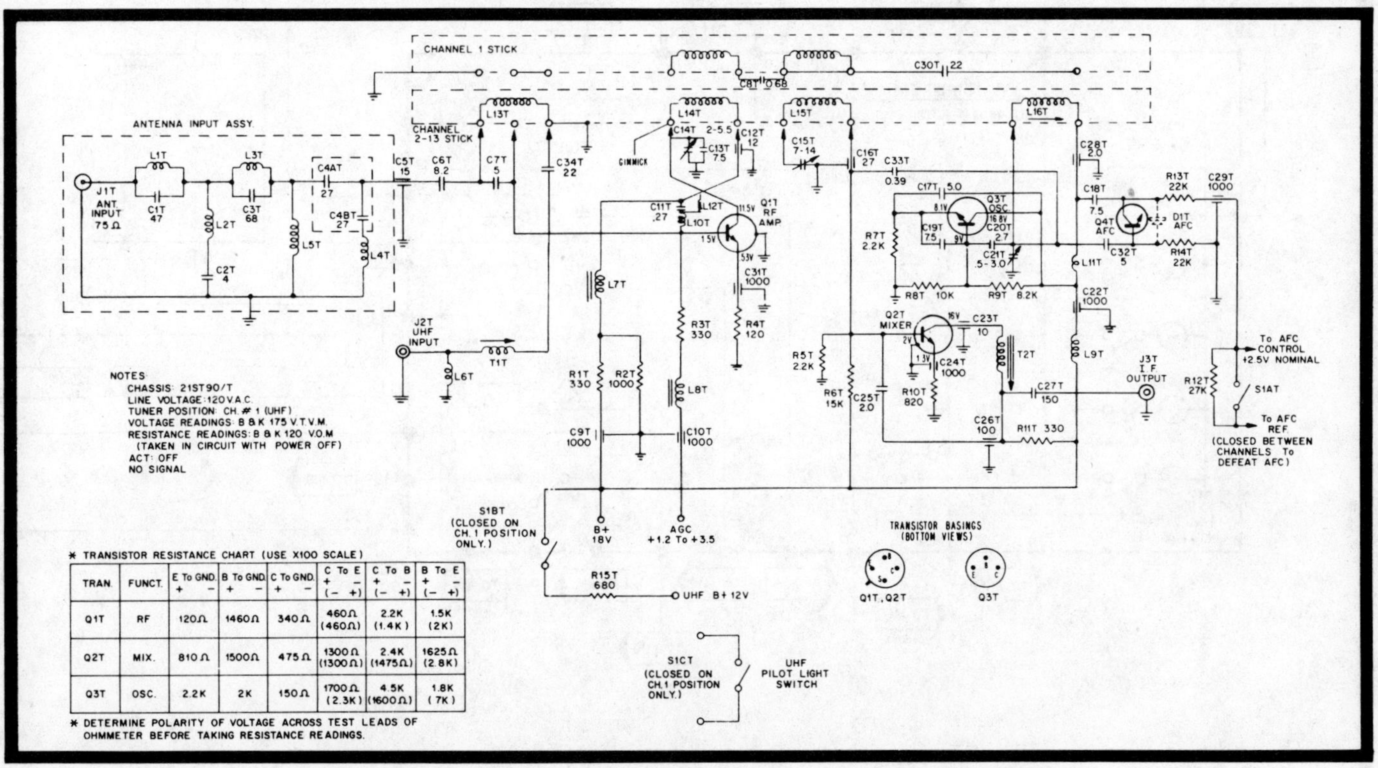

✱ TRANSISTOR RESISTANCE CHART (USE X100 SCALE)

TRAN.	FUNCT.	E To GND. + −	B To GND. + −	C To GND. + −	C To E + − (− +)	C To B + − (− +)	B To E + − (− +)
Q1T	RF	120Ω	1460Ω	340Ω	460Ω (460Ω)	2.2K (1.4K)	1.5K (2K)
Q2T	MIX.	810Ω	1500Ω	475Ω	1300Ω (1300Ω)	2.4K (1475Ω)	1625Ω (2.8K)
Q3T	OSC.	2.2K	2K	150Ω	1700Ω (2.3K)	4.5K (1600Ω)	1.8K (7K)

✱ DETERMINE POLARITY OF VOLTAGE ACROSS TEST LEADS OF OHMMETER BEFORE TAKING RESISTANCE READINGS.

Fig. 8-26. Philco-Ford VHF tuner TT182AD schematic.

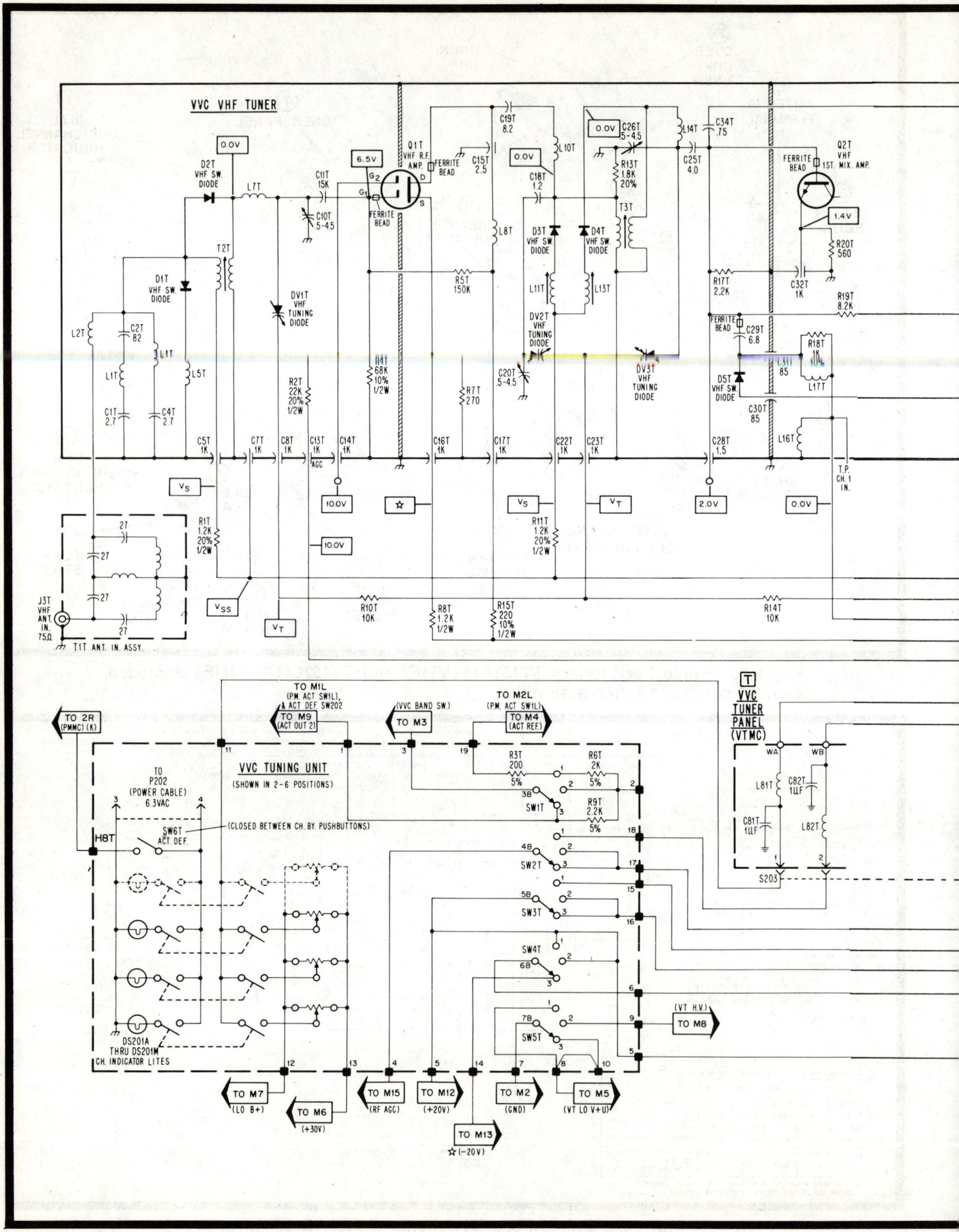
VVC VHF TUNER
Q1T VHF R.F. AMP.
Q2T VHF 1ST. MIX. AMP.
D2T VHF SW. DIODE
D1T VHF SW. DIODE
DV1T VHF TUNING DIODE
DV2T VHF TUNING DIODE
DV3T VHF TUNING DIODE
D3T VHF SW. DIODE
D4T VHF SW. DIODE
D5T VHF SW. DIODE
FERRITE BEAD
J3T VHF ANT. IN. 75Ω
T1T ANT. IN. ASSY.
T.P. CH. 1 IN.
VVC TUNING UNIT
(SHOWN IN 2-6 POSITIONS)
TO P202 (POWER CABLE) 6.3VAC
SW6T ACT. DEF.
(CLOSED BETWEEN CH. BY PUSHBUTTONS)
DS201A THRU DS201M CH. INDICATOR LITES
TO 2R (PMMC) (K)
TO M1L (P.M. ACT SW1L), & ACT DEF. SW202
TO M9 (ACT OUT 2)
(VVC BAND SW.) TO M3
TO M2L (P.M. ACT SW1L)
TO M4 (ACT REF.)
(VT H.V.) TO M8
TO M7 (LO B+)
TO M6 (+30V)
TO M15 (RF AGC)
TO M12 (+20V)
TO M13 ☆(-20V)
TO M2 (GND)
TO M5 (VT LO V+U)
VVC TUNER PANEL (VTMC)
S203

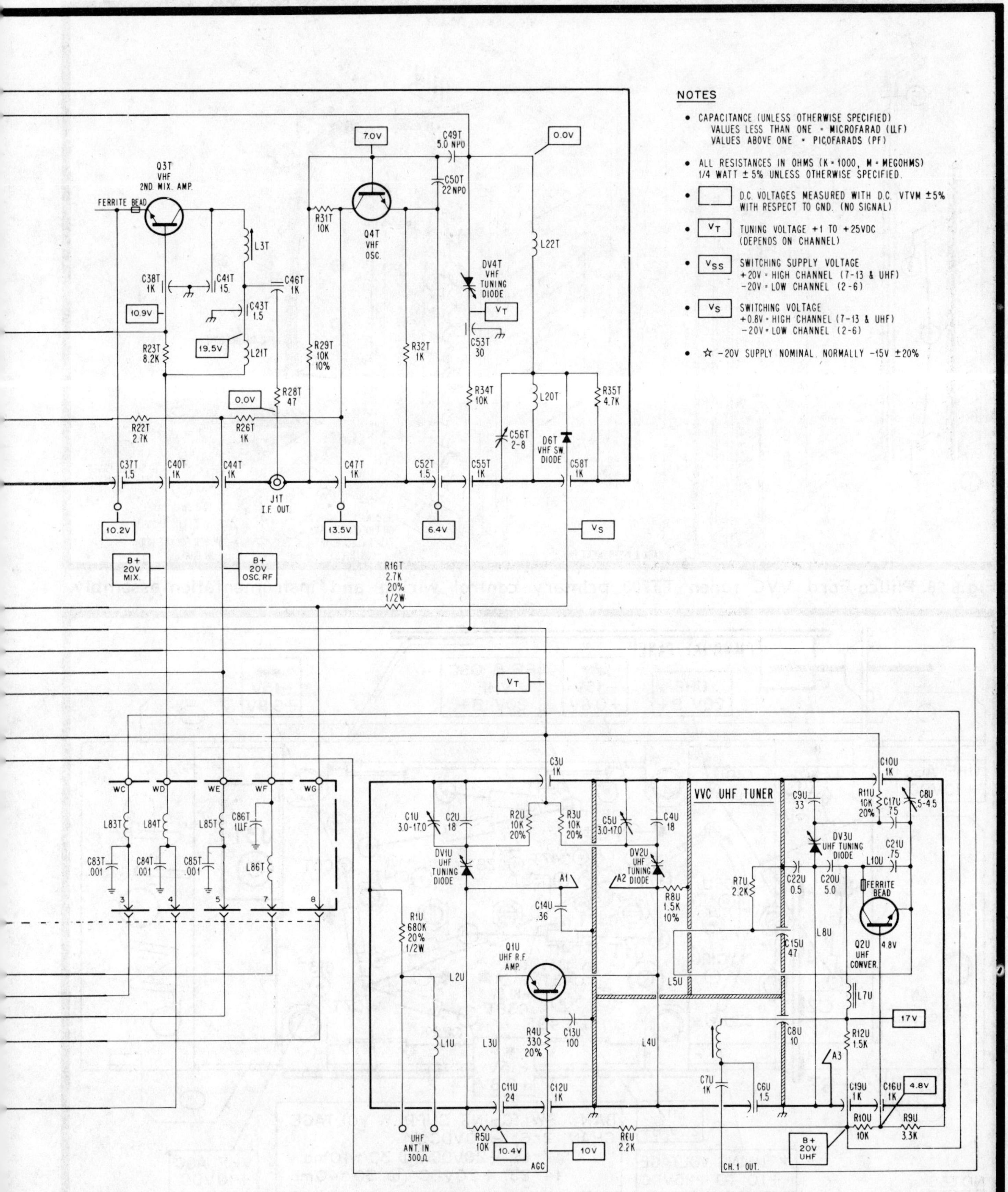

Fig. 8-27. Philco-Ford VVC tuner TT700 schematic.

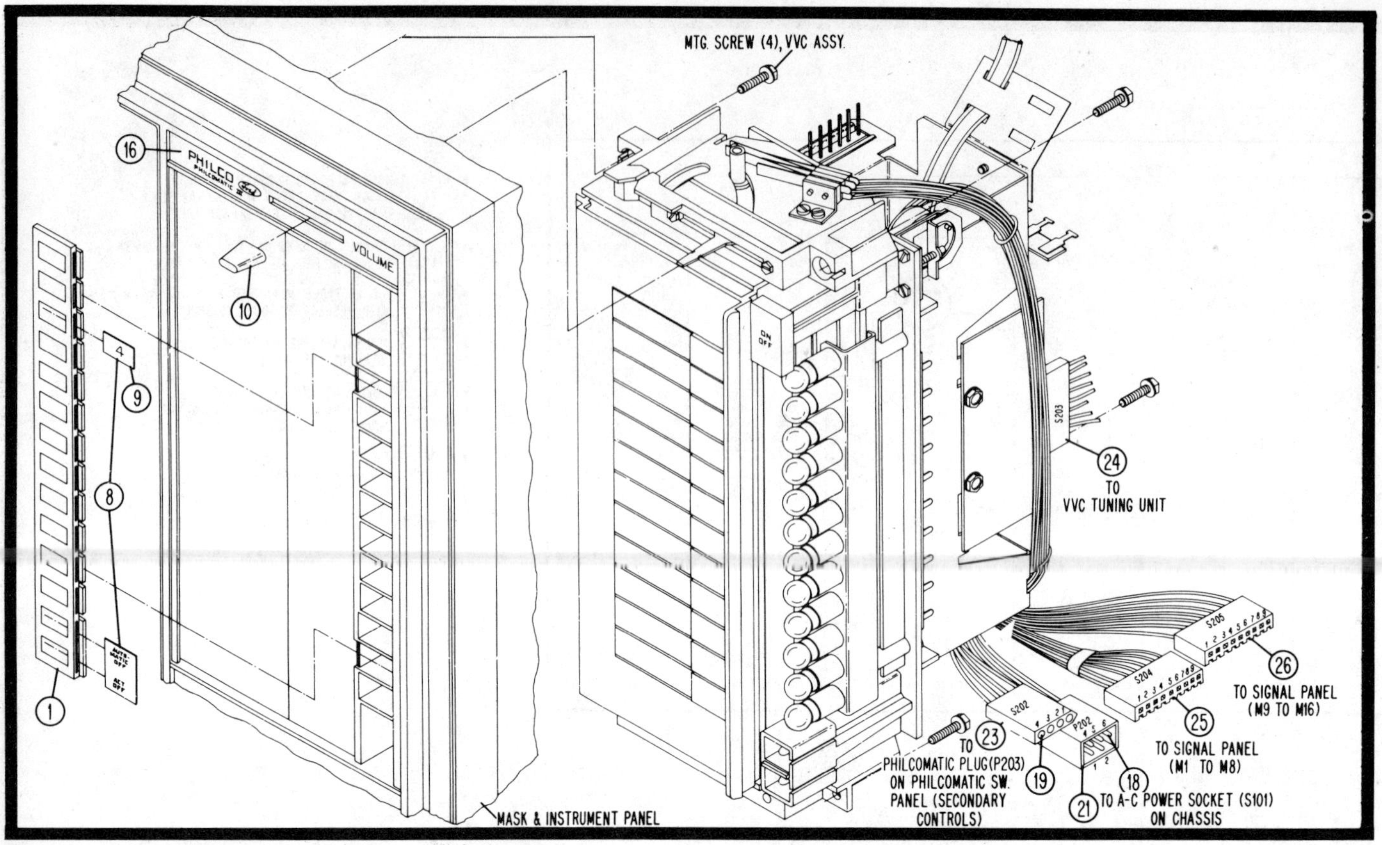

Fig.8-28. Philco-Ford VVC tuner TT700 primary control wiring and instrumentation assembly.

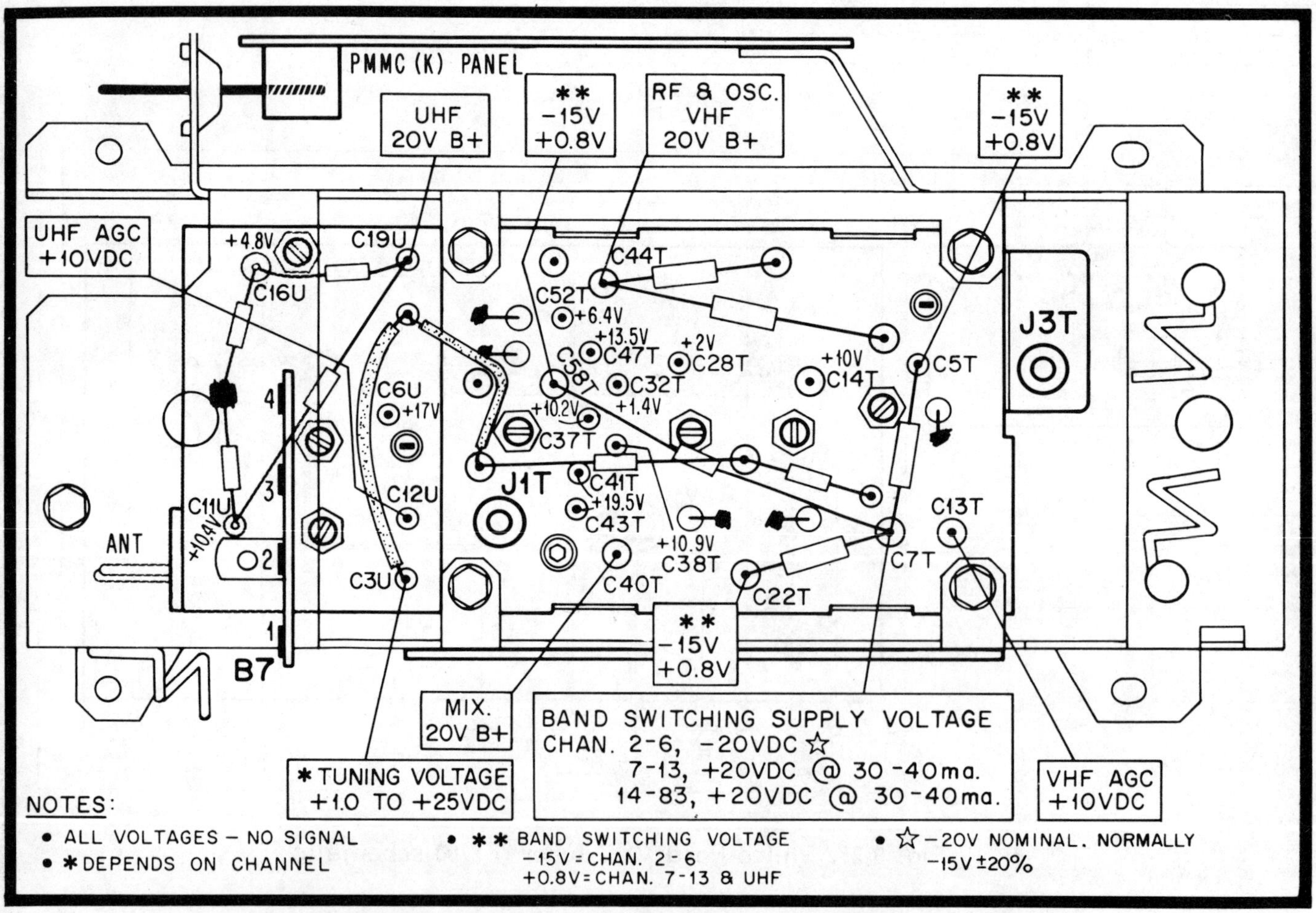

Fig. 8-29. Philco-Ford VVC tuner TT700 voltage checkpoints and parts layout.

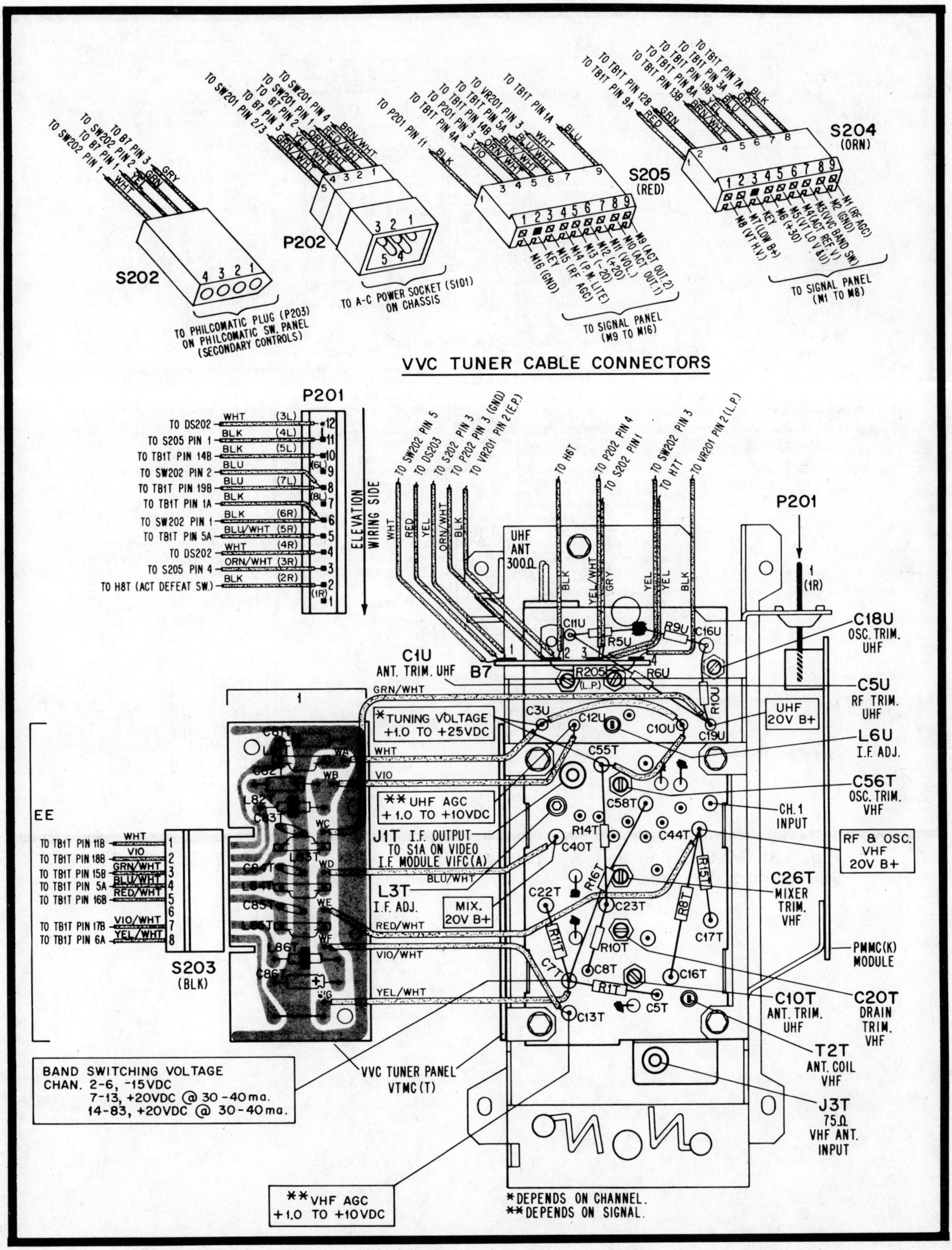

Fig. 8-30. Top of tuner layout, cable connections, and interconnect wiring for Philco-Ford VVC tuner TT700 (3CS91 and 3CY91 chassis).

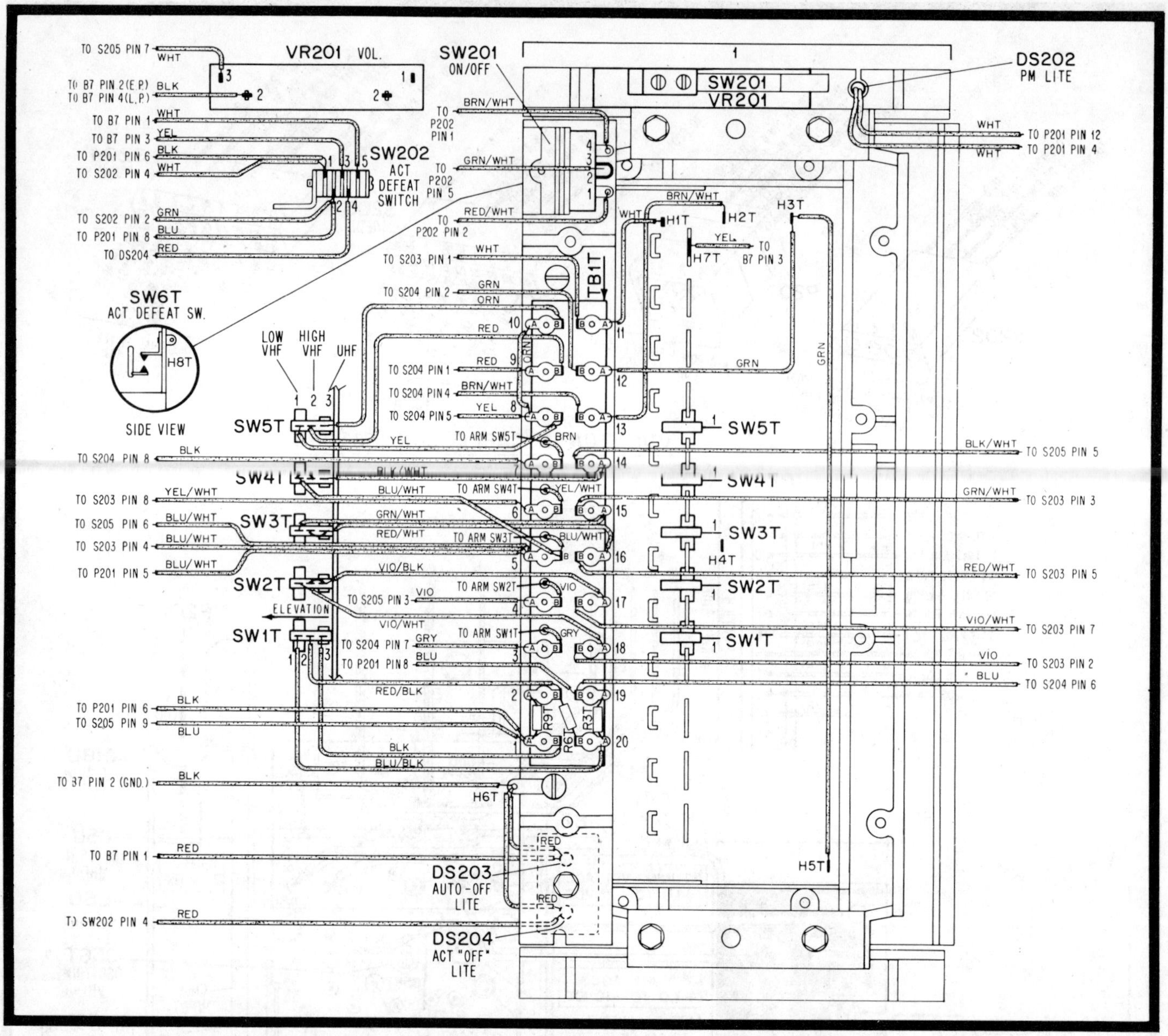

Fig. 8-31. VVC tuning unit and primary control wiring (TT700), Philco-Ford 3CS91 and 3CY91 chassis.

Chapter 9

RCA

To completely cover all RCA tuner models produced during the 1960s and 1970s, we would have to devote more pages to this chapter than we include in this entire manual. As a result, we're incorporating but a representative number of them. In the cross-reference table of Fig. 9-1, you can see which RCA chassis is applicable to each of the tuners covered here. The slash in the KRK number separates the VHF tuners (left of slash) from the UHF tuners. You'll find that some RCA chassis use one combination of UHF/VHF tuners, others use different combinations. The KRK 132 UHF tuner, for example, is mated with the KRK 140 VHF tuner in chassis CTC 38 and 39, but it's mated with the KRK 142 in chassis CTC 40; and the same basic tuner was mated in 1968 with KRK 131 (chassis CTC 35) or KRK 143 (chassis CTC 27X).

EQUIPMENT CONSIDERATIONS

The test equipment specified below, or its equivalent, is essential to properly perform the alignment procedures which are outlined on the following pages. Use of equipment which does not meet these requirements may result in inability to properly align the tuners.

A warmup period of at least fifteen minutes should be allowed for proper stabilization of such equipment as signal and sweep generators.

It is essential that the bias values as specified are maintained during alignment to insure the proper results.

The alignment pads and the input head are designed for correct matching of the equipment to the circuits involved. Failure to use proper matching will result in responses which cannot be depended upon as

Chassis	VHF / UHF Tuners	Chassis	VHF / UHF Tuners
CTC 22	KRK 128 / 122		
CTC 22	KRK 167 / 168 *	CTC 53	KRK 182 / 186
CTC 27X	KRK 143 / 132	CTC 54	KRK 155 / 160
CTC 28	KRK 131 / 132	CTC 55	KRK 140 / 156
CTC 31	KRK 128 / 138	CTC 55	KRK 140 / 185
CTC 35	KRK 131 / 132	CTC 55	KRK 144 / 152
CTC 36	KRK 128 / 122	CTC 59	KRK 165 / 156
CTC 36	KRK 128 / 122 / 143 / 156		
CTC 36X	KRK 143 / 151	KCS 156	KRK 127 / 120
CTC 38	KRK 140 / 142	KCS 157	KRK 129 / 122
CTC 38	KRK 144 / 138	KCS 165	KRK 137 / 122
CTC 38X	KRK 140 / 132	KCS 168	KRK 137 / 122
CTC 39	KRK 140 / 132	KCS 168	KRK 137 / 152
CTC 39	KRK 173 / 132	KCS 168	KRK 177 / 152
CTC 39	KRK 173 / 170	KCS 169	KRK 146 / 147 *
CTC 39	KRK 173 / 132 / 170 *	KCS 169	KRK 146 / 150
CTC 40	KRK 142 / 132	KCS 171	KRK 123 / 122
CTC 41	KRK 140 / 156 / 144 / 152	KCS 171	KRK 148 / 152
CTC 43	KRK 140 / 133 / 152 / 156	KCS 172	KRK 149 / 152
CTC 44	KRK 172 / 132	KCS 173	KRK 127 / 120 or 122
CTC 46	KRK 165 / 132	KCS 174	KRK 133 / 120
CTC 46	KRK 165 / 156	KCS 174	KRK 149 / 152
CTC 46	KRK 165 / 170	KCS 175	KRK 133 / 120
CTC 49	KRK 165 / 156	KCS 176	KRK 157 / 150
CTC 50	KRK 173 / 132	KCS 177	KRK 157 / 150
CTC 51	KRK 181 / 156	KCS 179	KRK 133 / 120
CTC 51	KRK 181 / 152	KCS 179	KRK 149 / 152
CTC 51	KRK 182 / 186	KCS 183	KRK 149 / 152
CTC 52	KRK 140 / 156	KCS 184	KRK 161 / 152
CTC 52	KRK 182 / 186	KCS 185	KRK 157 / 150
CTC 52	KRK 188 / 185	KCS 186	KRK 161 / 152
CTC 53	KRK 144 / 186	KCS 188	KRK 137 / 152
CTC 53	KRK 181 / 185		

Fig. 9-1. RCA tuner and chassis cross reference.

representing the true operation of the receiver. The pads should be constructed as compactly as possible, and all unshielded leads at the end of the test equipment cables should be as short as possible, preferably not in excess of one inch long.

In many instances a small ceramic capacitor of approximately 1000 pF connected from the oscilloscope probe to ground will eliminate stray pickup of unwanted signals. If used, make sure the capacitor does not affect the shape of the response being observed.

Horizontal Interference

Interference from the horizontal sweep circuits of the receivers may appear on the response curves, making it difficult to observe a clearly defined trace. It is recommended, therefore, that the sweep circuits be disabled during alignment.

Signal Overload

Use of excessive signal from the sweep generator can cause overloading of the receiver circuits. To determine that this condition is not present and that the response is a true representation, turn the sweep generator output to zero. Gradually increase the output until a response is obtained. Further increase of the sweep output should not change the configuration of the response except in amplitude. If the response changes, such as flattening at the top or dropping below the base line at the bottom, decrease the sweep output to restore the proper configuration. The oscilloscope gain should be run as high as possible to maintain a usable pattern with the peak-to-peak values specified, thus requiring a lower output from the sweep generator and less chance of overload.

Insertion of markers from the signal generator should not cause distortion of the response. The markers should be kept as small as possible and still remain visible. Generally, loose coupling of the generator cable to the sweep cable will provide ample marker information without distortion of the response.

To properly align UHF/VHF tuners, it is recommended that the following test equipment be available:

Bias supply—RCA WG-307B or equivalent.

Oscilloscope—WO-91B or equivalent.

Electronic voltmeter—A voltmeter with a 5V dc scale is required.

Heterodyne frequency meter—With crystal calibrator if the signal generator is not crystal controlled.

Marker generator—To provide the following frequencies with crystal accuracy (RCA WR99A or equivalent) at 100 mV.

Intermediate frequencies:
4.5, 39.25, 47.25 MHz

Radio frequencies:
55 to 258 MHz

VHF sweep generator—Meeting the following requirements (RCA WR69A or equivalent):

(a) Frequency ranges:
35 to 90 MHz, to 12 MHz sweep width
170 to 225 MHz, 12 MHz sweep width

(b) Output adjustable with at least 100 mV maximum.

(c) Output constant on all ranges.

(d) "Flat" output on all attenuator positions.

UHF alignment fixture—The schematic layout is designed to include the mechanical assembly indicating the relative size of the compartmented areas. High-quality components should be used throughout and shield covers employed to completely enclose the assembly.

UHF sweep generator—With a frequency range of 470 to 890 MHz (RCA type WR86A or equivalent).

UHF signal generator—To provide frequencies between 461 and 931 MHz with crystal accuracy.

Oscilloscope preamplifier—Required for alignment of UHF tuner only.

PADS & FIXTURES

In building test pads, the precautions in construction should be followed in the paragraph "Equipment Terminations." Good quality disc capacitors and 5 percent composition resistors are required. Figures 9-2A through G and Fig. 9-3A and B are schematics of the pads and fixtures required for proper alignment of the tuners covered in this chapter. Several of these pads are available through RCA Parts and Accessories. Each pad is housed in a compact plastic case equipped with a self-holding connector. The shielded connecting cable is provided with the proper connector for the instrument involved.

The sweep-matching pads, shown in Fig. 9-2C, are more reliable when terminating the output with the resistor leads rather than attaching clips. The leads may be inserted in the hollow terminals of the tuner balun and thus eliminate the possibility of shunt capacitance appearing across the balun due to the attaching clips. If attaching clips are desired, the original 300-ohm line from the antenna terminals to the tuner should be left connected to the tuner and the matching pad connected to the antenna binding posts.

KRK 123 VHF RF ALIGNMENT

Connect test equipment as shown, then proceed with alignment.

Bias—Apply −1.5 volts to rf agc terminal, C3.

Oscilloscope—Connect to tuner i-f output at C16. Use direct probe with loading fixture (Fig. 9-2F). Calibrate for 0.5 volt peak-to-peak sensitivity. (See Figs. 9-4 through 9-6.)

Sweep generator—Connect to receiver antenna terminals through matching pad (Fig. 9-2C).

Generator sweep signal must be coupled to tuner through matching balun. Balun is located on receiver antenna terminal bracket assembly.

Marker generator—Couple rf output cable loosely to sweep output cable to provide markers.

VTVM—Take injection voltage reading at mixer grid test point in series with 100K resistor. Do not leave connected during alignment. Injection voltage should read —1.5 to —4.0 volts at no signal. Replace V2 if not within limits.

Miscellaneous—Detune L30, the i-f mixer plate coil to reduce its effect on the trace (core out of coil).

B+ supply—If tuner is to be aligned while removed from chassis, connect +140V to C22.

Filament supply—If tuner is to be aligned while removed from chassis, connect appropriate supply to C4 and adjust for 450 mA load.

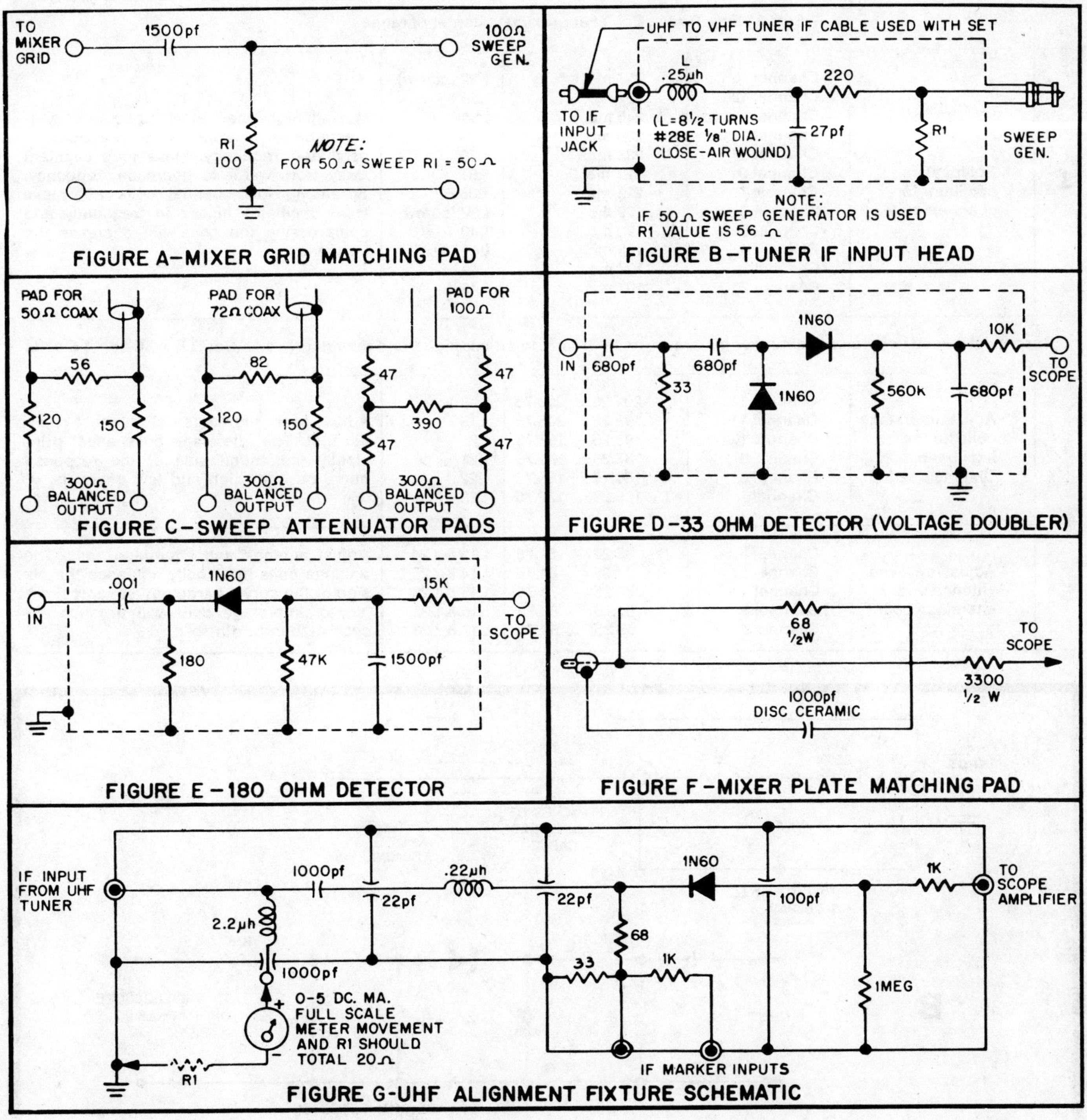

Fig. 9-2. Matching pads and accessories.

STEP		SWEEP GENERATOR	MARKER GENERATOR	ADJUST	REMARKS
1	Establish curve on Channel 13	Channel 13	211.25 mc. 215.75 mc.	L5 L6	Knife L5 & L6 for optimum curve. L5 Positions marker. L6 adjusts for maximum mid-channel gain.
Bias tuner to cut off; —10 to —5 volts at R F AGC terminal. Maximum sweep output and "scope" vertical gain may be required to observe correct neutralization.					
2	Adjust neutralization	Channel 10	Not used this step	C20 "Gimmick"	Move copper leaf (C20) in relation to body of C9 for least mid-channel response.
Set L31 Fine Tuning Coil at approximate center of range.					
3	Set local oscillator for channels 13-2	Channel 13 Channel 12 Channel 11 Channel 10 Channel 9 Channel 8 Channel 7 Channel 6 Channel 5 Channel 4 Channel 3 Channel 2	257 mc. 251 mc. 245 mc. 239 mc. 233 mc. 227 mc. 221 mc. 129 mc. 123 mc. 113 mc. 107 mc. 101 mc.	L32 (screw) L33 L34 L35 L36 L37 L38 L39 (screw) L40 L41 L42 L43	Moving high channel stamped loop segments closer to wafer switch will increase oscillator frequency. Move loop segment away from wafer to decrease frequency. Spreading low channel coils will move local oscillator higher in frequency and compressing the coils will decrease the frequency.
Reconnect sweep generator and marker generator, as in Step 1, to adjust high band interstage coils. Set R F AGC to—1.5 volts					
4	Adjust interstage coils for the high band Channels 12 to 7	Channel 12 Channel 11 Channel 10 Channel 9 Channel 8 Channel 7	205.25 209.75 199.25 203.75 193.25 197.75 187.25 191.75 181.25 185.75 175.25 179.75	L18 L19 L20 L21 L22 L23	Adjust each interstage coil, highest channel first. The interstage coils affect principally the sound side of the response curve on both high and low channels (Fig. 9-7).
5	Adjust low band antenna and interstage coils	Channel 6 Channel 5 Channel 4 Channel 3 Channel 2	83.25 87.75 77.25 81.75 67.25 71.75 61.75 65.75 55.25 59.75	L13 & L24 L14 & L25 L15 & L26 L16 & L27 L17 & L28	Adjust antenna and interstage coils. The antenna coils principally influence the pix side of the curve. Large adjustments (knifing) of interstage coils will produce excessive bandwidth.

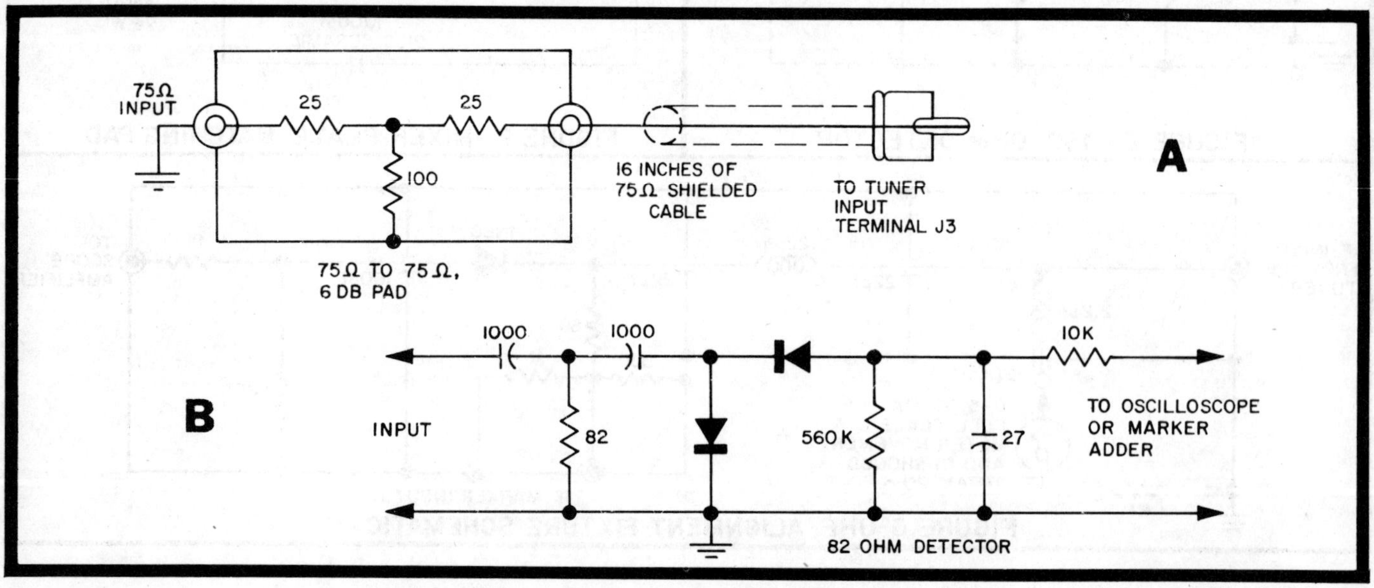

Fig. 9-3. Matching pad and detector schematic.

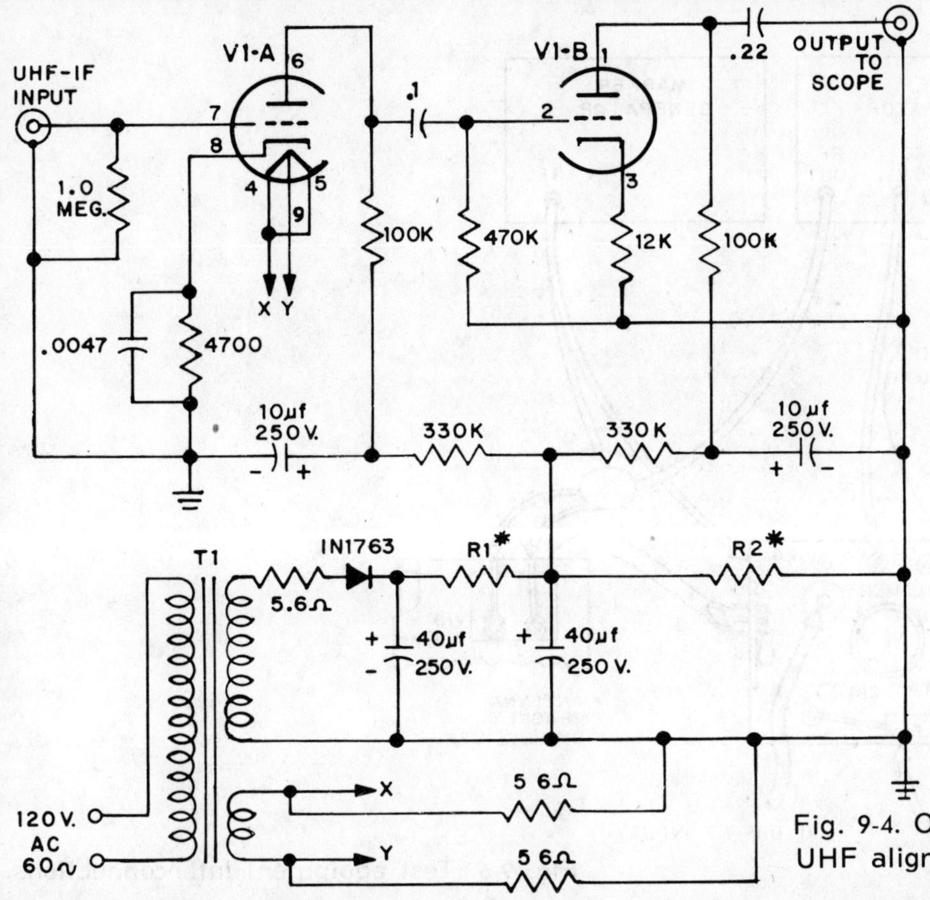

V1—12AX7 or 7025 (7025 preferable)

All resistors 1/2 W unless otherwise noted.

T1—Merit P3046 or equivalent

Sec: 150 VAC @ 25ma DC.

Fil: 6.3 VAC @ 0.5 A.

R1 6.8K 2W,

R2 100K 2W

T1—Stancor PS8415 or equivalent

Sec: 125V @ 15ma DC.

Fil: 6.3 VAC @ 0.6 A.

R1 4.7K, 2W,

R2 47K, 2W

Fig. 9-4. Oscilloscope preamplifier for UHF alignment.

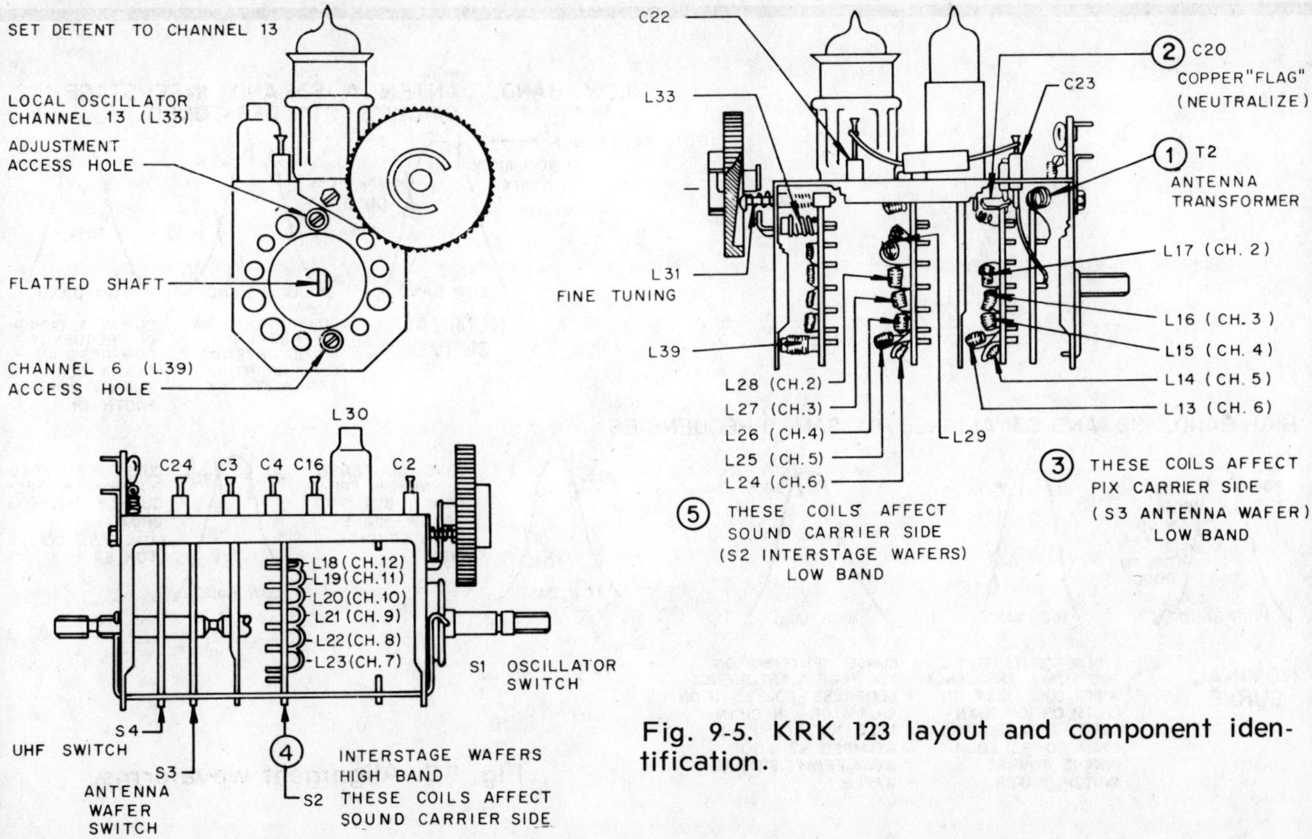

Fig. 9-5. KRK 123 layout and component identification.

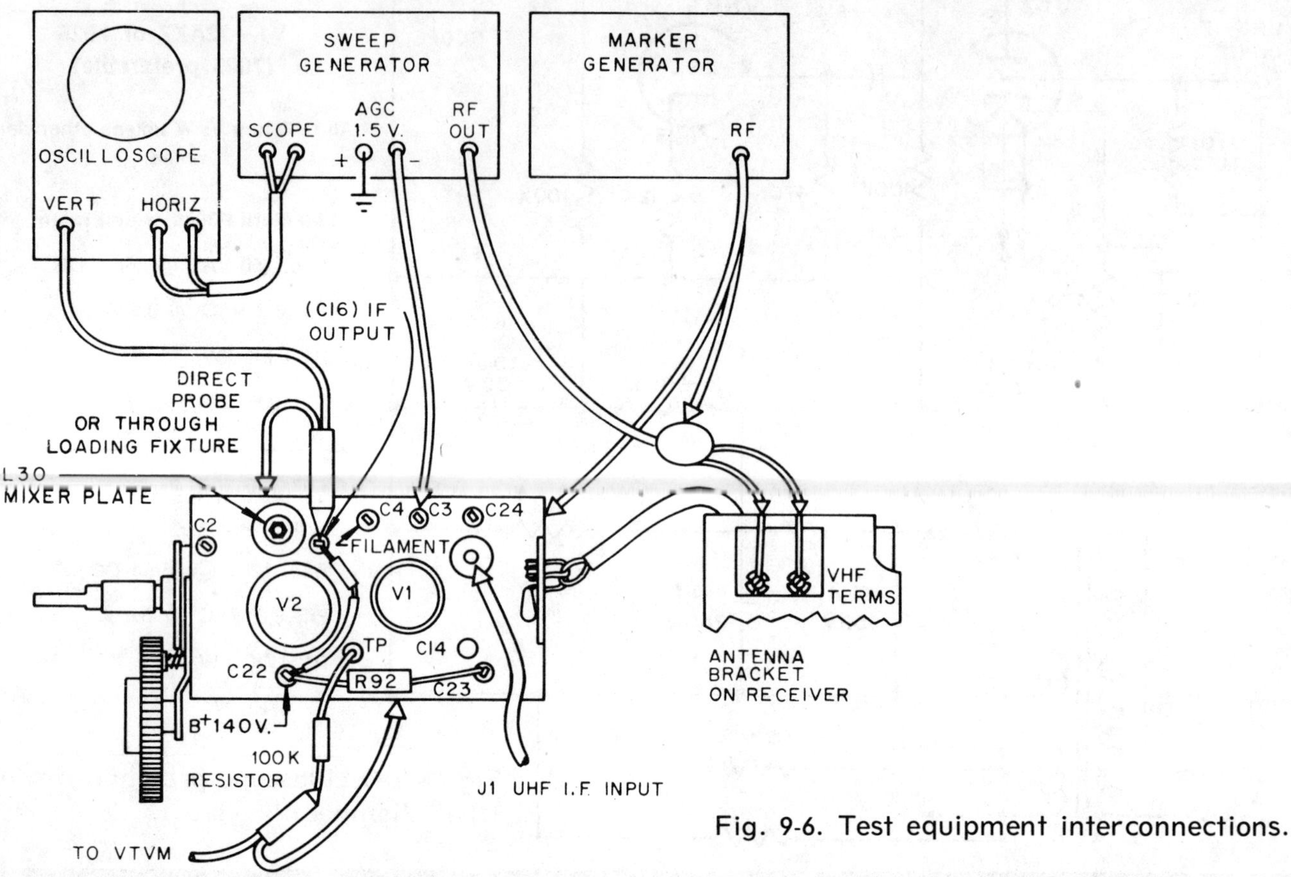

Fig. 9-6. Test equipment interconnections.

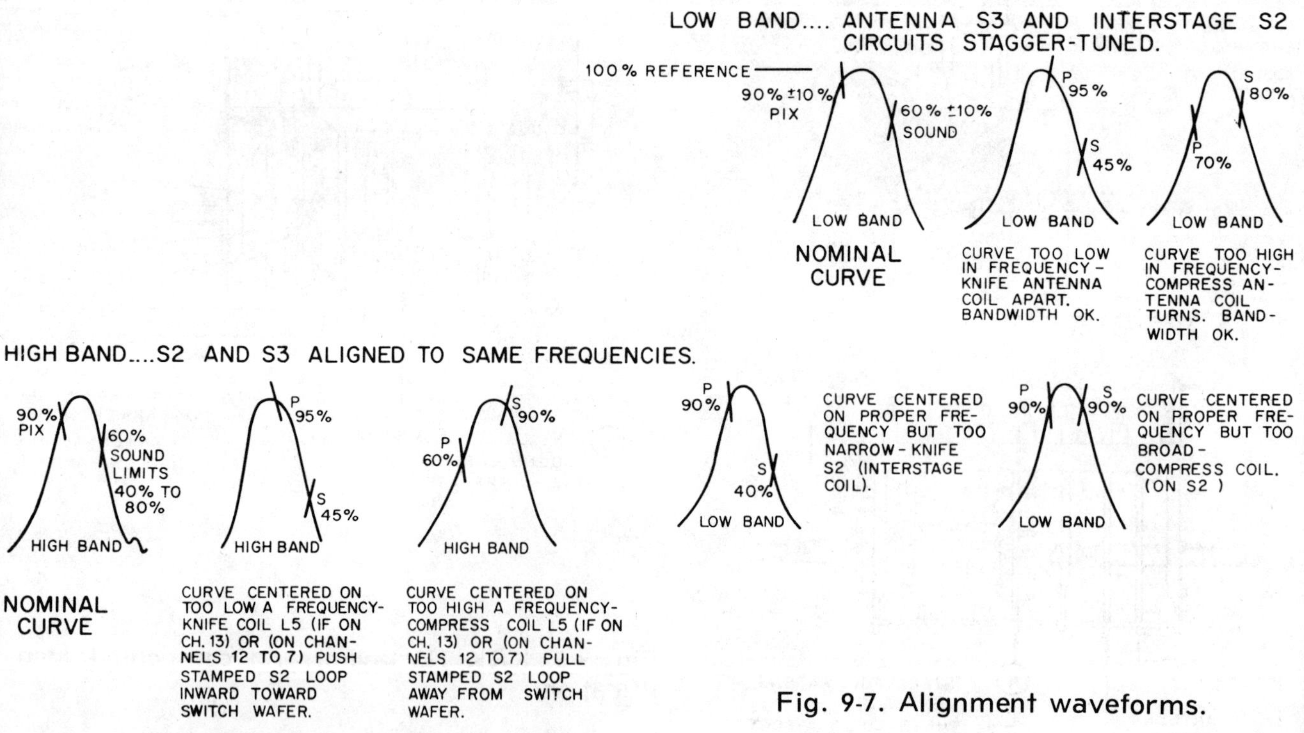

Fig. 9-7. Alignment waveforms.

Channel Number	Picture Carrier Freq. MHz	Sound Carrier Freq. MHz	Receiver R-F Osc. Freq. MHz (Zero Beat)	Channel Number	Picture Carrier Freq. MHz	Sound Carrier Freq. MHz	Receiver R-F Osc. Freq. MHz (Zero Beat)	Channel Number	Picture Carrier Freq. MHz	Sound Carrier Freq. MHz	Receiver R-F Osc. Freq. MHz (Zero Beat)
VHF TELEVISION FREQUENCIES				28	555.25	559.75	601	57	729.25	733.75	775
2	55.25	59.75	101	29	561.25	565.75	607	58	735.25	739.75	781
3	61.25	65.75	107	30	567.25	571.75	613	59	741.25	745.75	787
4	67.25	71.75	113	31	573.25	577.75	619	60	747.25	751.75	793
5	77.25	81.75	123	32	579.25	583.75	625	61	753.25	757.75	799
6	83.25	87.75	129	33	585.25	589.75	631	62	759.25	763.75	805
7	175.25	179.75	221	34	591.25	595.75	637	63	765.25	769.75	811
8	181.25	185.75	227	35	597.25	601.75	643	64	771.25	775.75	817
9	187.25	191.75	233	36	603.25	607.75	649	65	777.25	781.75	823
10	193.25	197.75	239	37	609.25	613.75	655	66	783.25	787.75	829
11	199.25	203.75	245	38	615.25	619.75	661	67	789.25	793.75	835
12	205.25	209.75	251	39	621.25	625.75	667	68	795.25	799.75	841
13	211.25	215.75	257	40	627.25	631.75	673	69	801.25	805.75	847
				41	633.25	637.75	679	70	807.25	811.75	853
UHF TELEVISION FREQUENCIES				42	639.25	643.75	685	71	813.25	817.75	859
14	471.25	475.75	517	43	645.25	649.75	691	72	819.25	823.75	865
15	477.25	481.75	523	44	651.25	655.75	697	73	825.25	829.75	871
16	483.25	487.75	529	45	657.25	661.75	703	74	831.25	835.75	877
17	489.25	493.75	535	46	663.25	667.75	709	75	837.25	841.75	883
18	495.25	499.75	541	47	669.25	673.75	715	76	843.25	847.75	889
19	501.25	505.75	547	48	675.25	679.75	721	77	849.25	853.75	895
20	507.25	511.75	553	49	681.25	685.75	727	78	855.25	859.75	901
21	513.25	517.75	559	50	687.25	691.75	733	79	861.25	865.75	907
22	519.25	523.75	565	51	693.25	697.75	739	80	867.25	871.75	913
23	525.25	529.75	571	52	699.25	703.75	745	81	873.25	877.75	919
24	531.25	535.75	577	53	705.25	709.75	751	82	879.25	883.75	925
25	537.25	541.75	583	54	711.25	715.75	757	83	885.25	889.75	931
26	543.25	547.75	589	55	717.25	721.75	763				
27	549.25	553.75	595	56	723.25	727.75	769				

Fig. 9-8. VHF / UHF Television Frequency Chart.

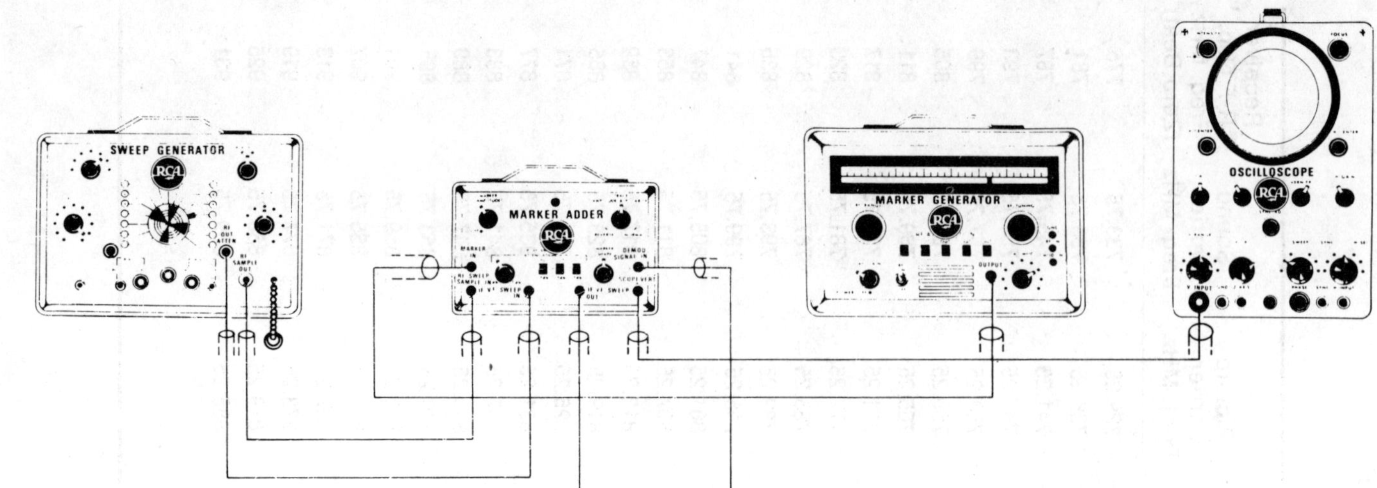

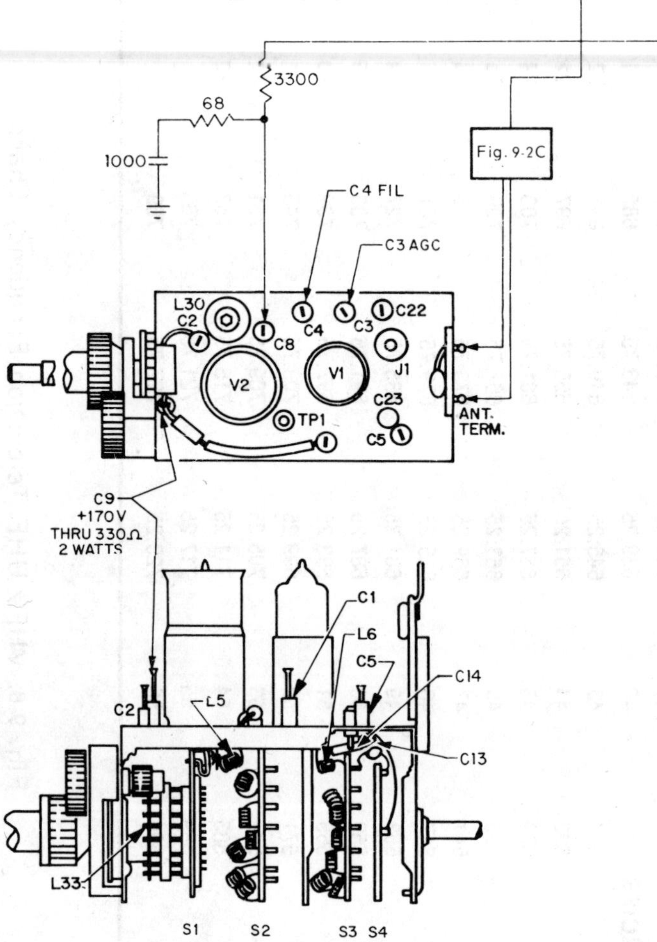

Fig. 9-9. KRK 127 tuner rf alignment.

INITIAL CONNECTIONS (Fig. 9-9).

Scope to C8 via matching pad.

Variable bias to C3.

Channel 10 sweep to antenna terminals, via pad of Fig. 9-2C.

Calibrate Scope for .1V P-P.

Tuner on Channel 10.

Markers (see frequency chart of Fig. 9-8).

OSCILLATOR INJECTION VOLTAGE CHECK (all channels).

VTVM via 100K resistor to mixer grid test point (TPI) should read −1.5 to −4 volts at no signal. Replace V2 if not within limits. Disconnect during alignment procedure.

STEP 1 NEUTRALIZATION

Bias tuner to cut-off −10 to −15 volts to C3.

Sweep output at maximum level.

Adjust C14 (tinned leaf adjacent to C13) either toward or away from body of C13 to achieve minimum mid-channel response.

STEP 2

Tuner and sweep on channel 13. Calibrate scope for .5V P-P.

−2 Volts bias to C3.

Adjust L6 for maximum mid-channel response, L5 to position markers (see response curves , Fig. 9-10).

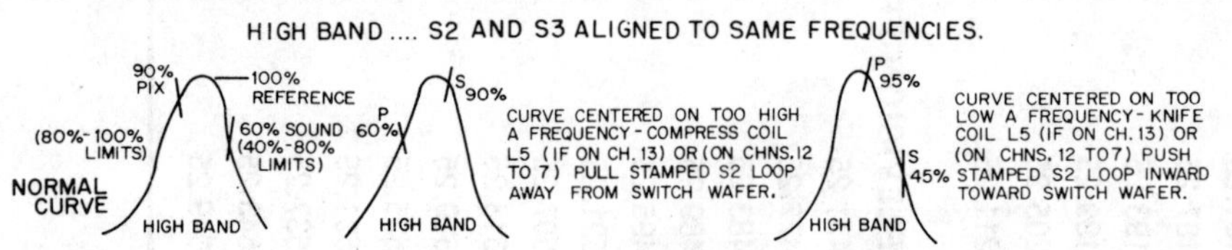

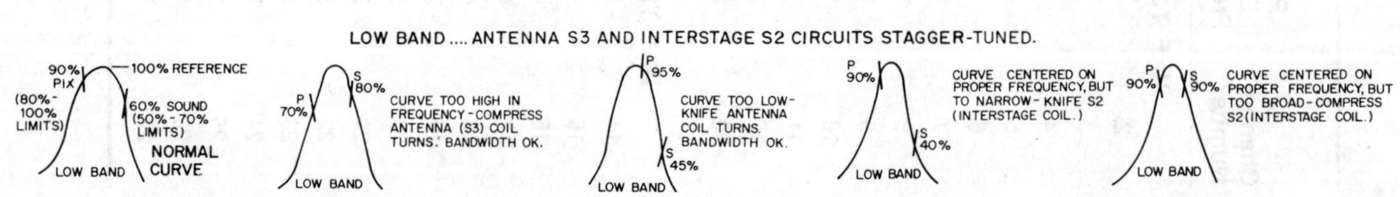

Fig. 9-10. KRK 127 VHF response curves.

STEP 3 OSCILLATOR ADJUSTMENT

Turn fine tuning maximum clockwise. Using zero beat method, adjust L32 to set local oscillator (261.5 MHz) 4.5 MHz above nominal frequency (257 MHz). Return fine tuning to nominal frequency (257 MHz) on Channel 13. (See Fig. 9-11.)

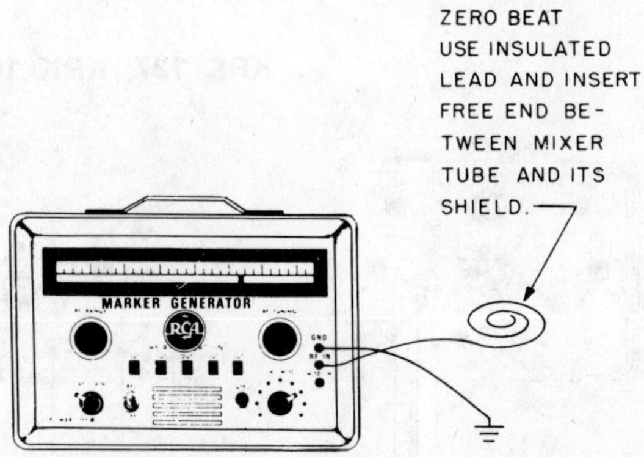

Fig. 9-11. Oscillator zero-beat.

STEP 4

Rotate tuner and sweep Channels 12 thru 7 in descending order, adjusting interstage stamped loops L18 thru L23 on S2 (RF). Interstage coils affect principally the sound side of response curve on both high and low Channels. (Refer to Fig. 9-12.)

STEP 5

Rotate tuner and sweep Channels 6 thru 2 adjusting L13 thru L17 on S3 (ant.) for correct response curve. Antenna coils principally influence the pix side of curve. Adjust interstage coils L24 thru L28 on S2 (RF) to obtain proper bandwidth.

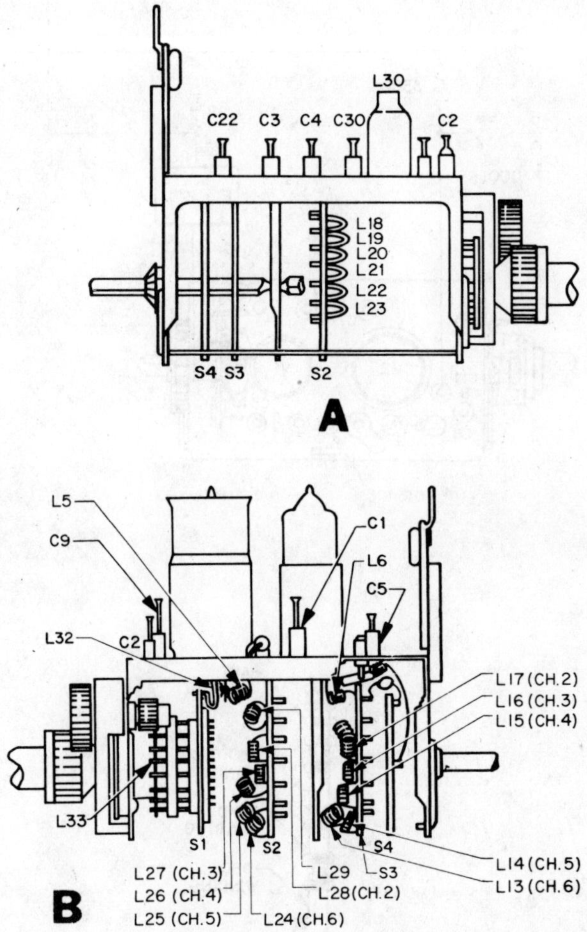

Fig. 9-12. Coil locations. A B

STEP 6

Tuner in UHF Position.

Apply IF sweep to J1 via IF input head. Adjust L29 for proper curve and maximum gain. (See Fig. 9-13.)

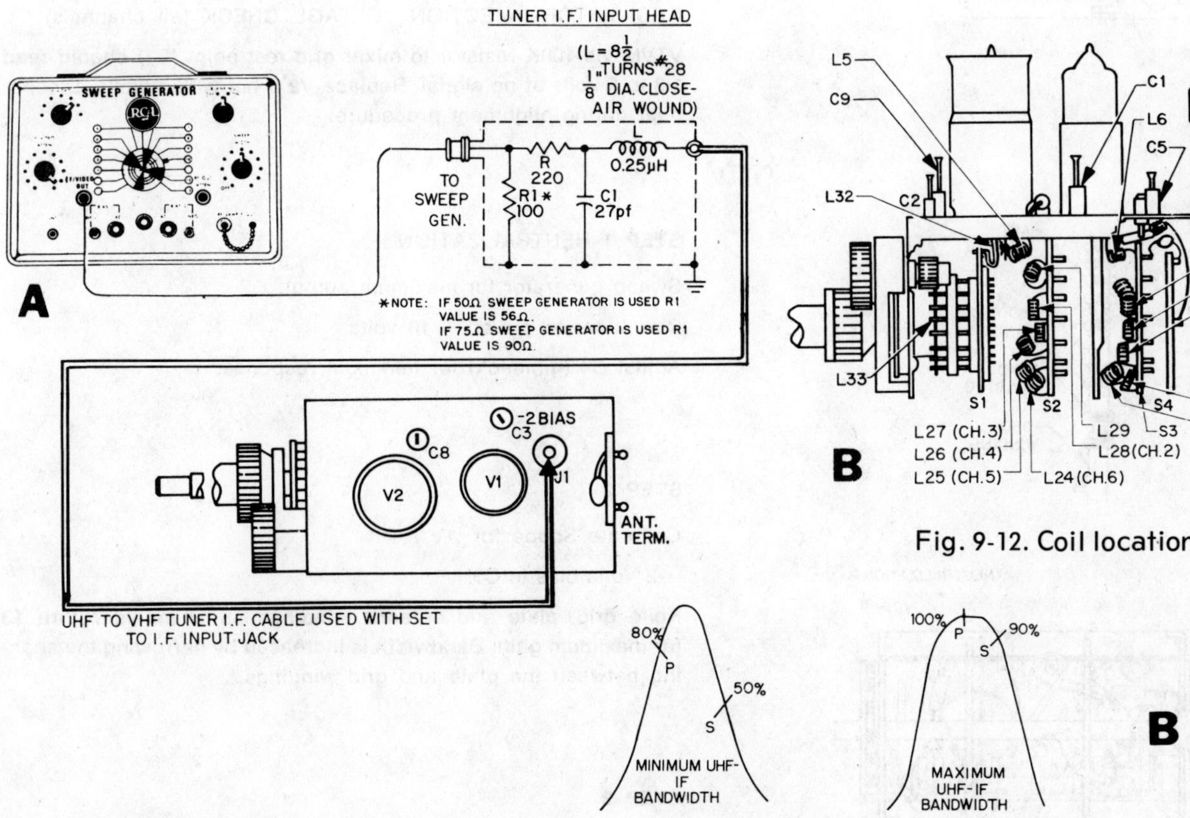

Fig. 9-13. Equipment connections (A) and response curves (B). (RCA.)

KRK 137, KRK 167 TUNER R F ALIGNMENT

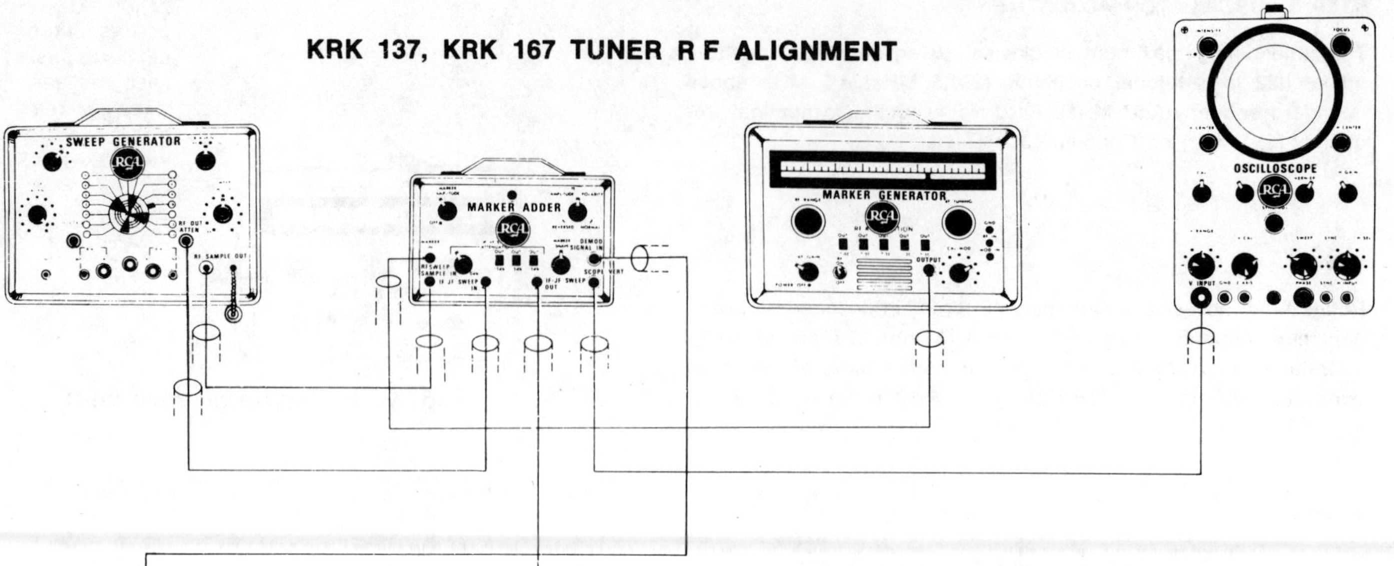

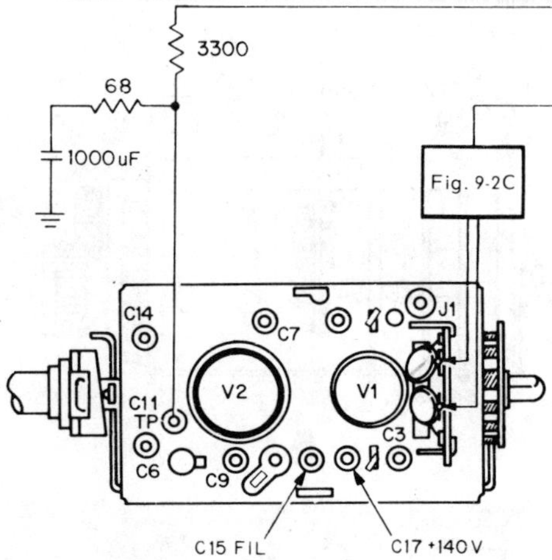

INITIAL CONNECTION (Fig. 9-14).

Scope to C11 via matching pad.

Variable bias to C3.

Channel 10 sweep to antenna terminals, via pad (Fig. 9-2C).

Tuner on Channel 10.

Calibrate Scope for .1V P-P.

Markers at (see frequency chart , Fig. 9-8)

OSCILLATOR INJECTION VOLTAGE CHECK (all channels).

VTVM via 100K resistor to mixer grid test point (C7) should read 1.8 to 5 volts at no signal. Replace V2 if not within limits. Disconnect during alignment procedure.

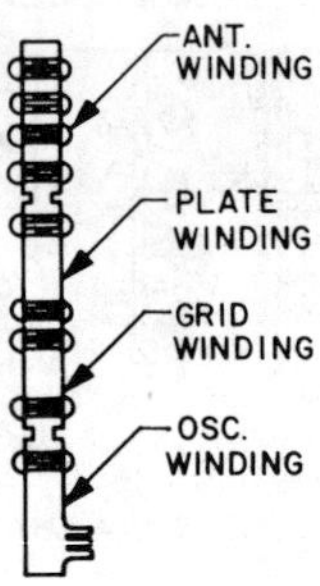

STEP 1 NEUTRALIZATION

Sweep generator for maximum output.

Bias tuner to cut-off —10 volts.

Adjust C4 (gimmick) for minimum response.

STEP 2

Calibrate Scope for .1V P-P.

—2 Volts bias to C3.

Knife grid, plate and antenna windings on channels 7 thru 13 for maximum gain. Bandwidth is increased by narrowing the spacing between the plate and grid windings.

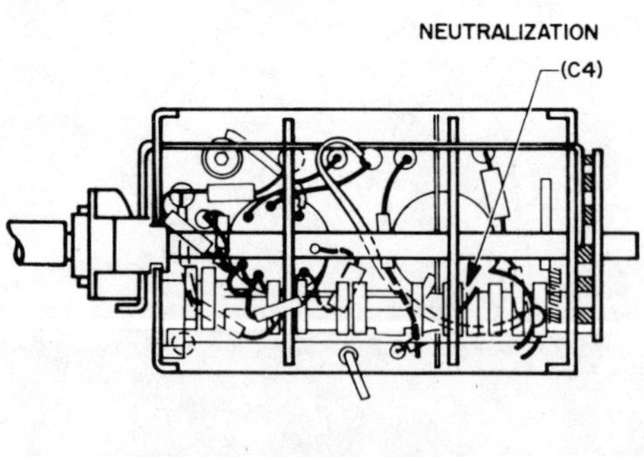

STEP 3

Knife grid, plate and antenna windings on channels 1 thru 6 for maximum gain. Bandwidth is increased by narrowing the spacing between the grid and coupling windings.

Fig. 9-14. KRK 137, KRK 167 tuner rf alignment.

STEP 4 OSCILLATOR ADJUSTMENT

Fine tuning to center of mechanical range. Using zero beat method, adjust oscillator screw on each channel.

STEP 5

Tuner in UHF Position.

Apply I F sweep to J1 via I F input head (Fig. 9-15).
Adjust slug rear of UHF strip to obtain response curve (Fig. 9-16).

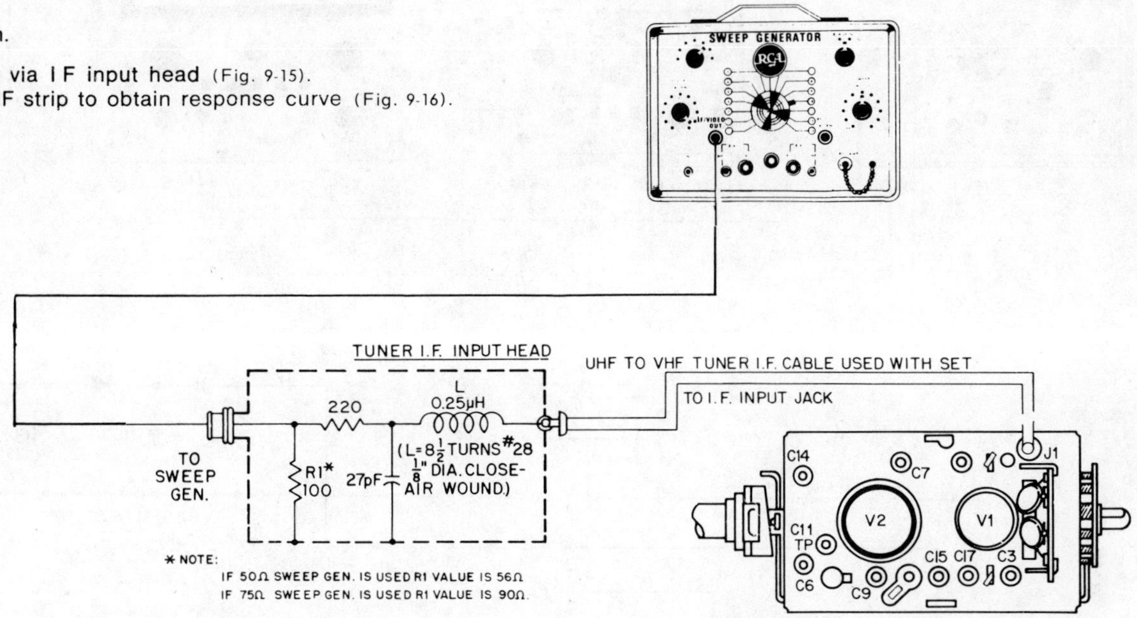

Fig. 9-15. Sweep generator alignment interconnection with matching network.

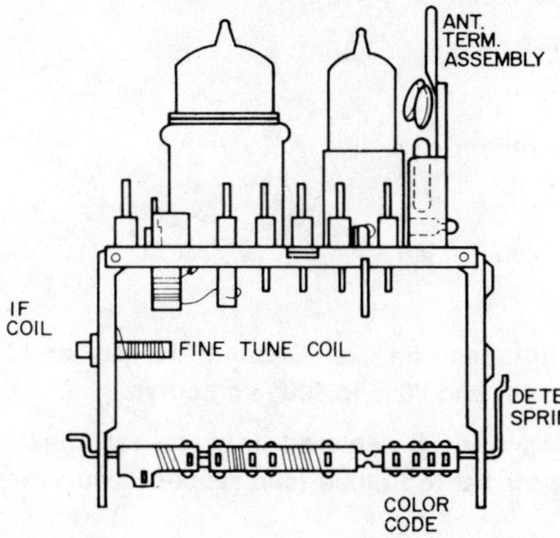

Fig. 9-16. Tuner alignment waveforms and adjustment points.

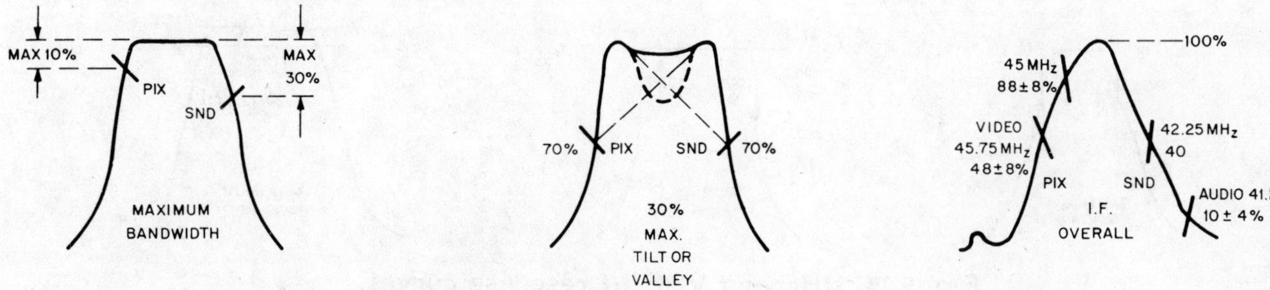

KRK 142 TUNER R F ALIGNMENT

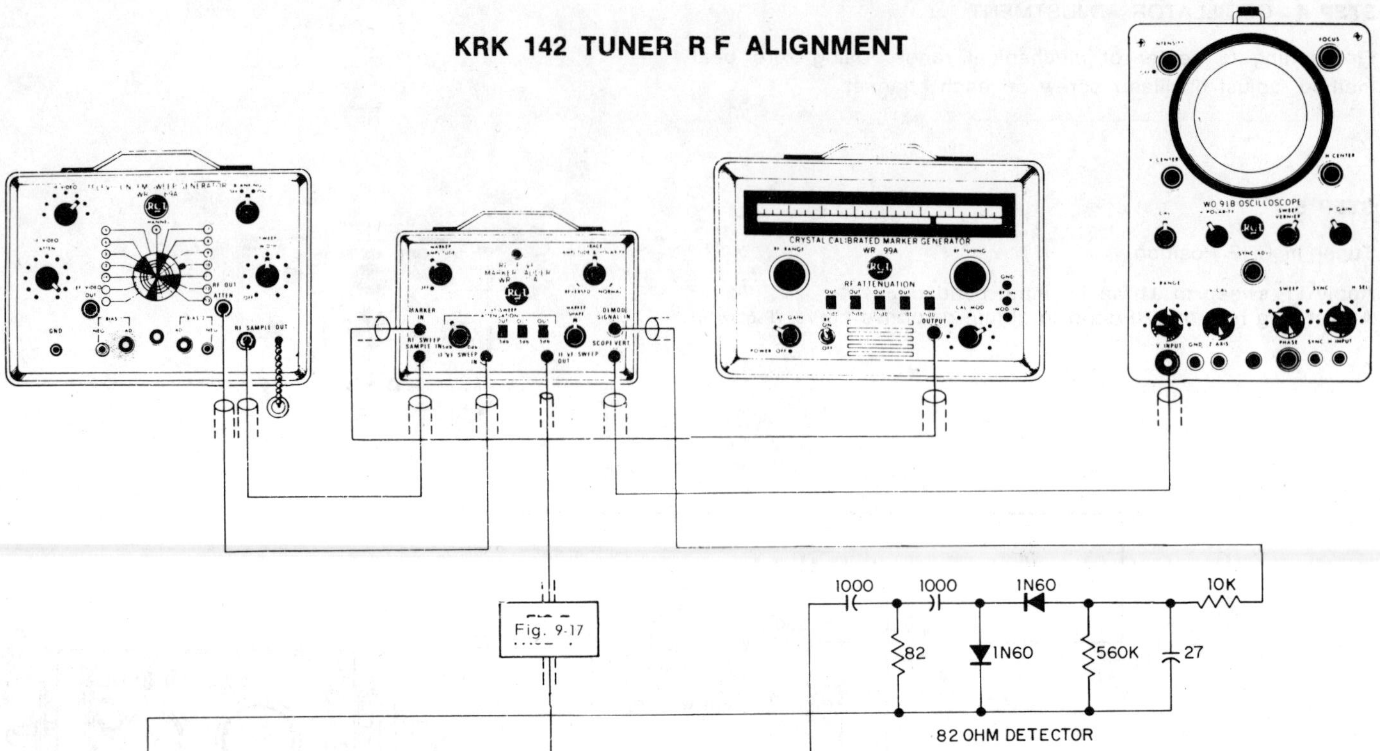

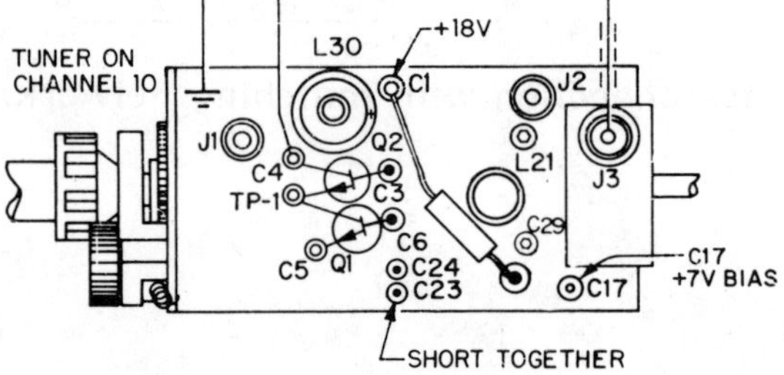

INITIAL CONNECTIONS (Fig. 9-17).

Scope to C4 via 82 ohm detector

+7V Bias to C17

Channel 10 sweep to J3 via T pad (Fig. 9-3A).

Calibrate Scope for 0.3V P-P.

Tuner on Channel 10.

Markers at (see frequency chart).

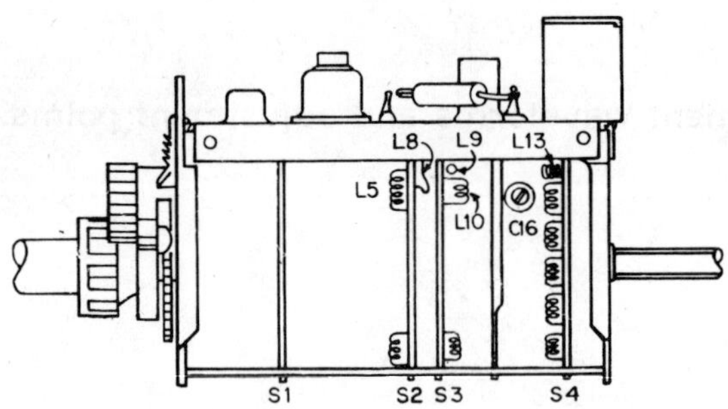

STEP 1

Adjust L5, L10 and C29 for centering on Channel 10, pix and sound markers at equal height and 80% to 100% on curve.

Adjust L9 for best coupling with L8 centered between switches. Then readjust L5 and L10 for flat top curve (see response curves of Fig. 9-18.

Fig. 9-17. KRK 142 tuner rf alignment.

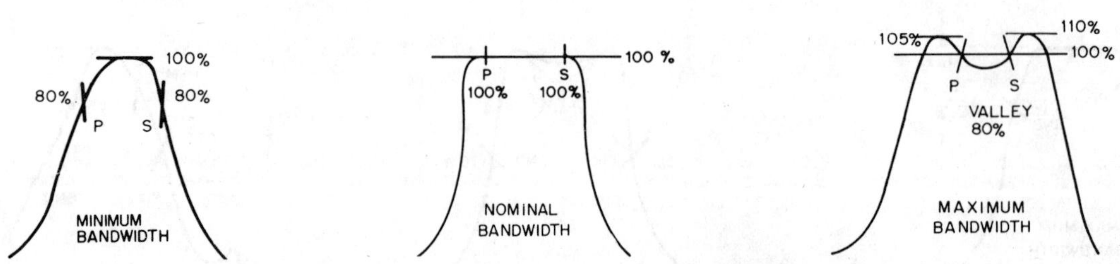

Fig. 9-18. UHF and VHF i-f response curves.

L.C.

STEP 2 OSCILLATOR ADJUSTMENT

+7V Bias to C17. Zero bias between C23 and C24.

Tuner and sweep Channel 13. Turn fine tuning maximum clockwise. Using zero beat method, adjust L3 to set local oscillator at (261.5 MHz) 4.5 MHz above nominal frequency (257 MHz). Return fine tuning to nominal frequency (257 MHz) on Channel 13.

STEP 3 OSCILLATOR ADJUTSMENT

Tuner and Sweep on Channel 10.

Adjust L5, L10, and C29 for centering on Channel 10. Adjust L8 and L9 if necessary for correct coupling. Check Channels 13 through 7 and retouch L5, L10, C29, L8 and L9 for best compromise on tracking and coupling on high channels. (See Fig. 9-19.)

STEP 4

Tuner and Sweep on Channel 6.

Adjust L20 for maximum output and L7 and L12 for proper response curve. Tuner and sweep on Channels 5 thru 2 in descending order and adjust coils L18 through L15 for flat topped curve on each channel.

STEP 5

Tuner in UHF Position. Apply IF sweep to J2 via IF in head. Adjust L21 for flat topped curve. (See Fig. 9-20.)

STEP 6 AFT CHECK

Apply ±8 volts between C23 and C24, oscillator frequency must be shifted ±1 MHz or more.

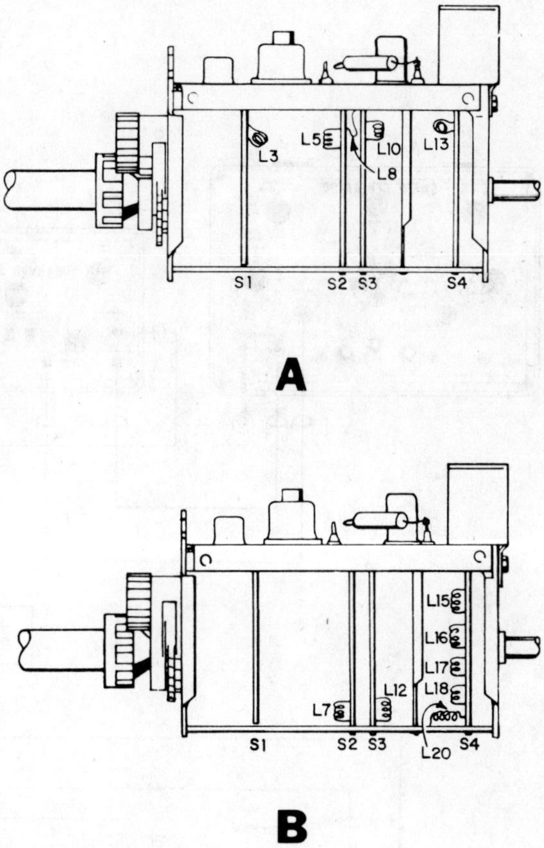

Fig. 9-19. Tuner layout.

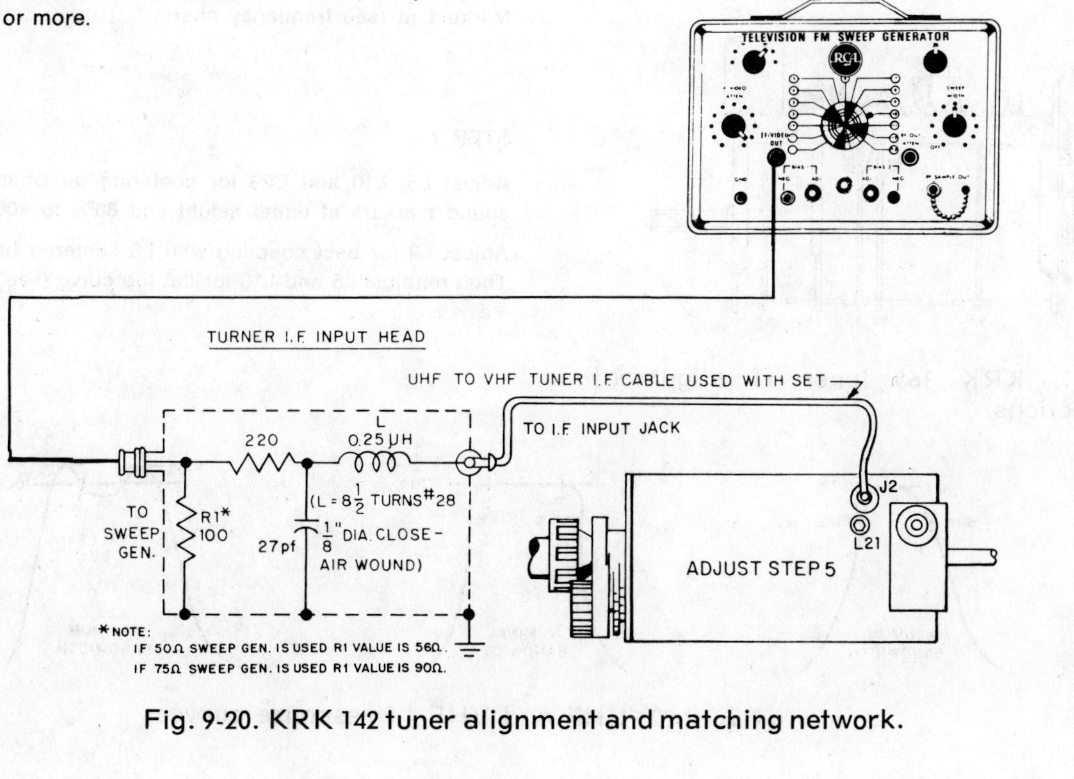

Fig. 9-20. KRK 142 tuner alignment and matching network.

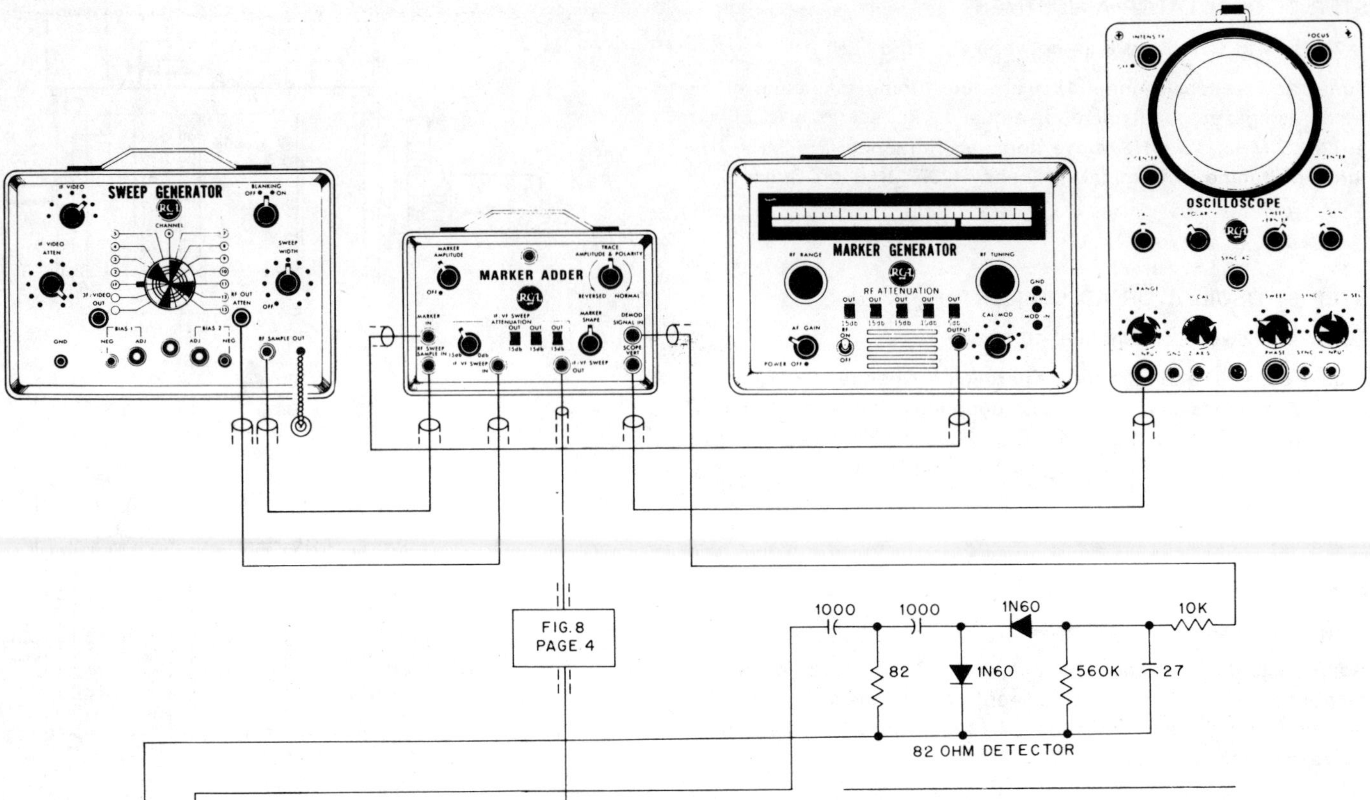

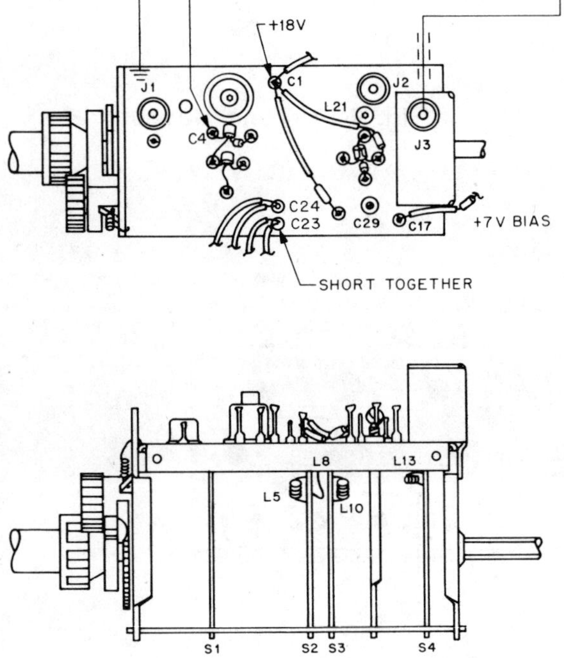

INITIAL CONNECTIONS (Fig. 9-21).

Scope to C4 via 82 ohm detector

+7V Bias to C17

Channel 10 sweep to J3 via T pad (Fig. 9-3A).

Calibrate Scope for .2V P-P.

Tuner on Channel 10.

Markers at (see frequency chart).

STEP 1

Adjust L5, L10 and C29 for centering on Channel 10, pix and sound markers at equal height and 80% to 100% on curve.

Adjust L9 for best coupling with L8 centered between switches. Then readjust L5 and L10 for flat top curve (see response curves, Fig. 9-22).

Fig. 9-21. KRK 165 tuner rf alignment interconnections.

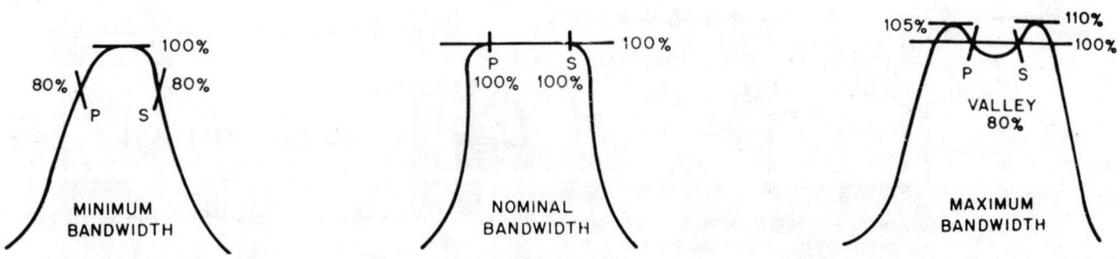

Fig. 9-22. UHF and VHF i-f response curves.

STEP 2 OSCILLATOR ADJUSTMENT

+7V Bias to C17. Zero bias between C23 and C24.

Tuner and sweep Channel 13. Turn fine tuning maximum clockwise. Using zero beat method, adjust L3 to set local oscillator at (261.5 MHz) 4.5 MHz above nominal frequency (257 MHz). Return fine tuning to nominal frequency (257 MHz) on Channel 13.

STEP 3

Tuner and Sweep on Channel 10.

Adjust L5, L10, and C29 for centering on Channel 10. Adjust L8 and L9 if necessary for correct coupling. Check Channels 13 through 7 and retouch L5, L10, C29, L8 and L9 for best compromise on tracking and coupling on high channels. (See Fig. 9-23.)

STEP 4

Tuner and Sweep on Channel 6.

Adjust L20 for maximum output and L7 and L12 for proper response curve. Tuner and sweep on Channels 5 thru 2 in descending order and adjust coils L18 through L15 for flat topped curve on each channel.

STEP 5

Tuner in UHF Position. Apply I F sweep to J2 via I F in head. Adjust L21 for flat topped curve. (See Fig. 9-24.)

STEP 6 A F T CHECK

Apply ±8 volts between C23 and C24. Oscillator frequency must be shifted ±1 MHz or more.

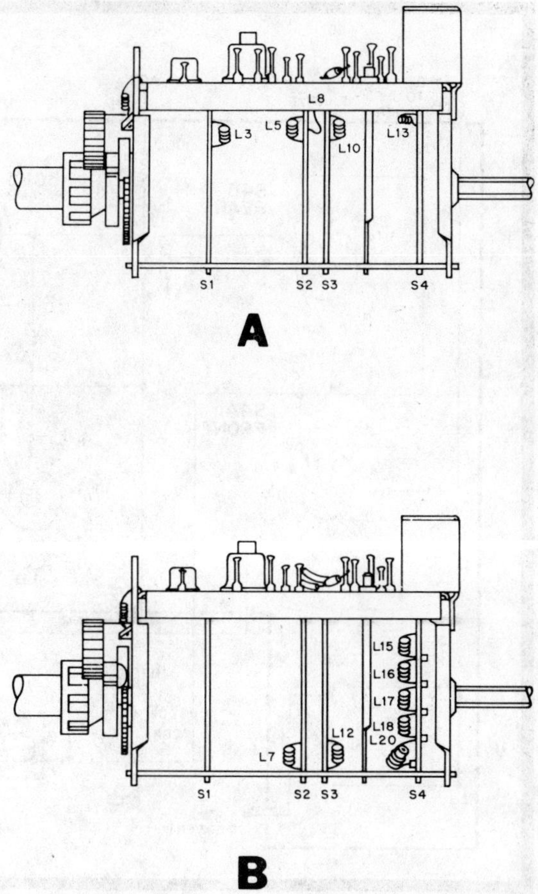

Fig. 9-23. KRK 165 tuner layout.

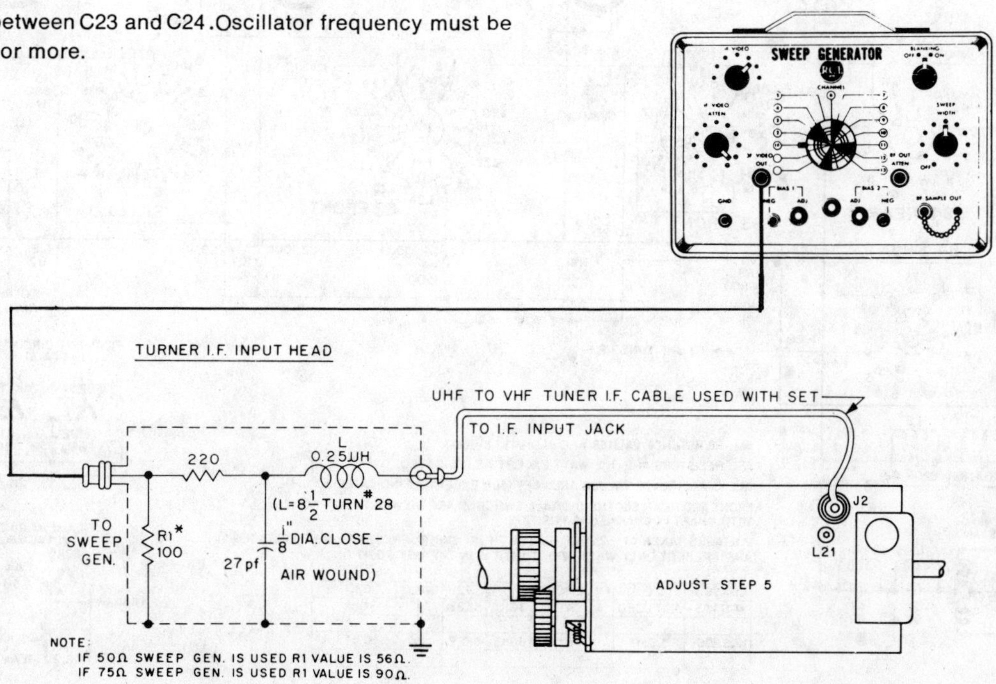

Fig. 9-24. KRK 165 tuner alignment and matching network.

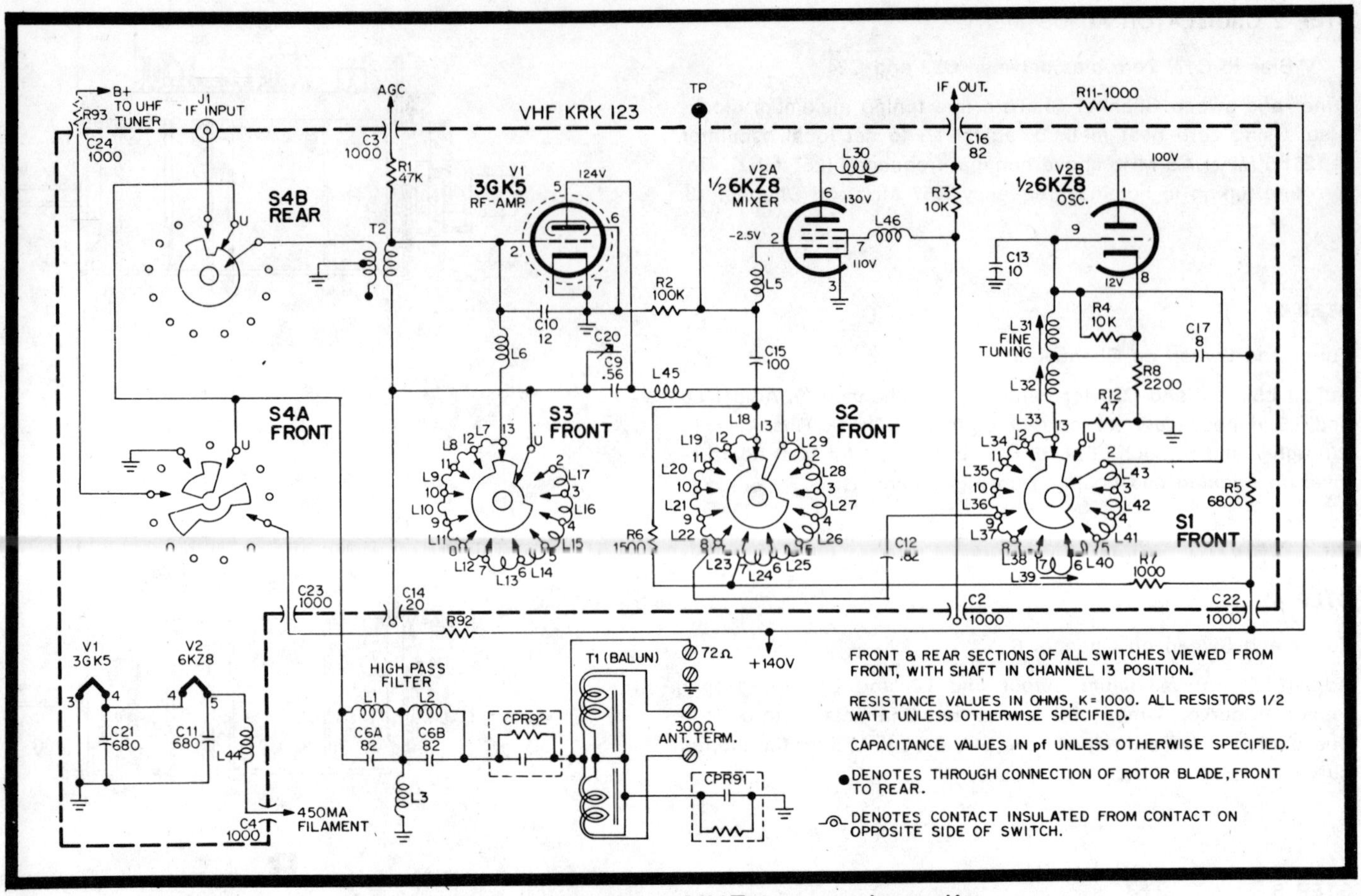

Fig. 9-25. KRK 123 VHF tuner schematic.

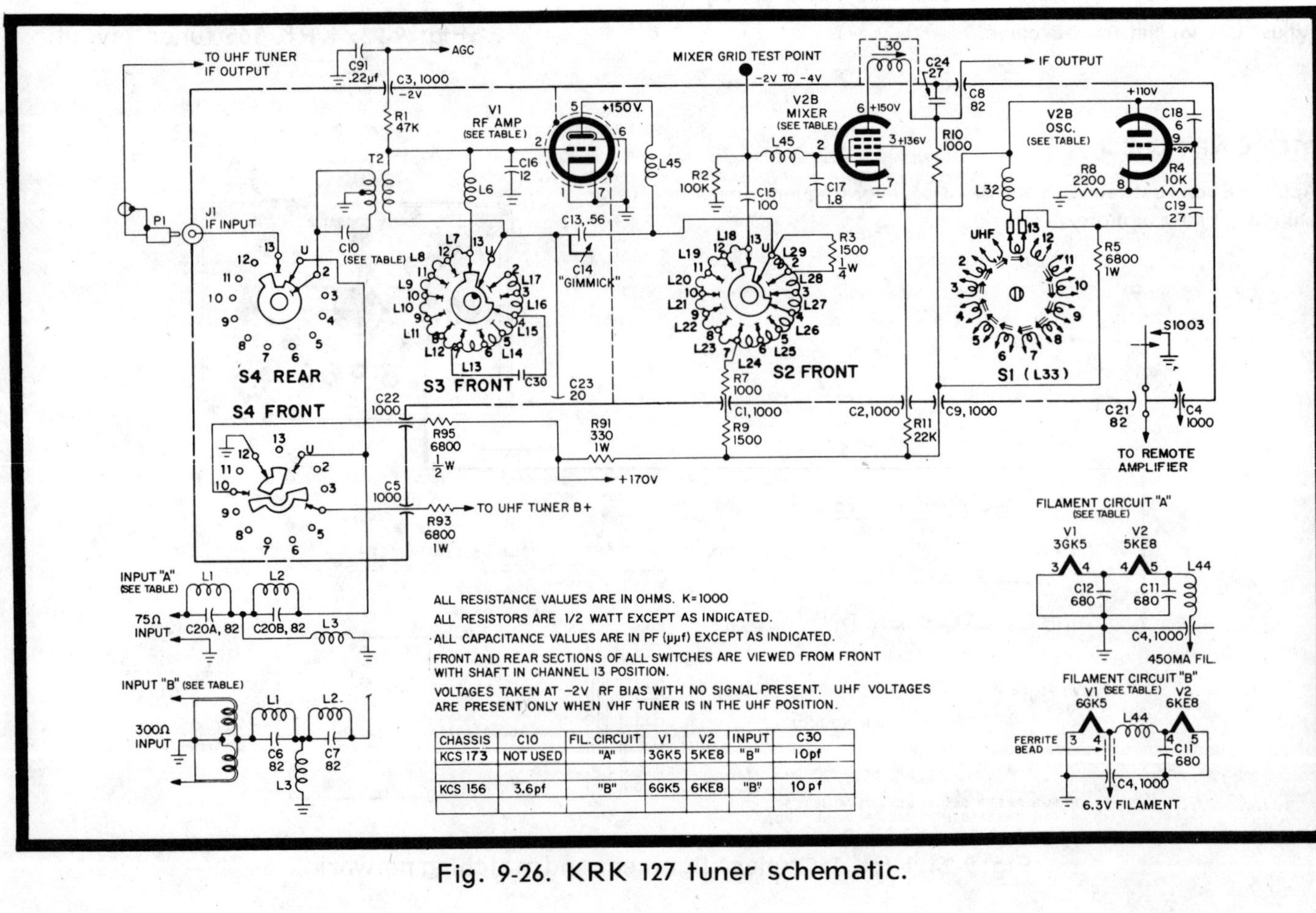

CHASSIS	C10	FIL. CIRCUIT	V1	V2	INPUT	C30
KCS 173	NOT USED	"A"	3GK5	5KE8	"B"	10pf
KCS 156	3.6pf	"B"	6GK5	6KE8	"B"	10 pf

Fig. 9-26. KRK 127 tuner schematic.

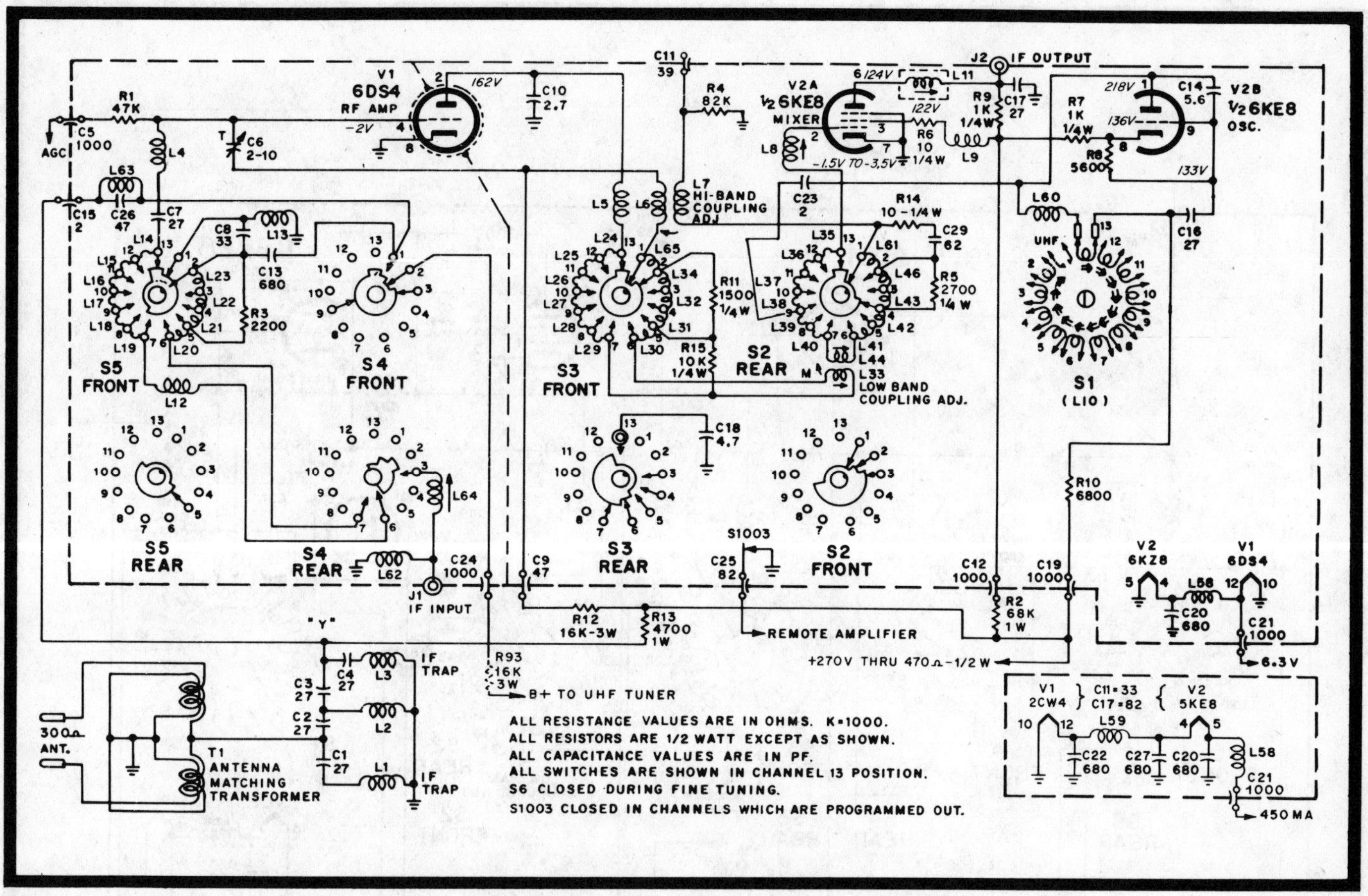

Fig. 9-27. KRK 128 VHF tuner schematic.

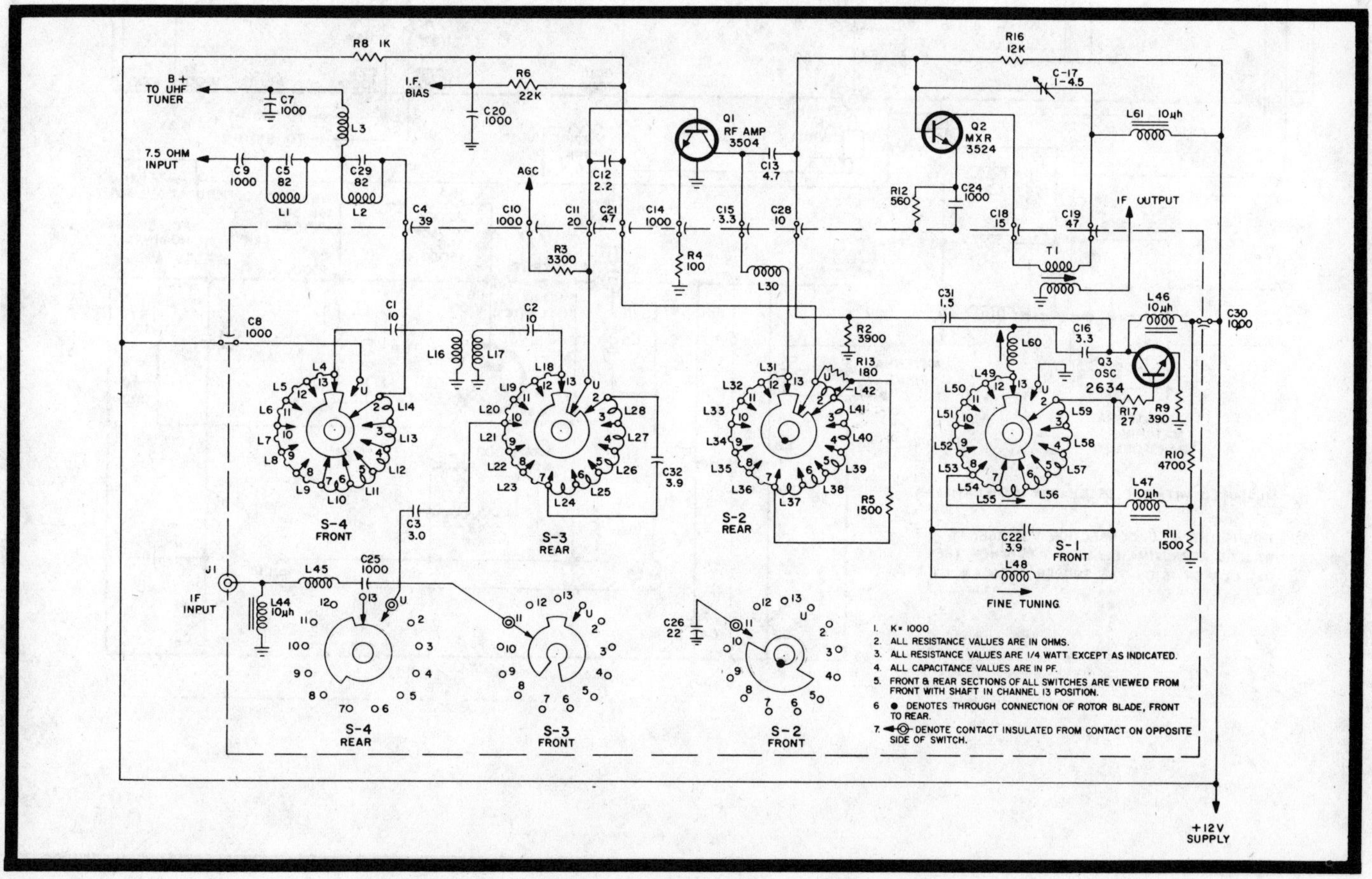

Fig. 9-28. KRK 129 VHF tuner schematic.

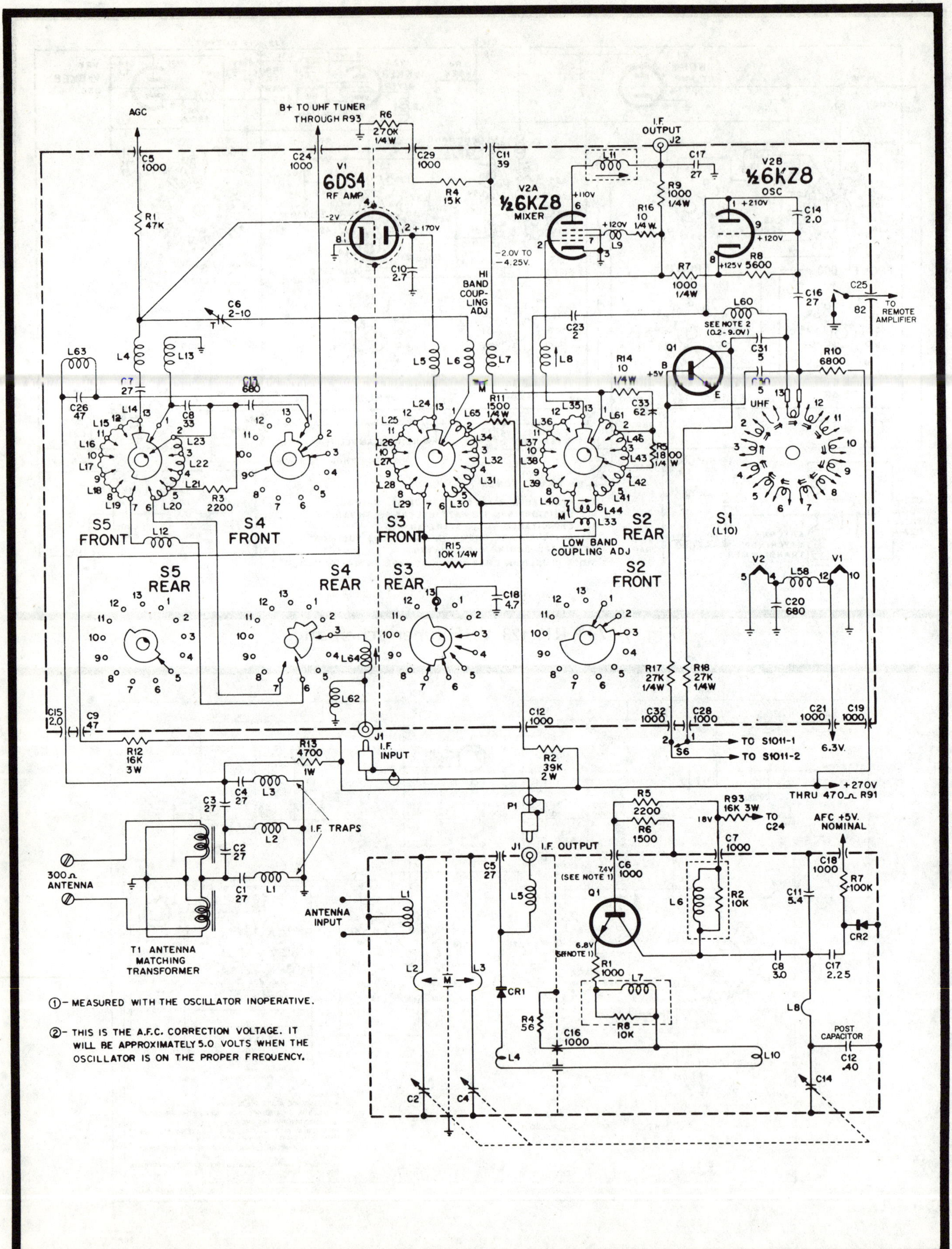

Fig. 9-29. KRK 131 / 132K schematic.

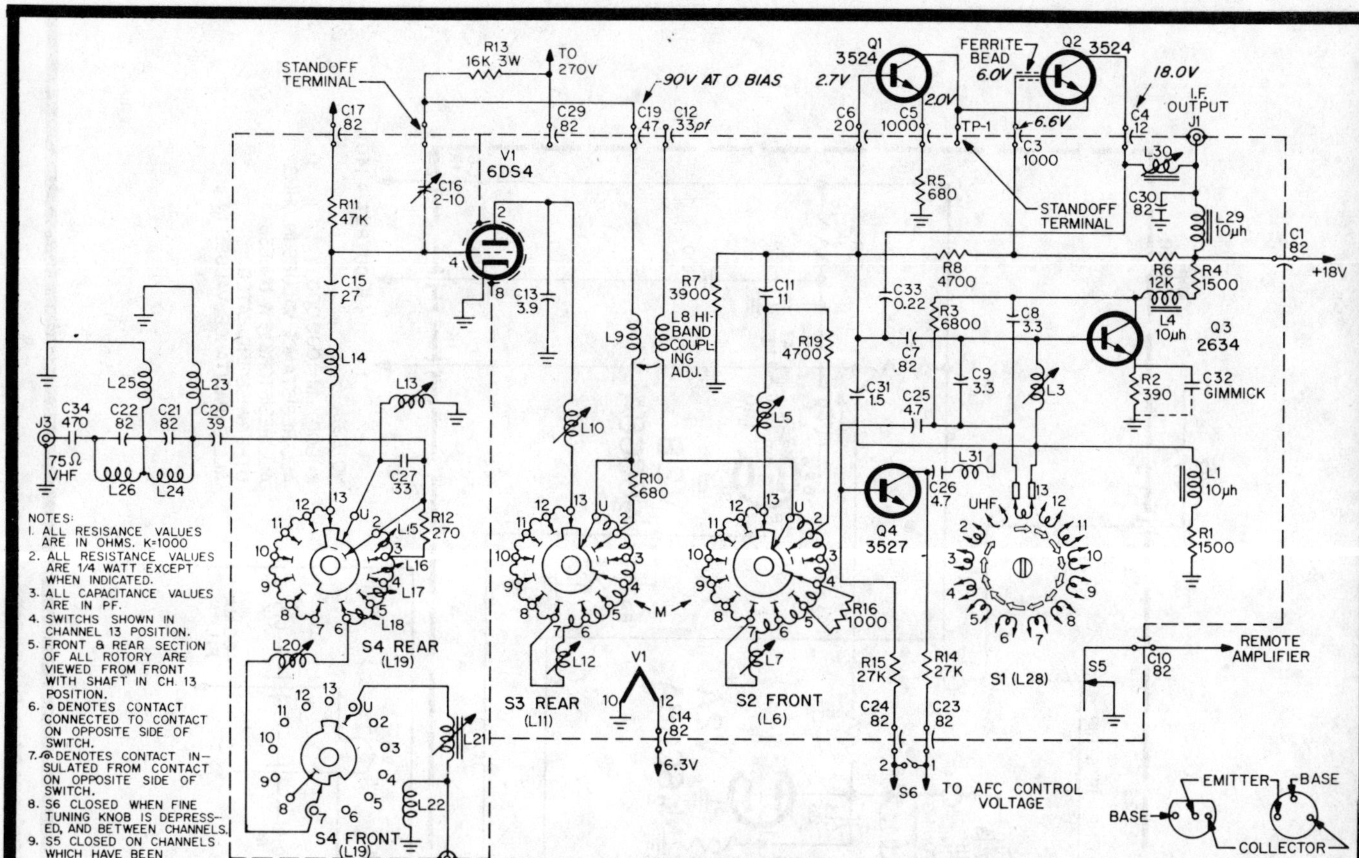

Fig. 9-30. KRK 140 VHF tuner schematic.

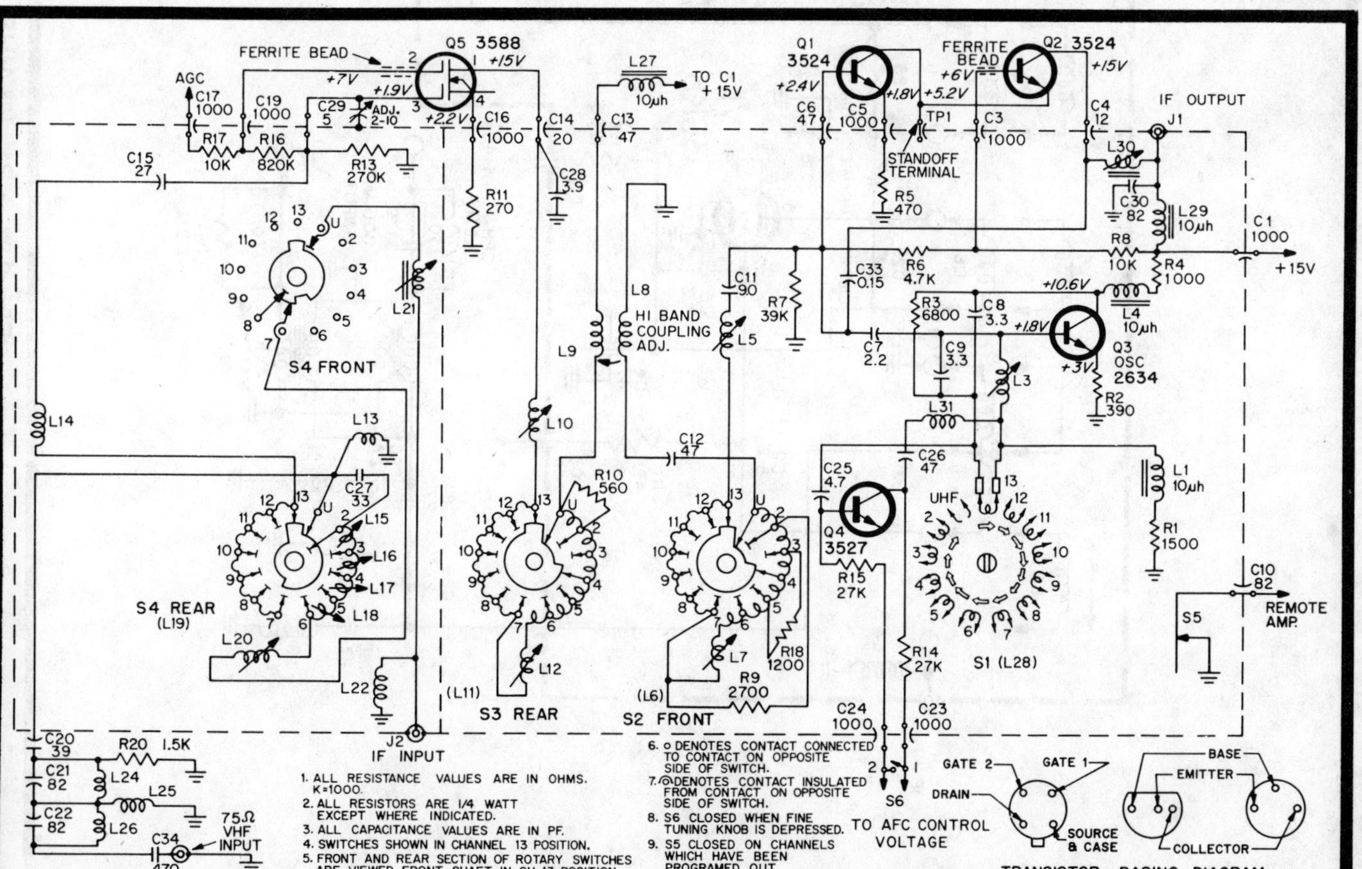

Fig. 9-31. KRK 142 VHF tuner schematic.

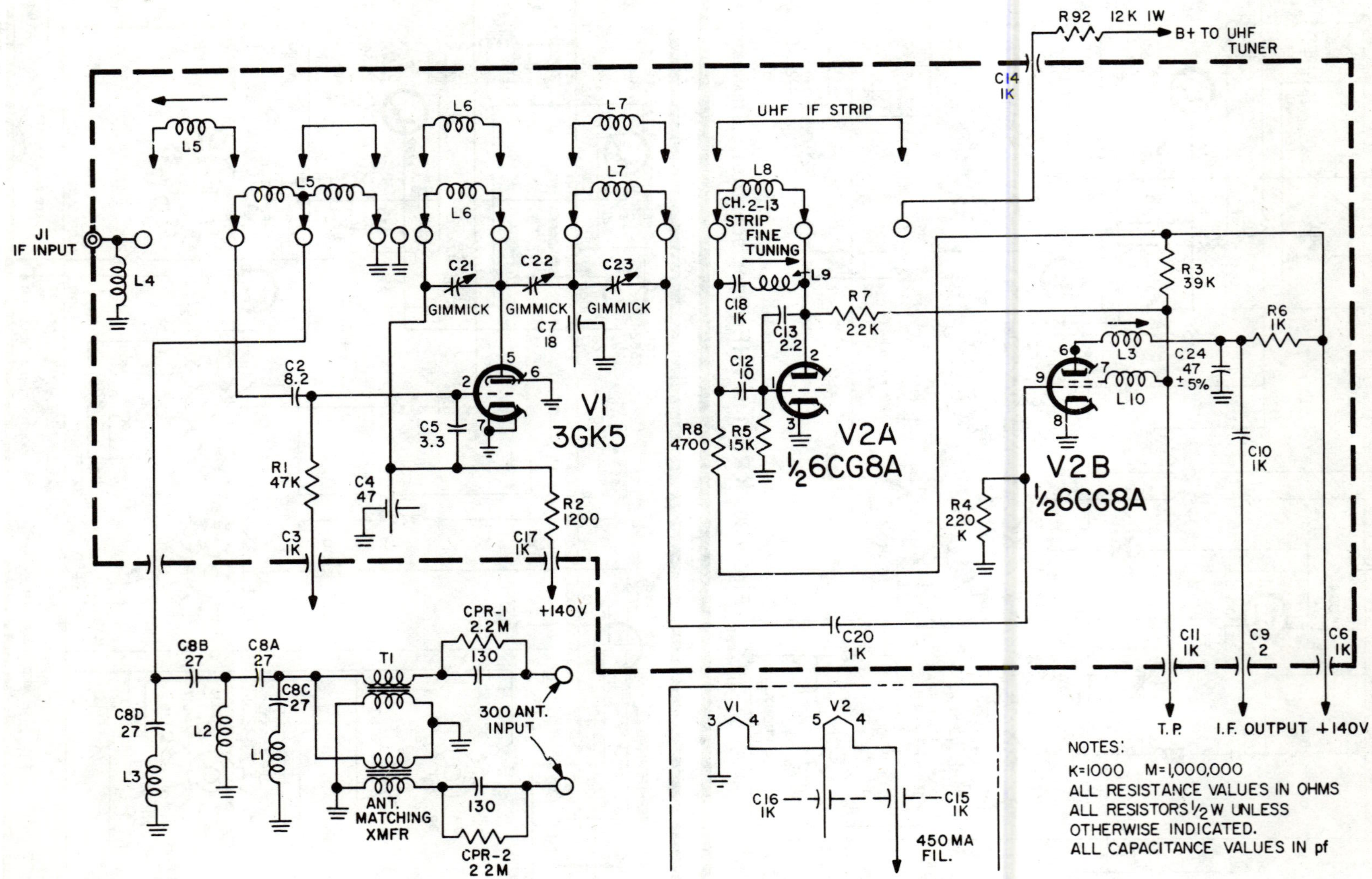

Fig. 9-32. KRK 137 VHF tuner schematic diagram.

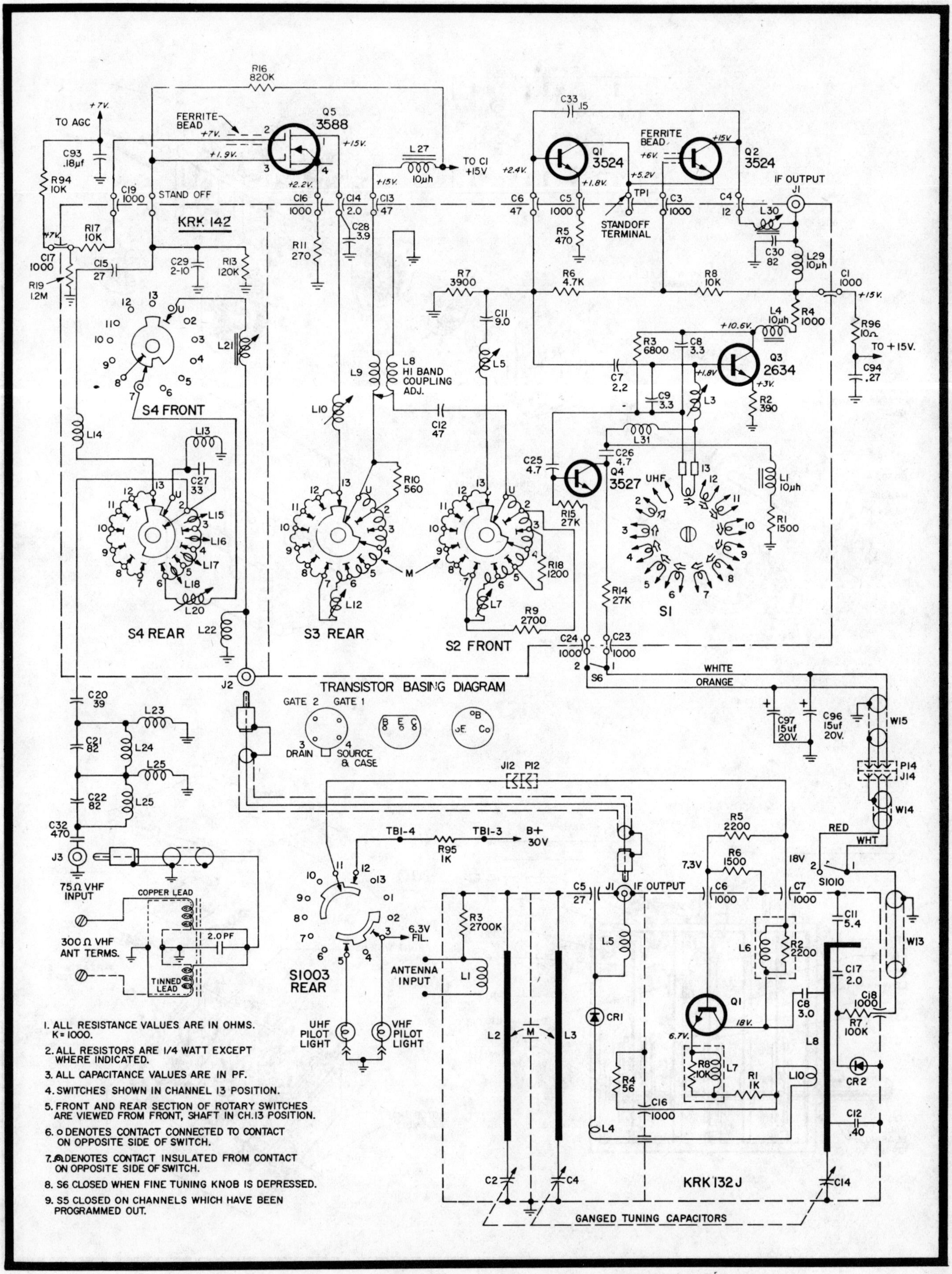

Fig. 9-33. KRK 142 / 132A VHF / UHF tuner schematic diagram.

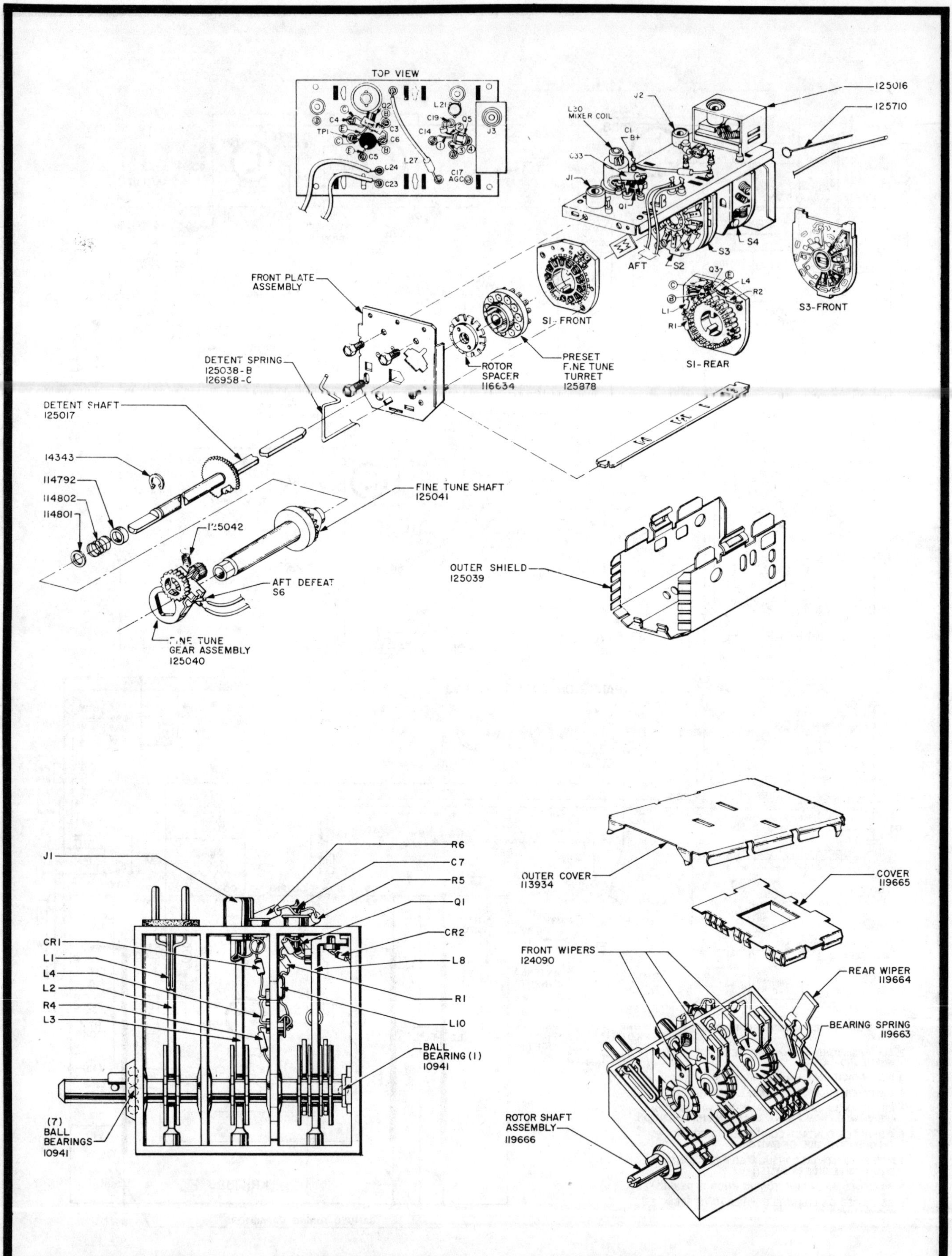

Fig. 9-34. KRK 142/132 VHF/UHF tuner parts identification.

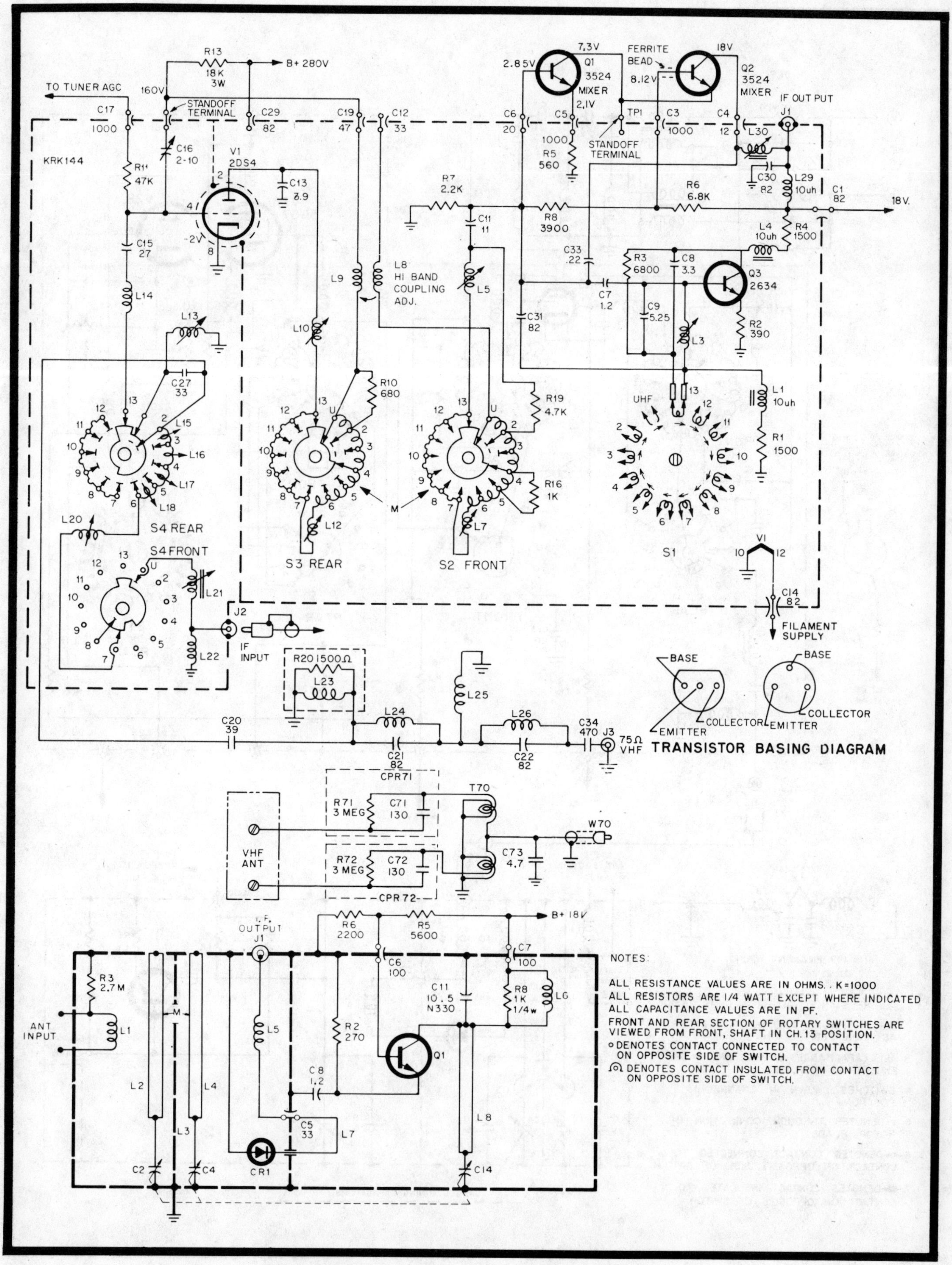

Fig. 9-35. KRK 144 / 186 tuner schematic.

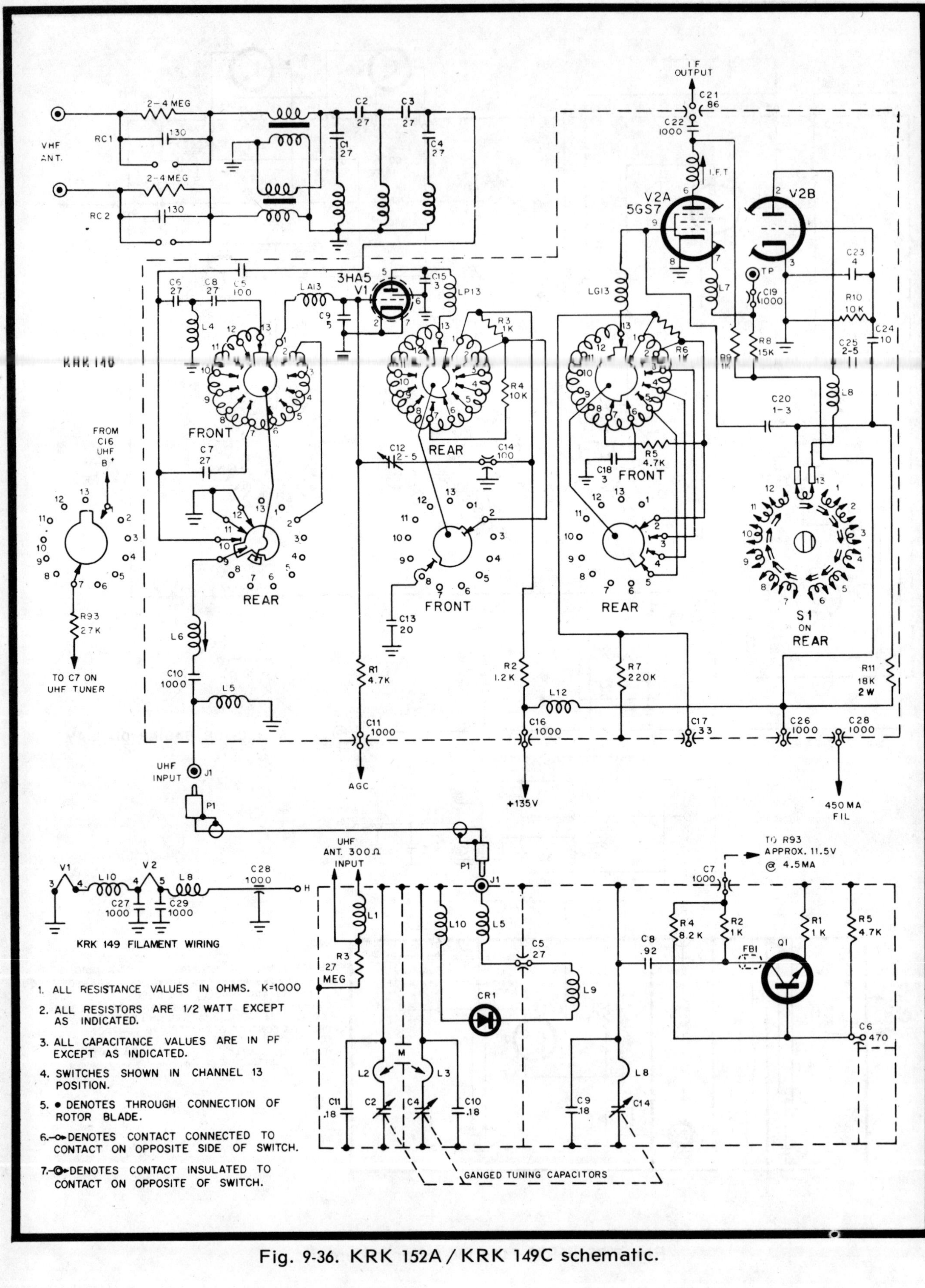

Fig. 9-36. KRK 152A / KRK 149C schematic.

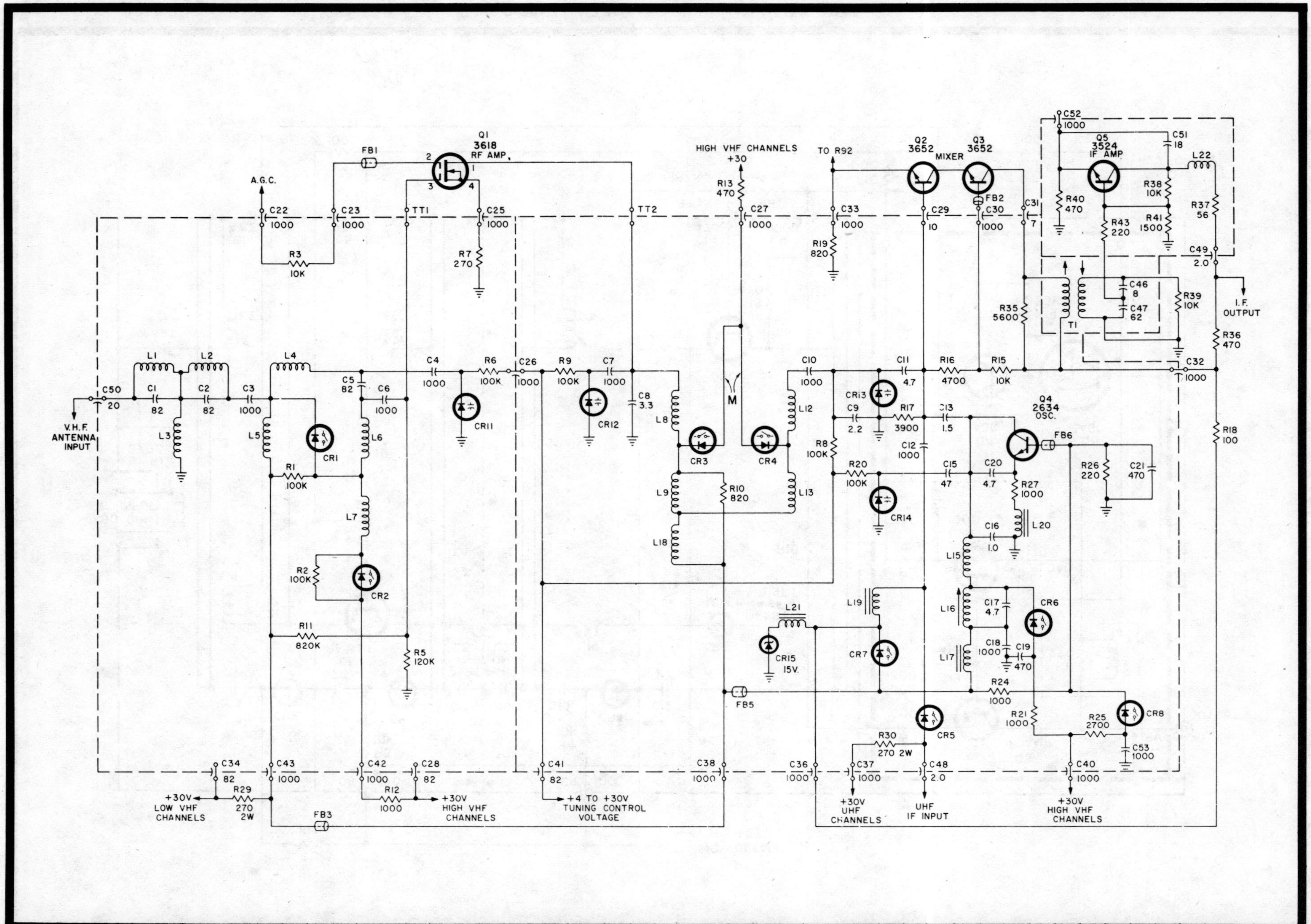

Fig. 9-37. VHF tuner schematic KRK 155B.

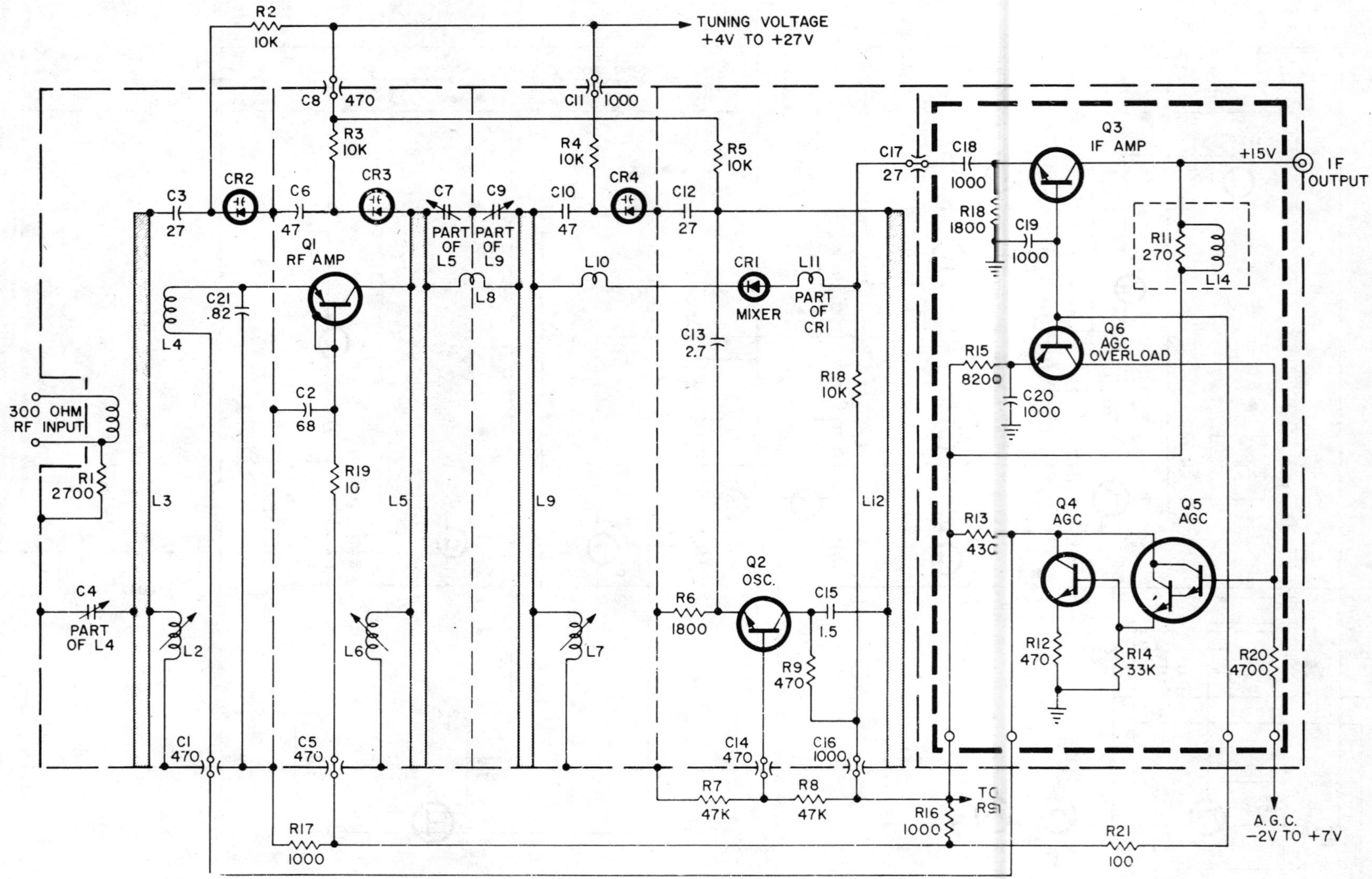

Fig. 9-38. UHF tuner schematic KRK 194A.

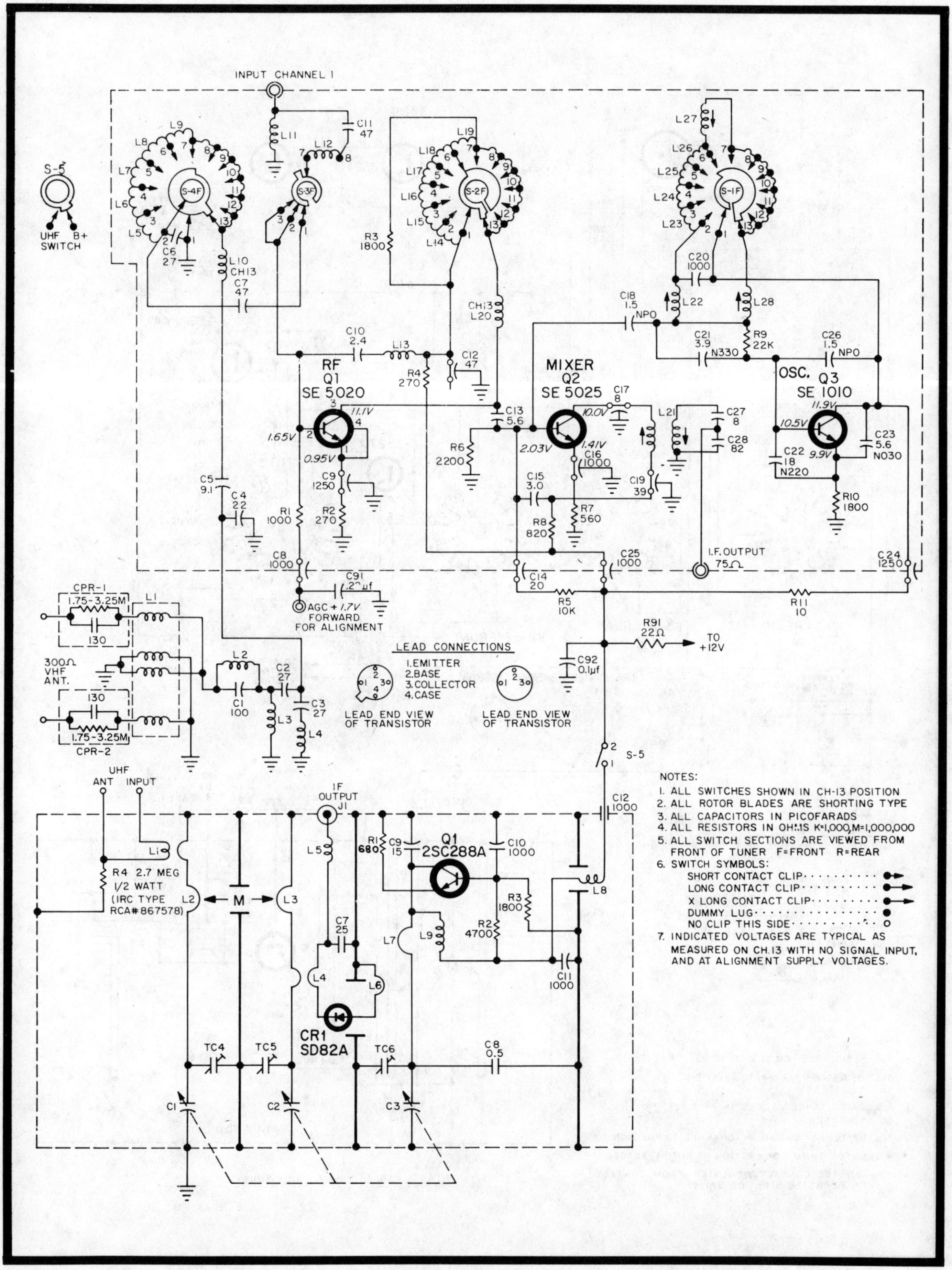

Fig. 9-39. KRK 157 / 150 VHF / UHF tuner schematic diagram.

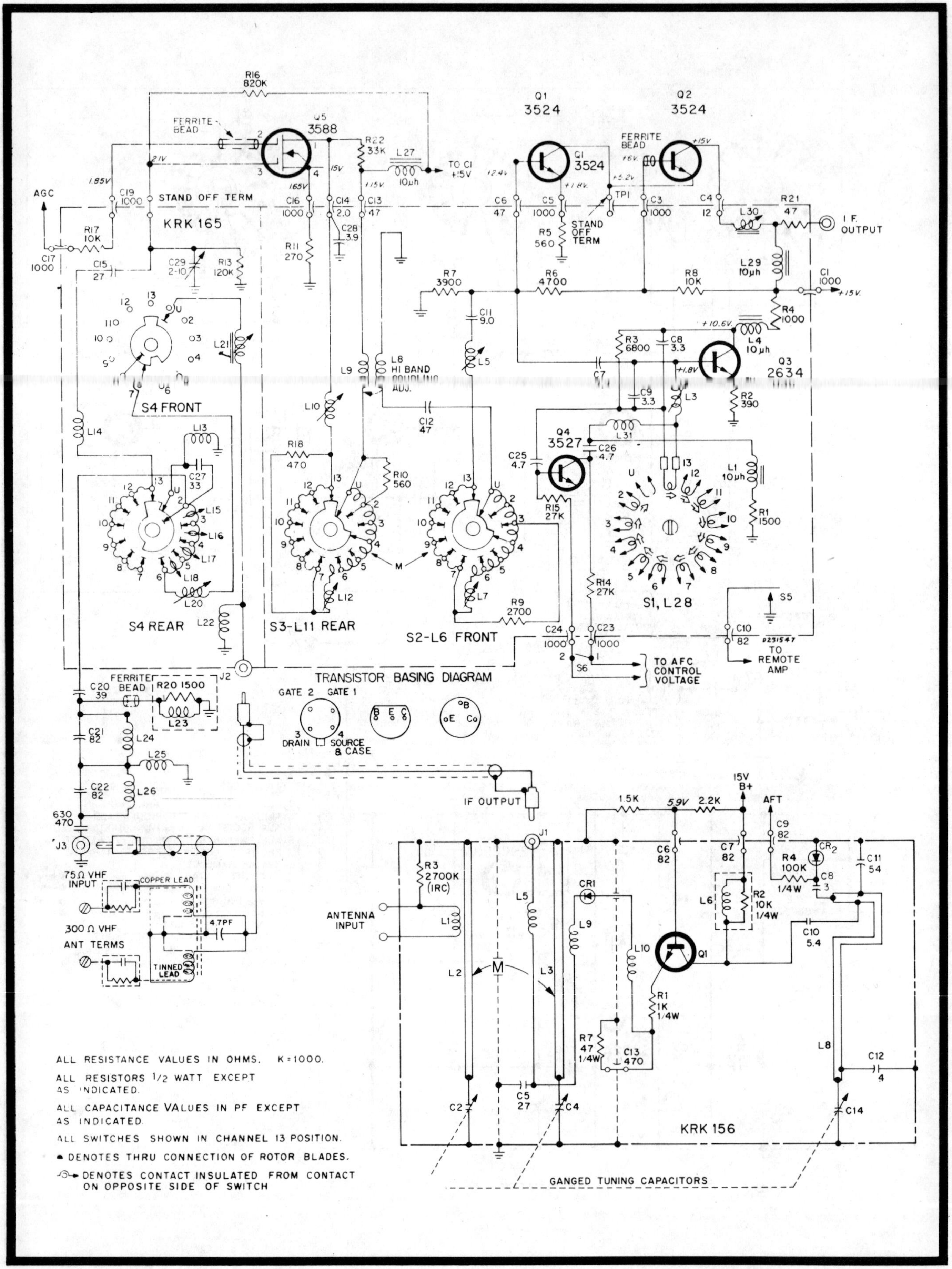

Fig. 9-40. KRK 165/156 VHF / UHF tuner schematic diagram.

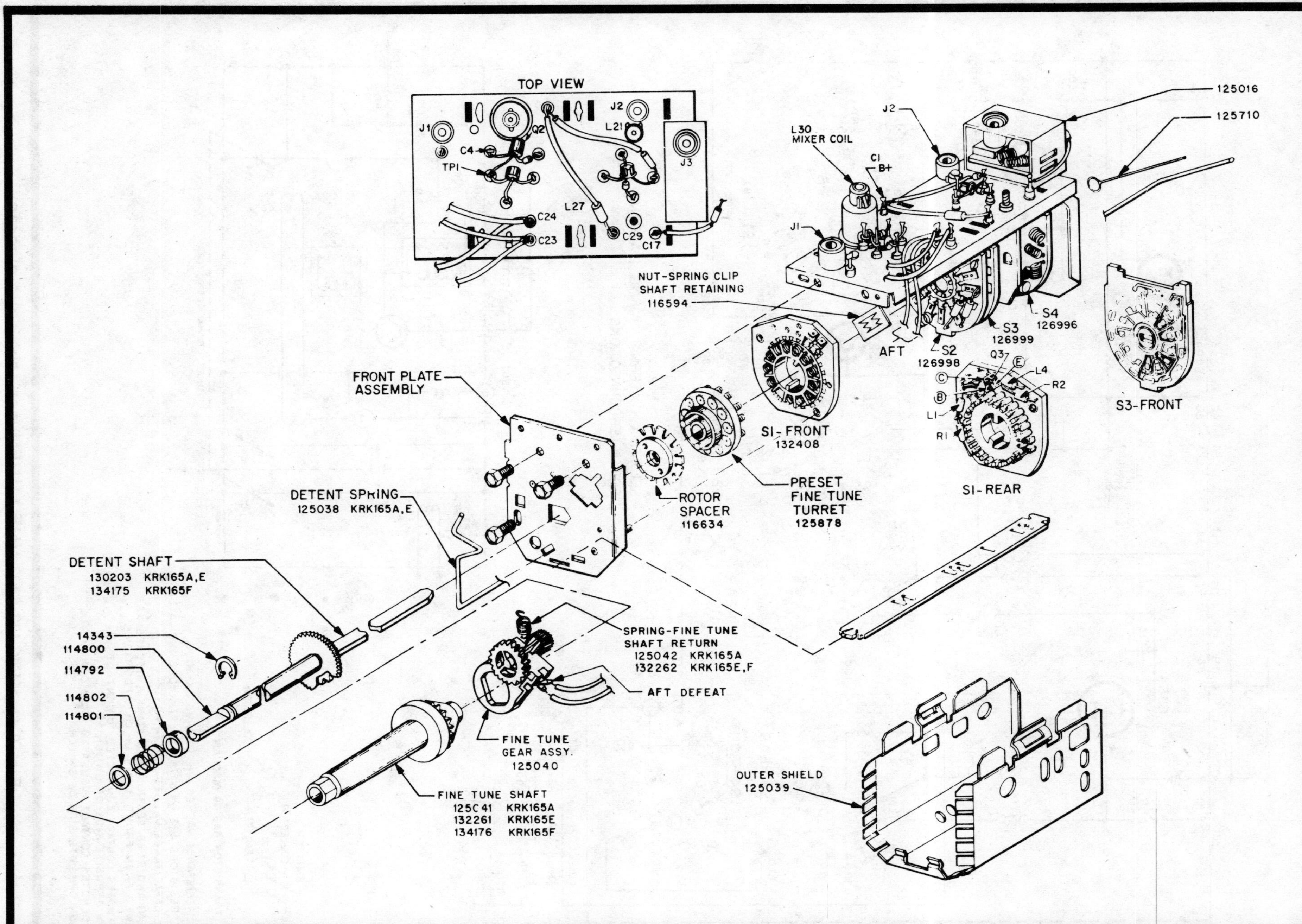

Fig. 9-41. KRK 165 / 156 VHF / UHF tuner exploded view.

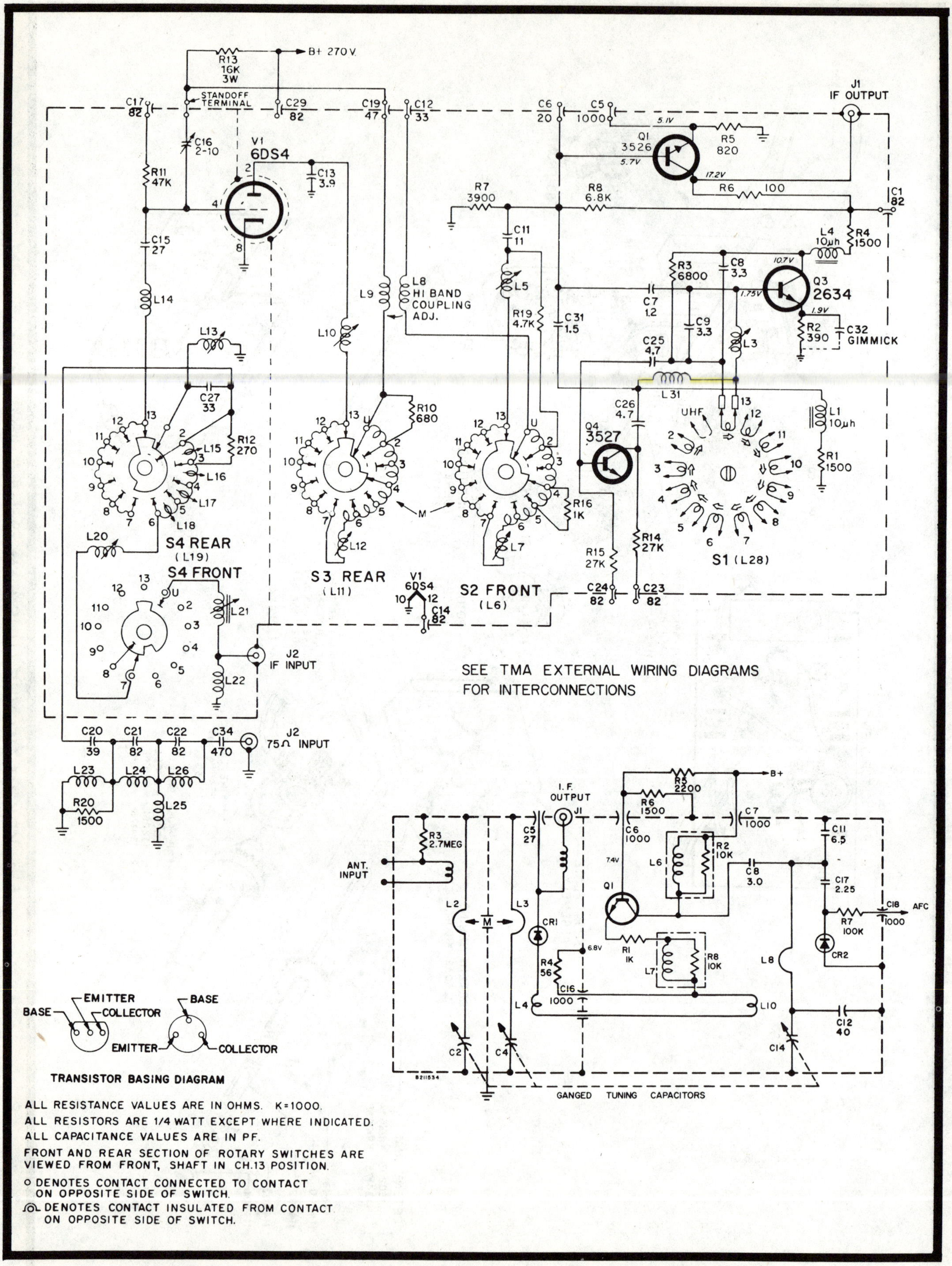

Fig. 9-42. KRK 173/132 VHF/UHF tuner schematic diagram.

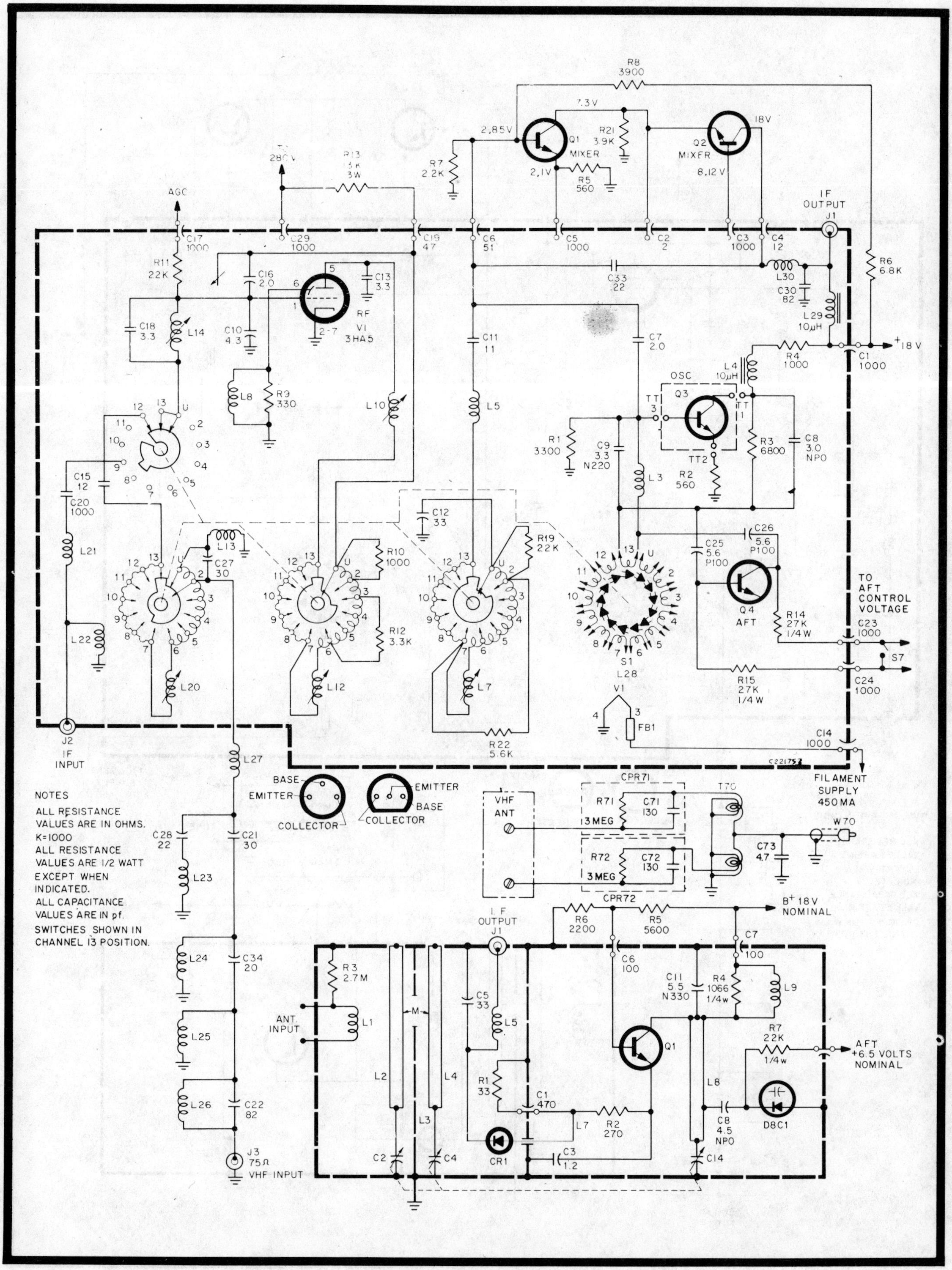

Fig. 9-43. KRK 181/185 tuner schematic.

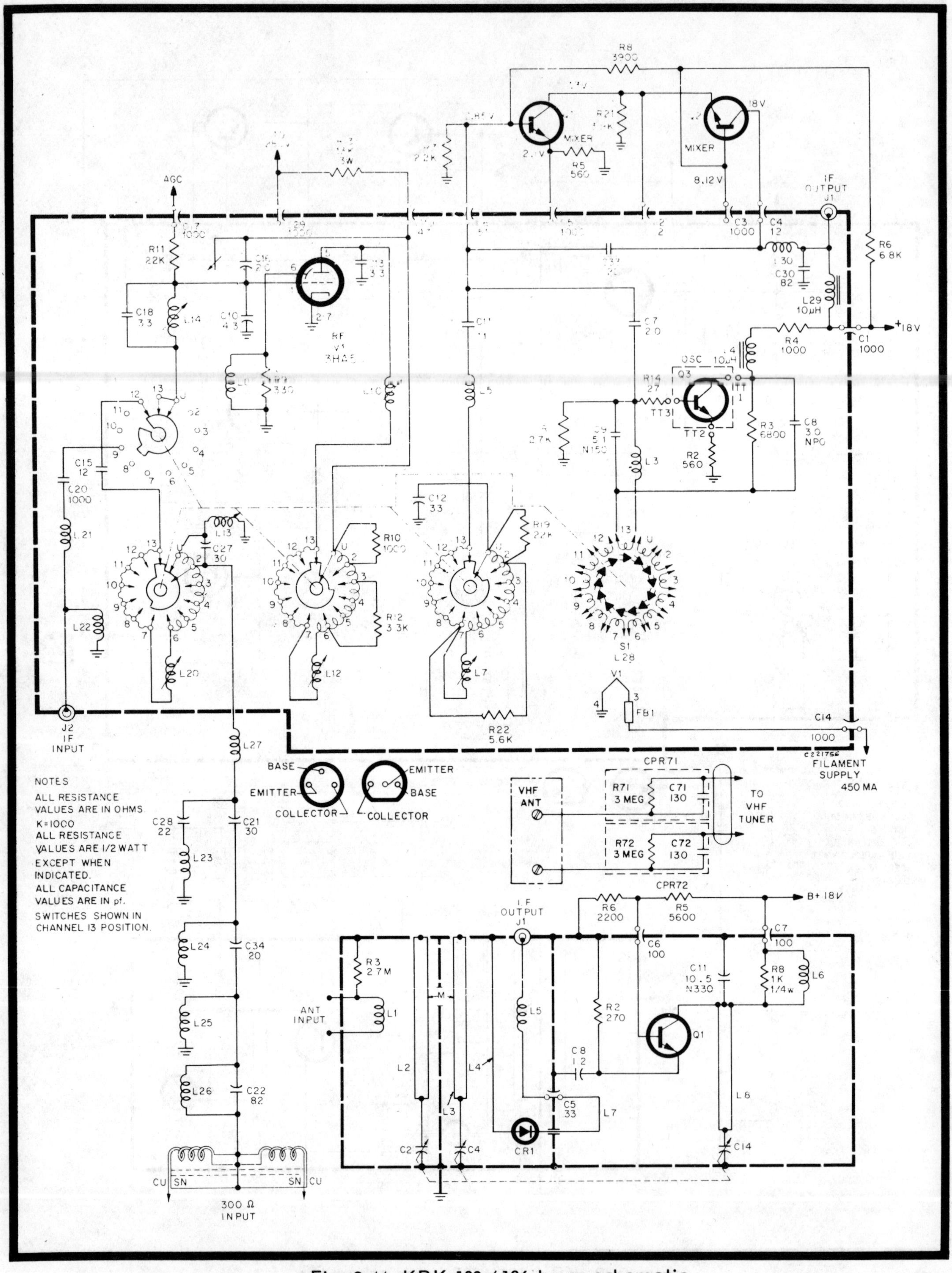

Fig. 9-44. KRK 182 / 186 tuner schematic.

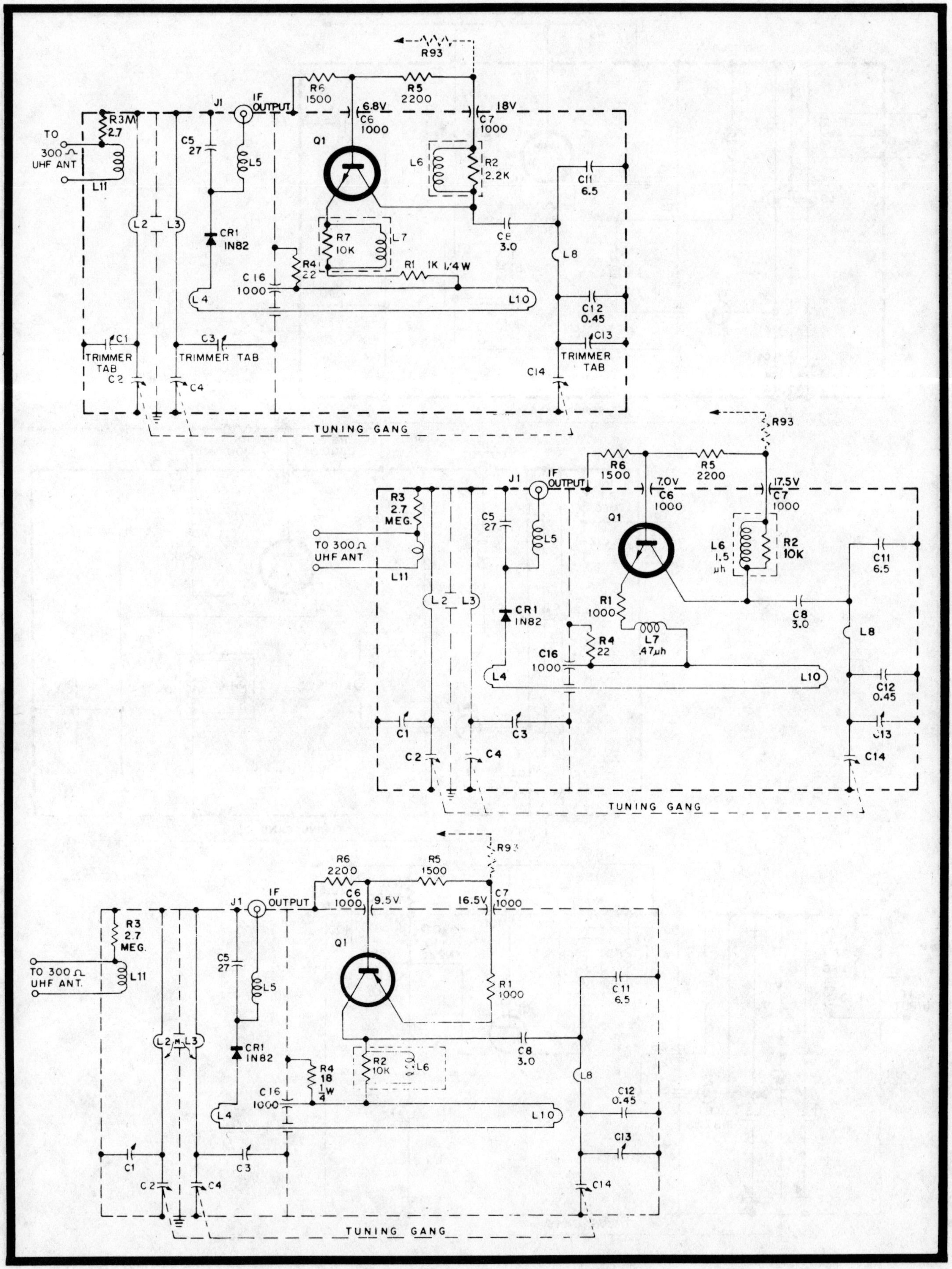

Fig. 9-45. KRK 120M, N, and P schematics.

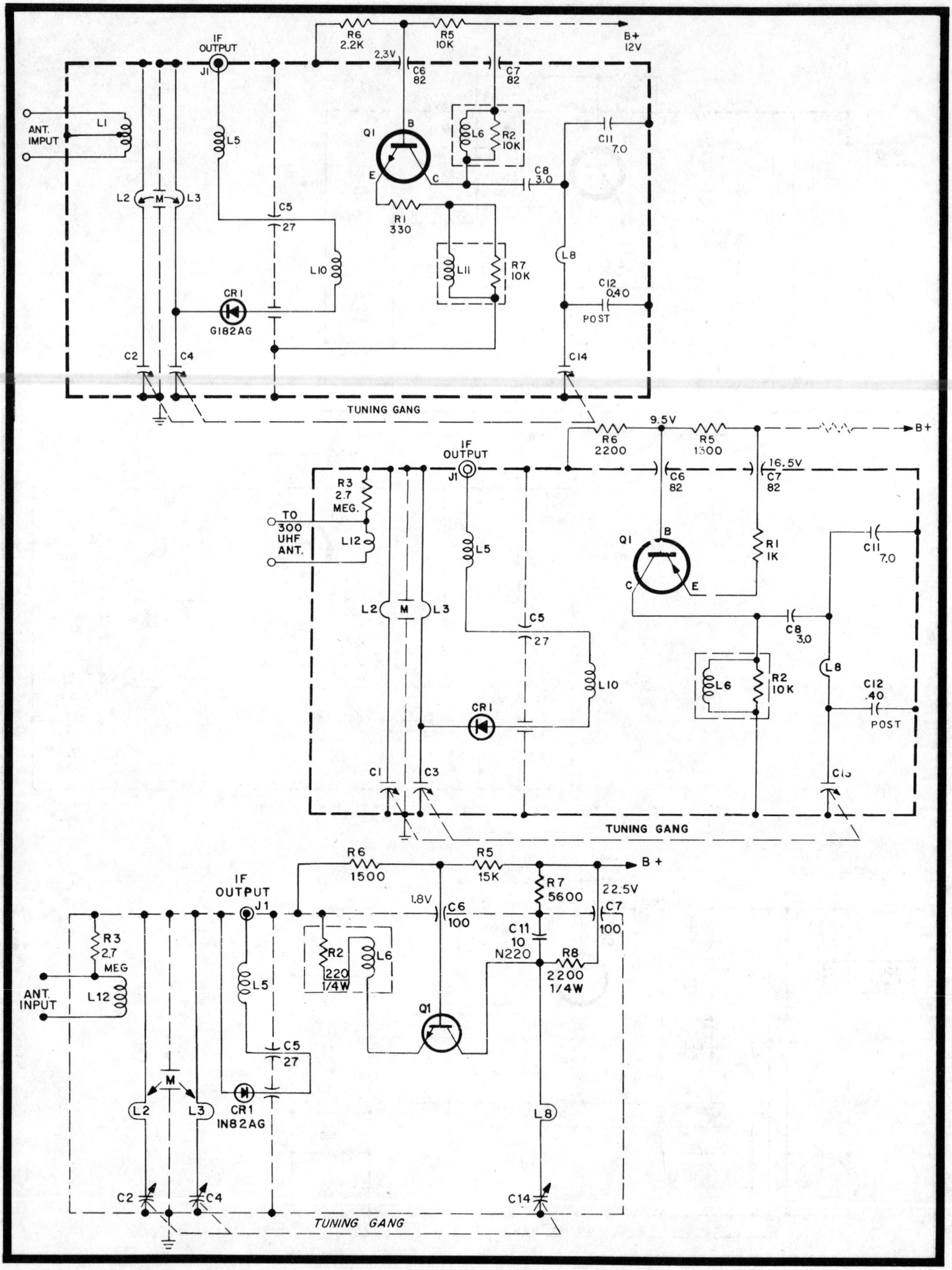

Fig. 9-46. UHF tuner schematic diagrams.

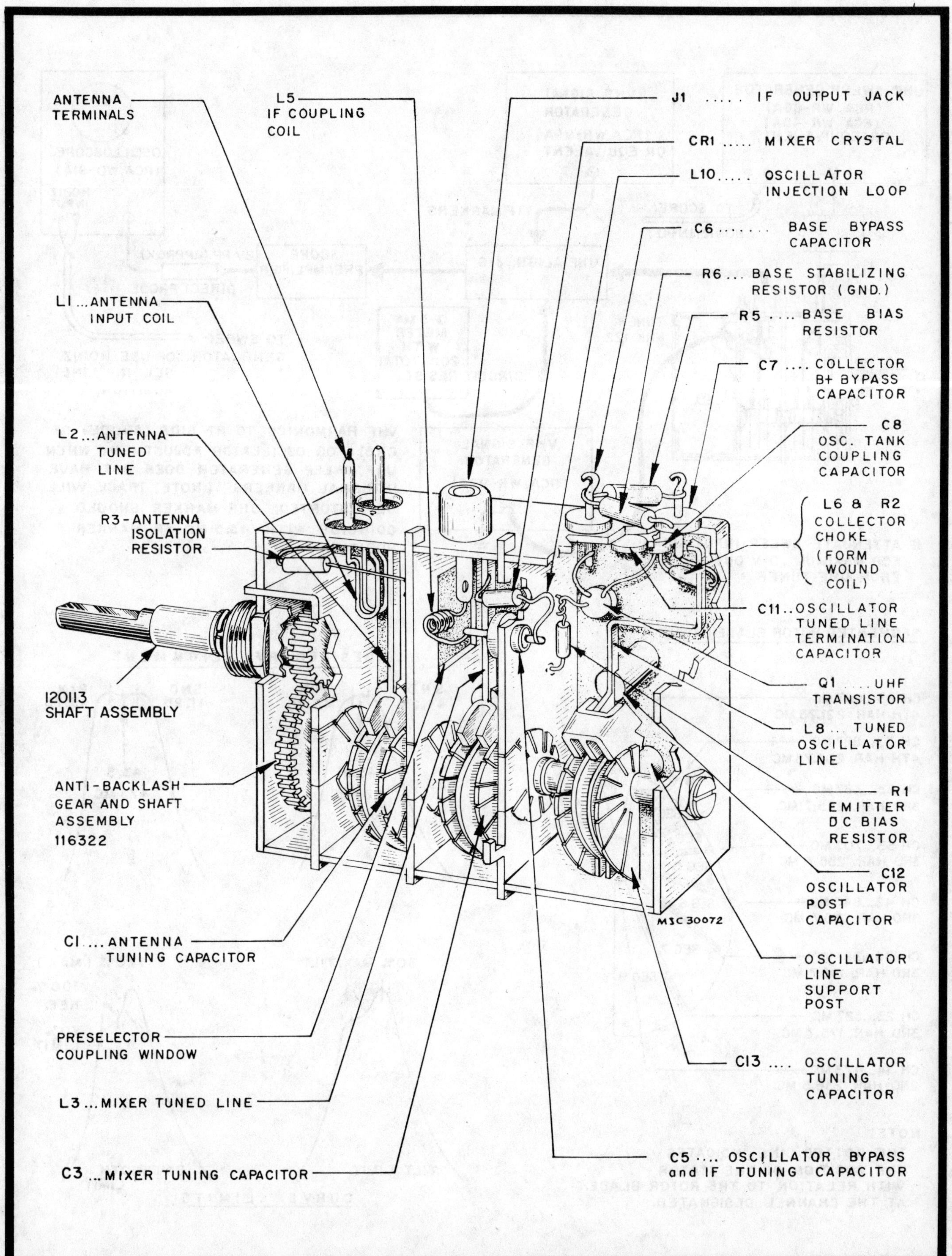

Fig. 9-47. KRK 122 A, B, AB parts identification.

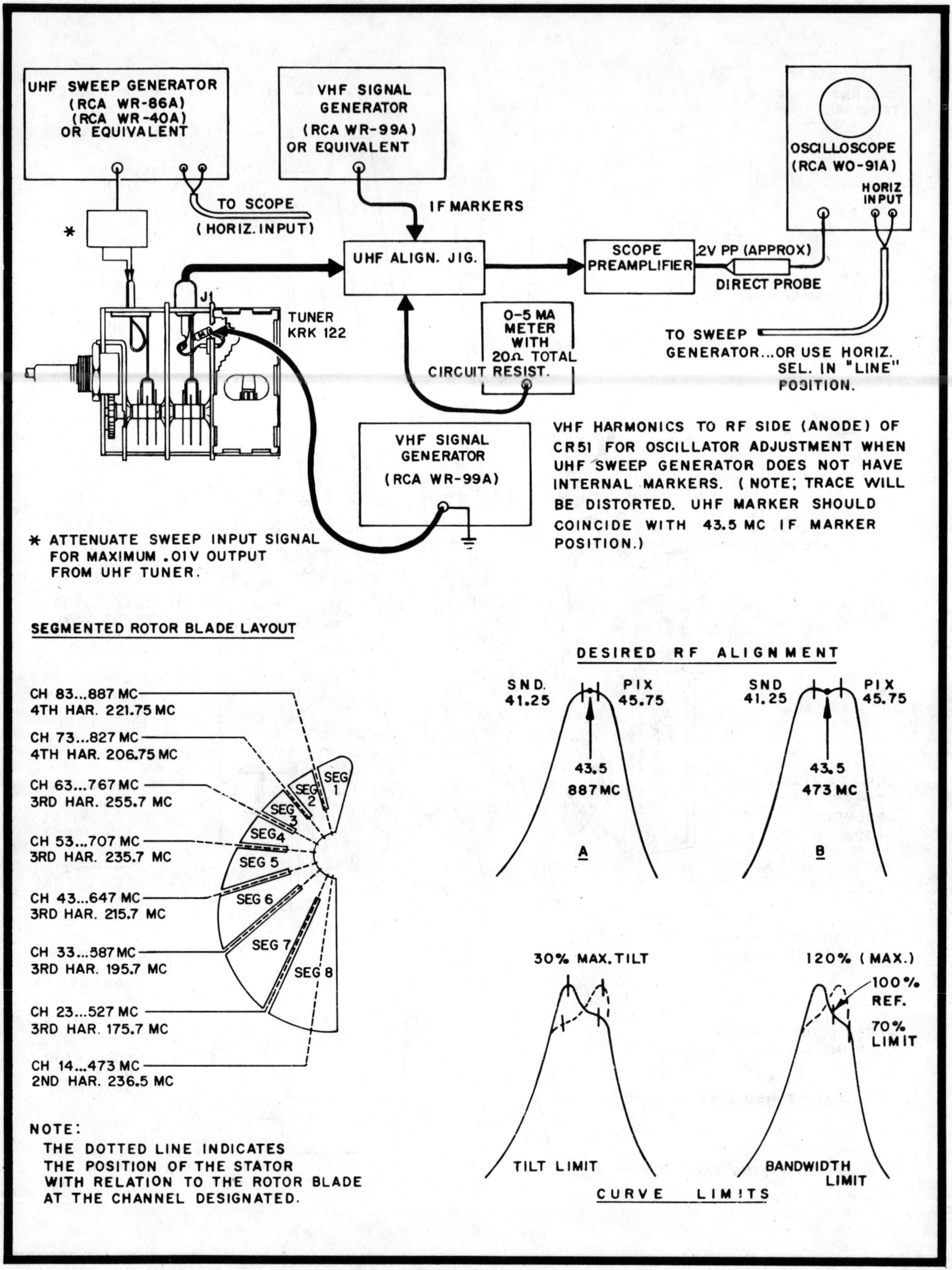

Fig. 9-48. KRK 122A, B, AB alignment drawing.

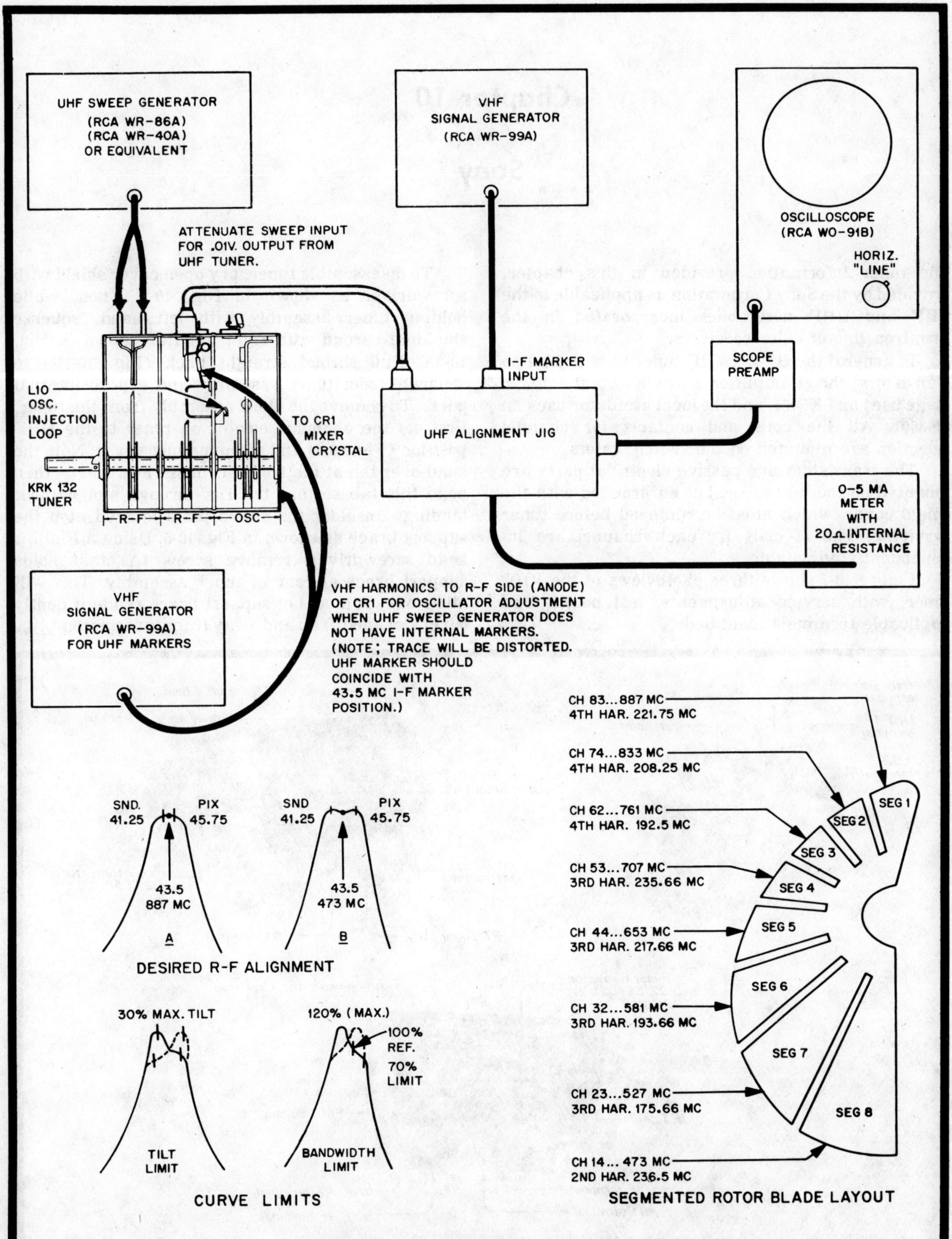

Fig. 9-49. KRK 132, 138 alignment setup.

Chapter 10

Sony

The tuner information provided in this chapter, provided by the Sony Corporation, is applicable to the VHF and UHF assemblies incorporated in the Trinitron line of color receivers.

In general the BT-602 VHF tuner uses three npn transistors; the rf amplifier is an SE5020, the mixer stage uses an SE5021, and the local oscillator uses an SE3001. All the coils and contacts for channel selection are mounted on the switch wafers.

The transistors and passive electrical parts are mounted on the tuner case but enshrouded with the shield cover, which must be removed before tuner servicing. The rf coils for each channel are independently adjustable.

Figure 10-1 shows three photoviews of the VHF tuner, with service adjustments, test points, and applicable terminals identified.

To disassemble tuner, pry open cover shield with screwdriver as shown in Fig. 10-2. Then, while holding tuner assembly with left hand, squeeze shielding shroud with right hand as shown in Fig. 10-3A. Pull shroud straight back (Fig. 10-3B) to separate from tuner assembly and expose internal parts. To remove the shaft assembly from the tuner, first set the channel selector on tuner to the UHF position (Fig. 10-4) and remove tuner cover. Note the bend-over tab at the front of tuner (Fig. 10-5), then bend this tab so that tab fits through slot without binding. Unsolder the shield plate located atop the support brace as shown in Fig. 10-5. Using a Phillips head screwdriver, remove screw (A) that holds support brace at rear of tuner assembly. This will allow easy removal of support brace (slide it gently out of slot and lift up and away from tuner assembly).

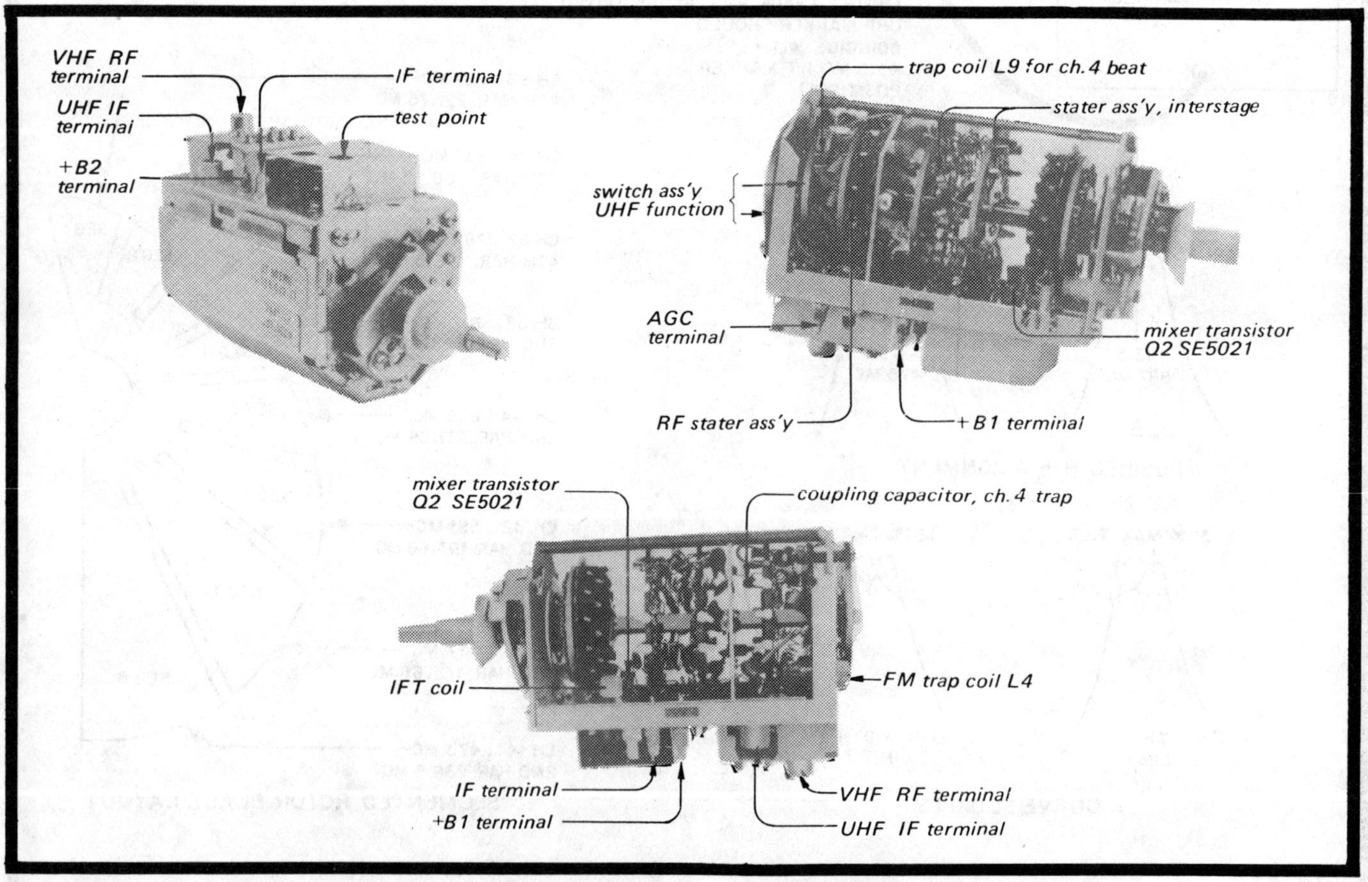

Fig. 10-1. Sony VHF tuner (Trinitron), internal and external views.

Fig. 10-2. Use small screwdriver and pry gently where shielding shroud clamps to tuner assembly.

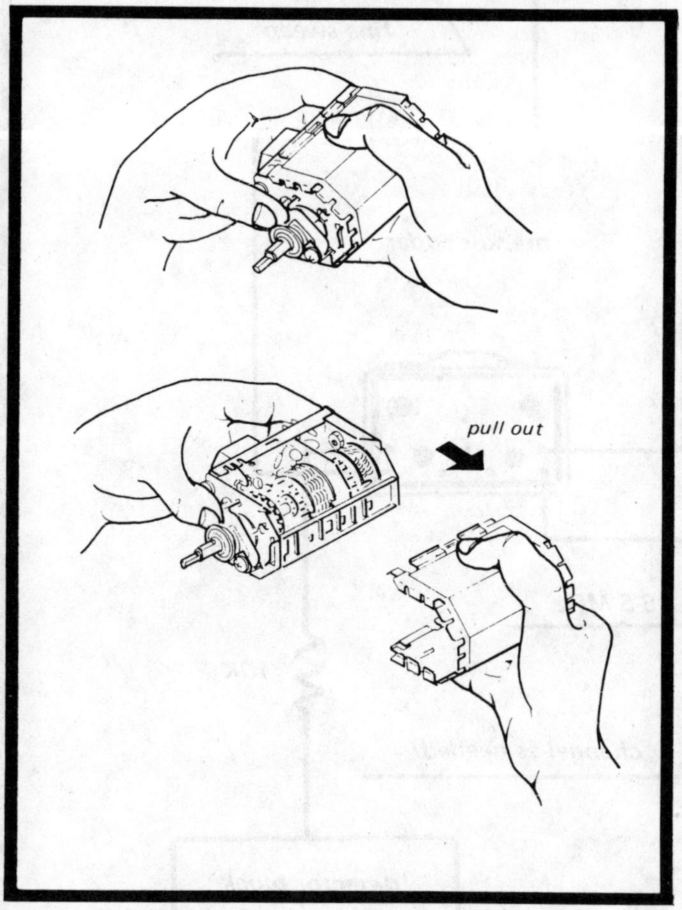

Fig. 10-3. Shield may be removed by squeezing shroud at sides (A) and pulling straight back (B).

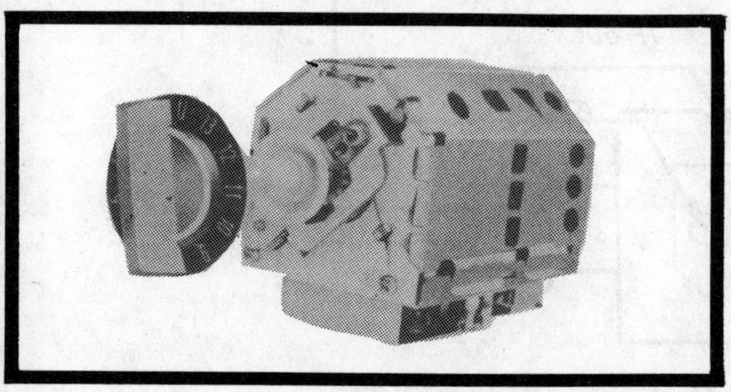

Fig. 10-4. Before attempting to remove the tuner shaft assembly, place tuner in UHF position as shown.

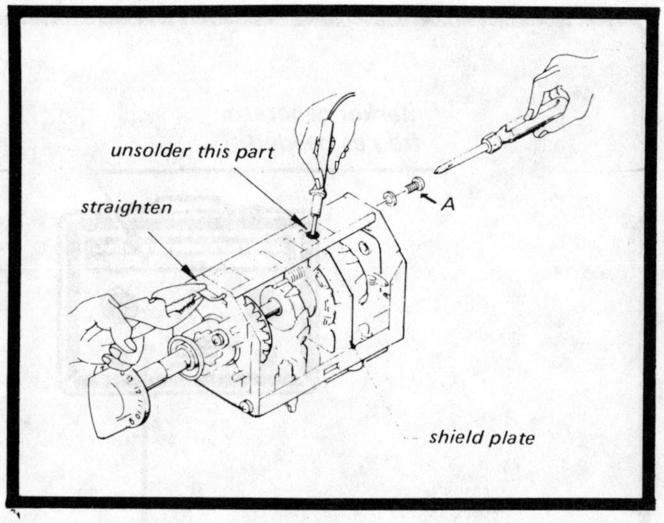

Fig. 10-5. Left-to-right sequence of steps required to remove tuner shaft assembly from tuner.

The last step is to remove the two screws at the front of tuner assembly (B1 and B2 of Fig. 10-6). The tuner shaft assembly will now readily pull out.

CIRCUIT ADJUSTMENTS

Equipment Needed

Sweep generator	— RCA WR-69A or equivalent
Marker generator	— RCA WR-99A or equivalent
Oscilloscope	
Regulated dc power supply	—0-30V 50 mA
VOM	— RCA WV-77E (K) or equivalent
Detector	—See Fig. 10-7.

Procedure

Connect all the equipment as shown in Fig. 10-8, and allow equipment to warm up for 20 minutes to stabilize.

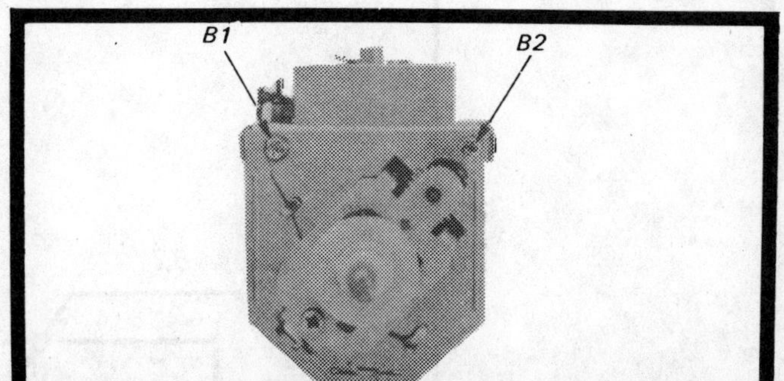

Fig. 10-6. Front-of-tuner screws that must be removed before separating shaft from tuner.

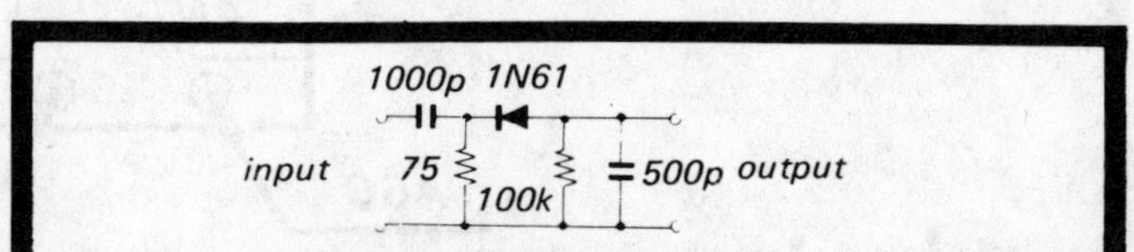

Fig. 10-7. Detector network.

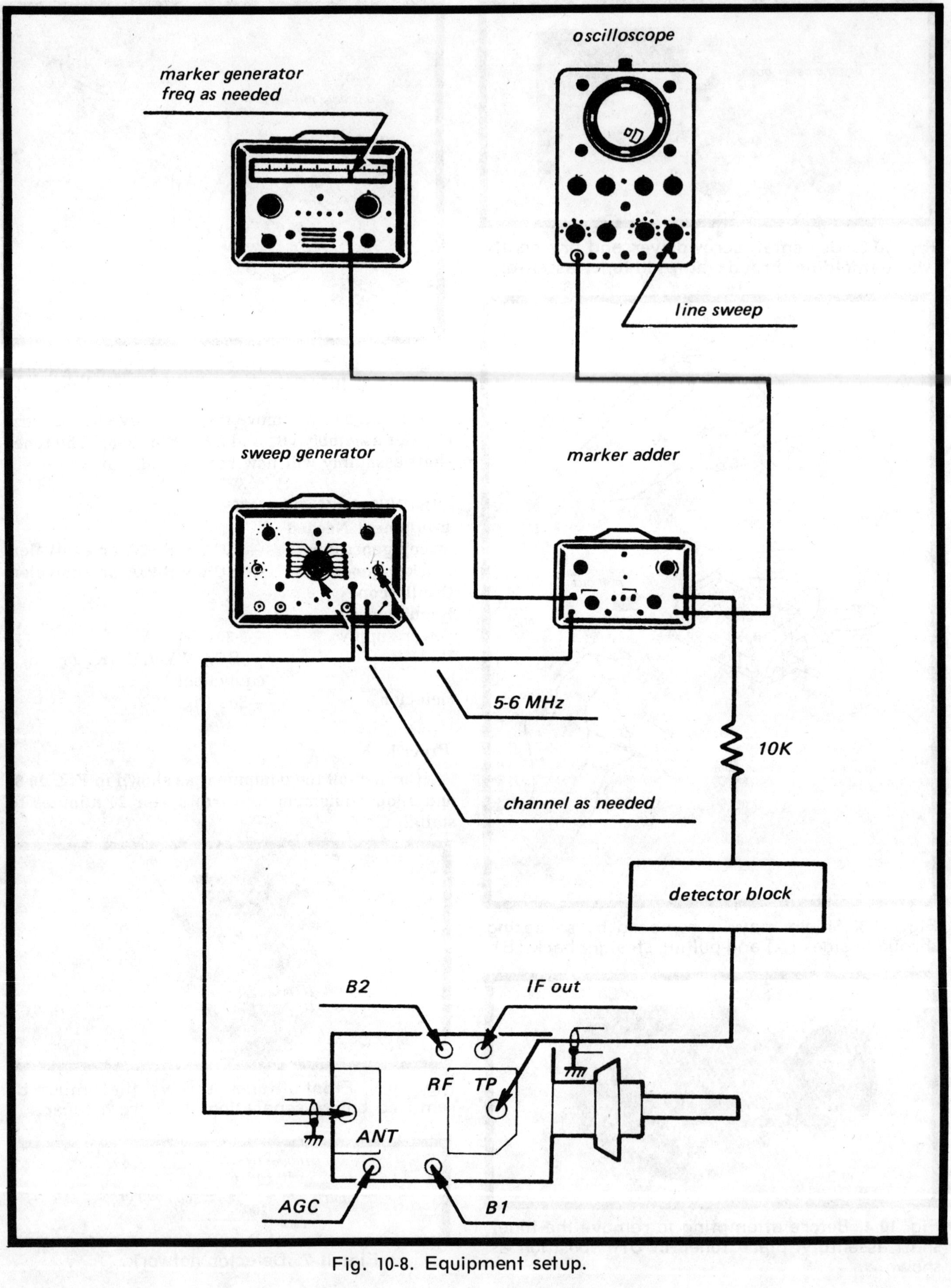

Fig. 10-8. Equipment setup.

STEP	CHANNEL	ADJUST	MARKER POINTS (MHz)		ADJUSTMENT PROCEDURE
			VIDEO	SOUND	
1	13	Lb 13, Lc 13	211.25	215.75	Adjust interstage coils Lb 13 and Lc 13 to position both markers (211.25 MHz and 215.75 MHz) at the peaks of the response curve. (See Fig. 10-9.) 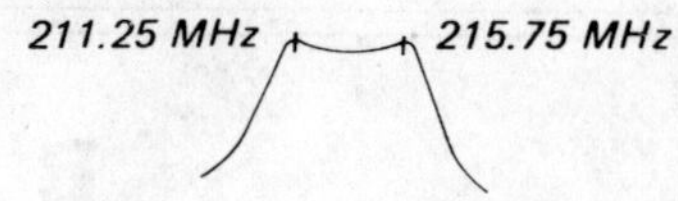 Fig. 10-9. Ch. 13 response curve.
2	12 11 10 9 8 7	Lb 12, Lc 12 Lb 11, Lc 11 Lb 10, Lc 10 Lb 9, Lc 9 Lb 8, Lc 8 Lb 7, Lc 7	205.25 199.25 193.25 187.25 181.25 175.25	209.75 203.75 197.75 191.75 185.75 179.75	Adjust each interstage coil same as described in the above procedure. However, pay attention to the differences in the markers at each channel.
3	6	Lb 6, Lc 6	83.25	87.75	Adjust interstage coils Lb 6 and Lc 6 to position the 83.25 MHz marker at the peak of response curve, and the 87.75 MHz marker at 6 dB (50%) below the video marker as shown in Fig. 10-10. 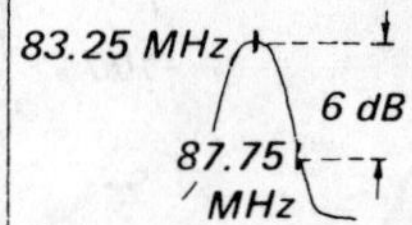 Fig. 10-10. Ch. 6 response curve.
4	4	Trap L 9	77.25		Sweep the sweep generator at 70 MHz. Adjust trap coil L 9 until the 77.25 MHz video marker appears at the dip in the curve as seen on the scope. (See Fig. 10-11.) 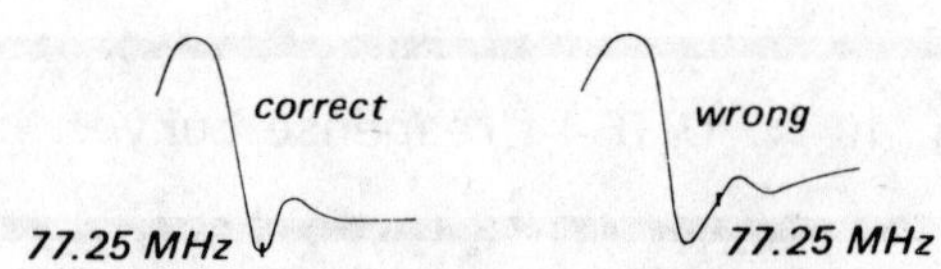 Fig. 10-11. Trap adjustment.
5	5	Lb 5, Lc 5	77.25	81.75	Adjust interstage coils Lb 5 and Lc 5 the same as in step 1. But pay attention to the differences in marker frequencies.
6	4	Lb 4, Lc 4	67.25	71.75	Adjust interstage coils Lb 4 and Lc 4 to position the 67.25 MHz marker at the peak of response curve, and the 71.75 MHz marker within 6 dB (50%) below the video marker as shown in Fig. 10-12. 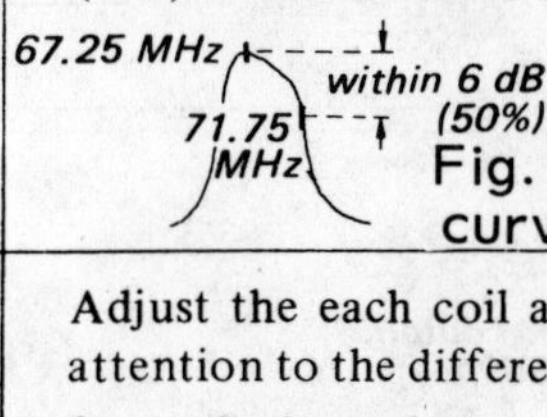 Fig. 10-12. Ch. 4 response curve.
7	3 2	Lb 3, Lc 3 Lb 2, Lc 2	61.25 55.25	65.75 59.75	Adjust the each coil as same as step 1. But pay attention to the difference in marker frequencies for each channel.

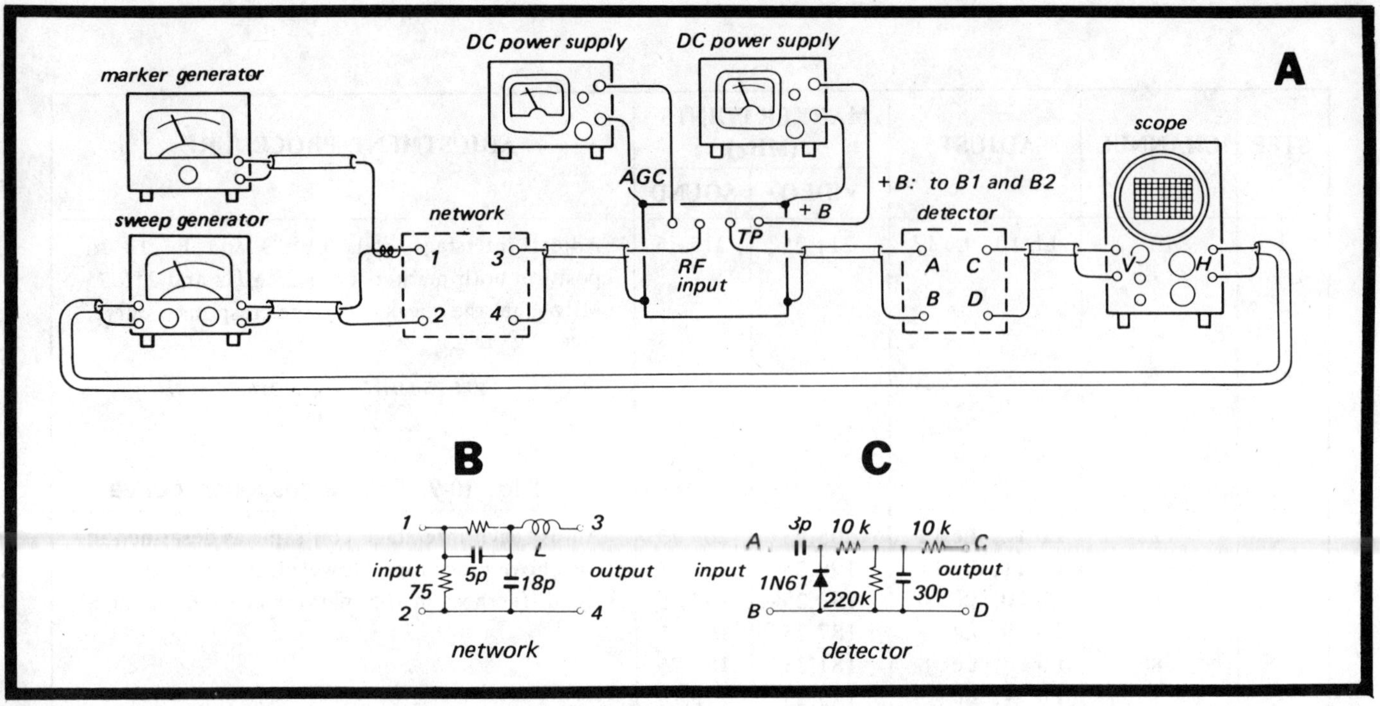

Fig. 10-13. Equipment setup (A). (B) (C)

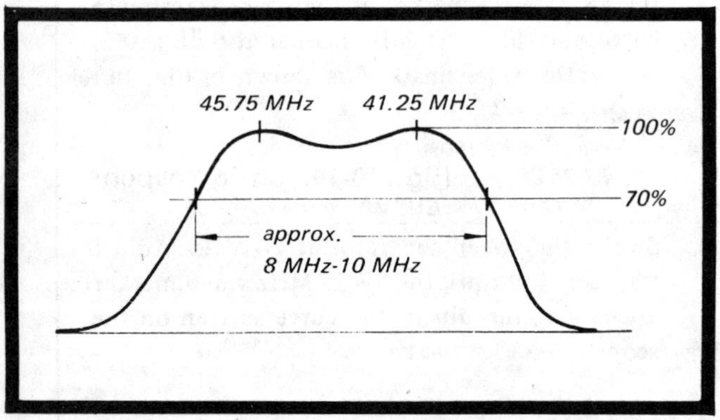

Fig. 10-14. UHF-i-f response curve.

Tear the shielding foil off the tuner cover, then perform the alignment steps (see Alignment Table) in the sequence described.

UHF Intermediate Frequency Adjustments

1. Connect a sweep generator to rf input terminal of tuner through the network as shown in Fig. 10-13.
2. Loosely couple the output of a marker generator to the output lead of the sweep generator as shown in Fig. 10-13A.
3. Connect scope to test point through detector as shown. (Detector circuit given in Fig. 10-13C.)

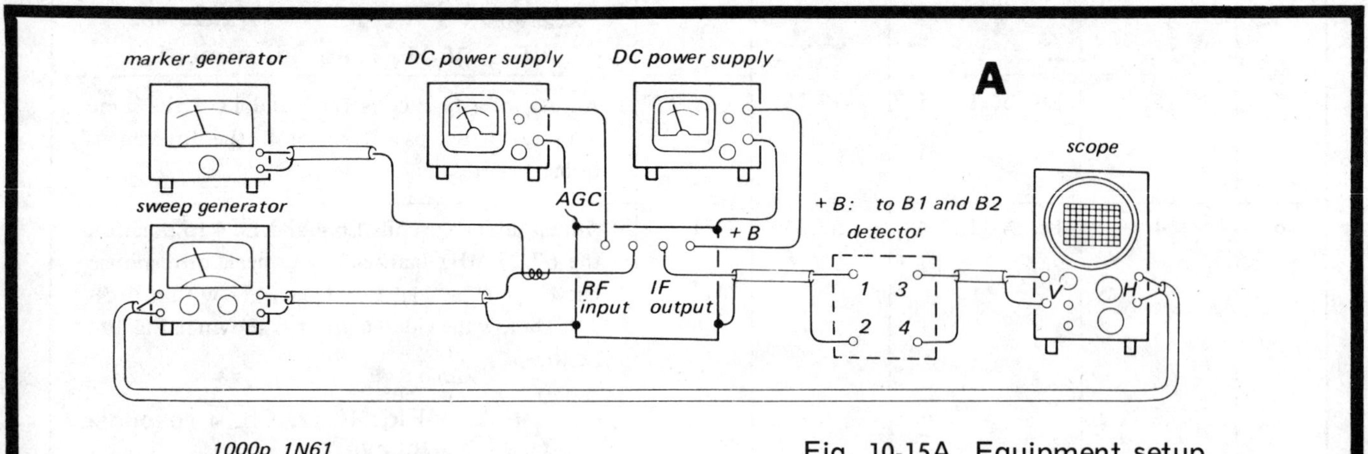

Fig. 10-15A. Equipment setup.

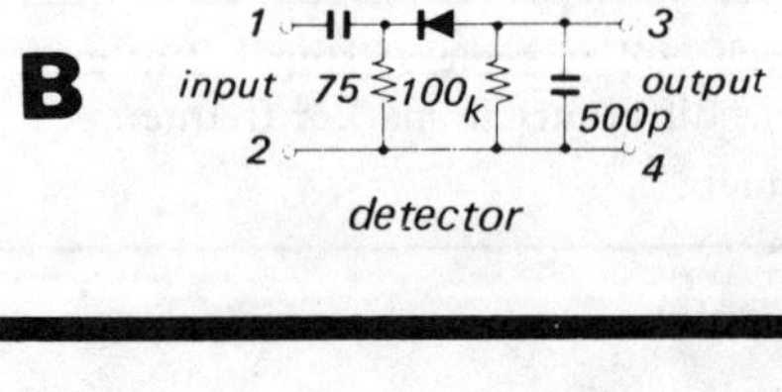

Fig. 10-15B. Detector.

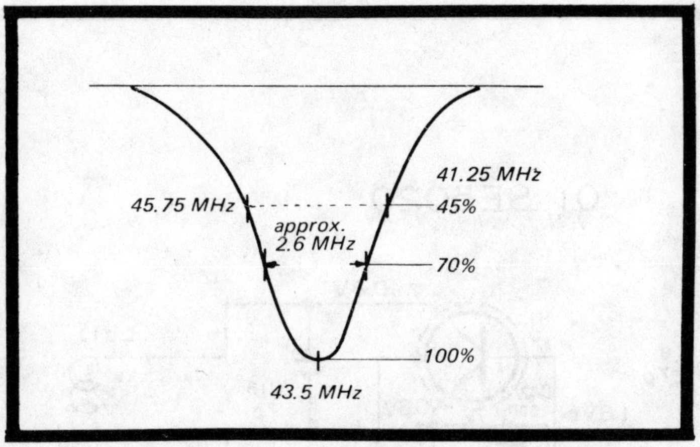

Fig. 10-16. Mixer output response curve.

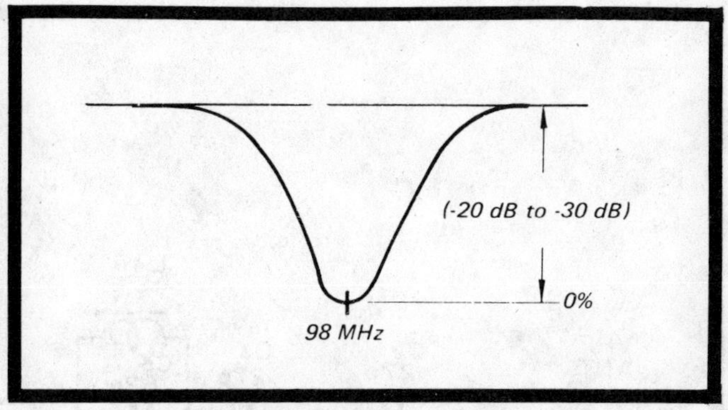

Fig. 10-18. Response curve for FM trap coil L4.

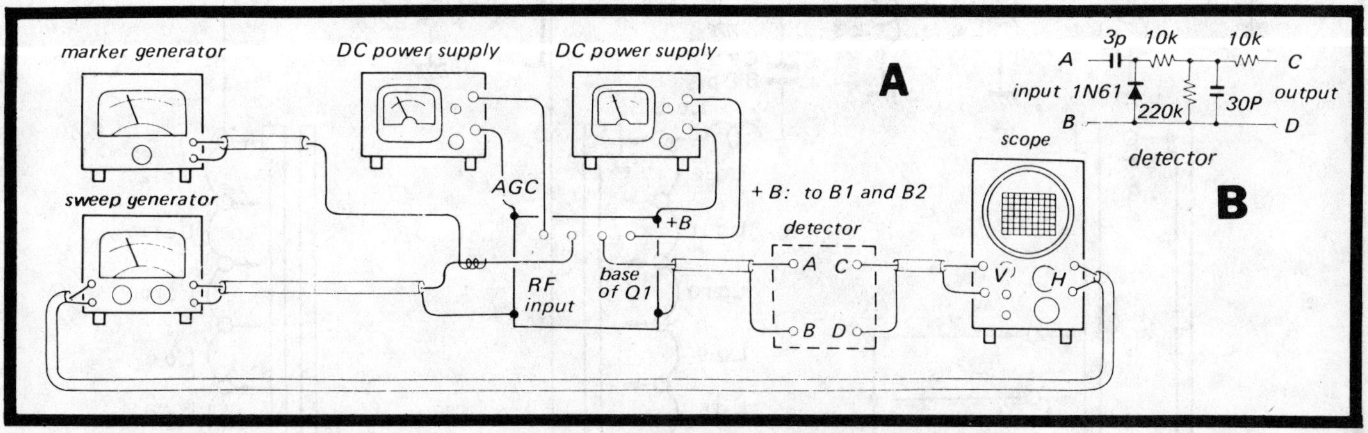

Fig. 10-17. Equipment setup (A), detector (B).

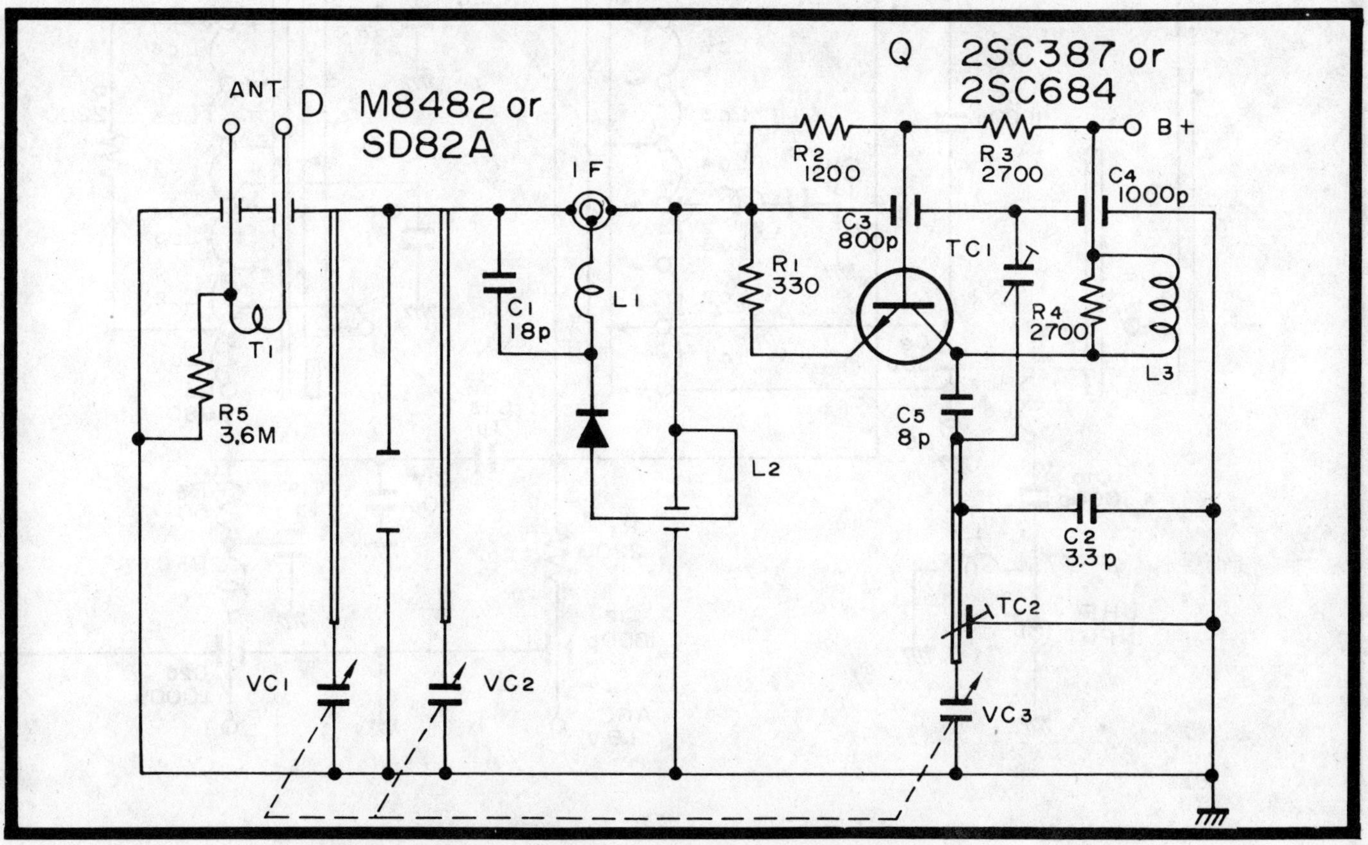

Fig. 10-19. Schematic diagram of UHF tuner.

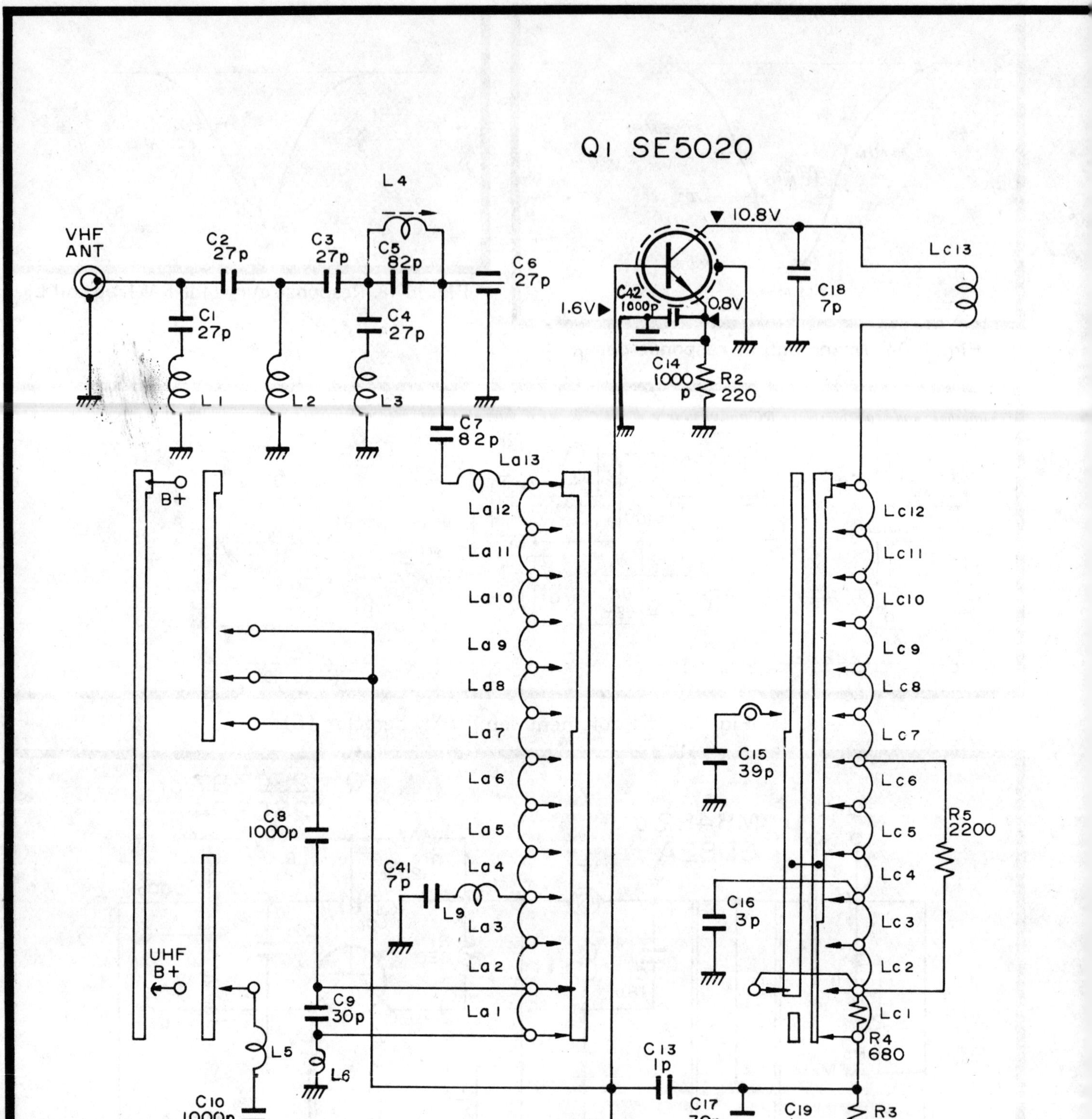
Q1 SE5020
VHF ANT
C2 27p
C3 27p
L4
C5 82p
C6 27p
C1 27p
C4 27p
L1
L2
L3
C7 82p
La13
La12
La11
La10
La9
La8
La7
La6
La5
La4
La3
La2
La1
B+
UHF B+
C8 1000p
C41 7p
L9
C9 30p
L5
L6
C10 1000p
L7
UHF INPUT
10.8V
1.6V
C42 1000p
0.8V
C14 1000 p
R2 220
C18 7p
Lc13
Lc12
Lc11
Lc10
Lc9
Lc8
Lc7
Lc6
Lc5
Lc4
Lc3
Lc2
Lc1
C15 39p
C16 3p
R5 2200
R4 680
C13 1p
C17 30p
C19 1p
R3 50
RFC1
R1 2200
C12 1000p
C20 1000p
AGC 1.6 V
+B1 11V

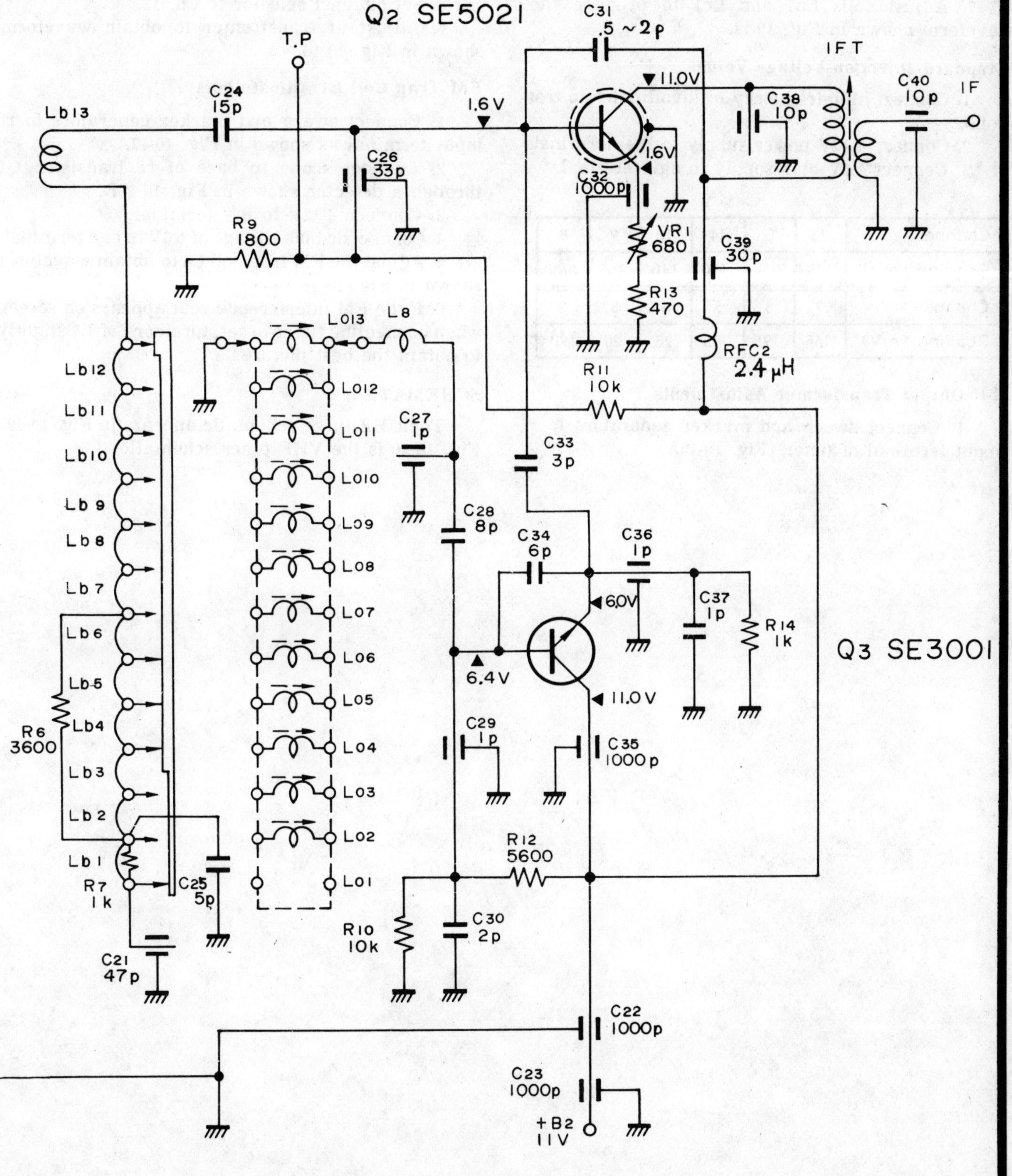

Fig. 10-20. Schematic diagram of VHF tuner.

4. Set the sweep generator so that the output level of about 70 dB is obtained.

5. Connect a dc power supply (+12V) to B+ terminal.

6. Connect dc bias supply (1.6V) to agc terminal.

7. Adjust coils Lb1 and Lc1 to produce the waveform shown in Fig. 10-14.

Standard Injection-Voltage Values

1. Connect high-frequency millivoltmeter to test point.

2. Connect +12V power supply to B+ terminal.

3. Connect 1.6V bias supply to agc terminal.

Channel	13	12	11	10	9	8
Reading (mV)	145	150	150	140	165	165
Channel	7	6	5	4	3	2
Reading (mV)	165	195	205	20	20	190

I-F Output Transformer Adjustments

1. Connect sweep and marker generators to rf input terminal of tuner (Fig. 10-15A).

2. Connect scope to i-f output terminal through detector of Fig. 10-15B. (Note this is not the same detector circuit used before.)

3. Connect +12V to B+ terminal.

4. Connect 1.6V bias supply to agc terminal.

5. Set channel selector to Ch. 13.

6. Adjust i-f transformer to obtain waveform shown in Fig. 10-16.

FM Trap Coil L4 Adjustments

1. Connect sweep and marker generators to rf input terminal as shown in Fig. 10-17A.

2. Connect scope to base of rf transistor Q1 through a detector shown in Fig. 10-17B.

3. Connect +12V to B+ terminal.

4. Connect dc bias voltage of 1.6V to agc terminal.

5. Adjust the FM trap coil L4 to obtain waveform shown in Fig. 10-18.

6. If the FM interference beat appears on screen when connecting tuner to set, turn core of L4 slightly to obtain the best picture.

SCHEMATICS

The UHF tuner schematic appears in Fig. 10-19; Fig. 10-20 is the VHF tuner schematic.

Chapter 11

Sylvania

This chapter contains representative alignment instructions for Sylvania tuners, plus schematics and part-location diagrams for a wide variety of Sylvania tuner models for which the alignment procedures are applicable. The preliminary instructions are applicable to all Sylvania tuners and should be followed carefully before performing any alignment procedure.

The material in this chapter has been provided through the courtesy of GTE Sylvania.

PRELIMINARY INSTRUCTIONS

1. Keep all exposed leads as short as possible.
2. Use shielded leads between test equipment and chassis, and terminate all lead shields as close as possible to their repsective hot leads.
3. Use high scope gain and keep sweep generator output at lowest usable value; check at intervals for possible sweep generator overloading by temporarily varying signal input level and noting any change (excluding amplitude) in response curve shape.
4. Keep marker generator coupling to a minimum to avoid distortion of response curve.
5. For optimum receiver alignment, power line voltage should be maintained at 120V.
6. Use nonmetallic tools for all alignment adjustments.
7. Refer to frequency chart (Fig. 11-1) as necessary for proper equipment setup.

RF ALIGNMENT (TUBE-TYPE TUNER)

1. Rotate tuner, sweep generator, and rf marker generator through all channels; refer to alignment

CHAN.	PICTURE	SOUND	SWEEP GEN.
2	55.25MHz	59.75MHz	57MHz
3	61.25MHz	65.75MHz	63MHz
4	67.25MHz	71.75MHz	69MHz
5	77.25MHz	81.75MHz	79MHz
6	83.25MHz	87.75MHz	85MHz
7	175.25MHz	179.75MHz	177MHz
8	181.25MHz	185.75MHz	183MHz
9	187.25MHz	191.75MHz	189MHz
10	193.25MHz	197.75MHz	195MHz
11	199.25MHz	203.75MHz	201MHz
12	205.25MHz	209.75MHz	207MHz
13	211.25MHz	215.75MHz	213MHz

Fig. 11-1. Alignment frequency chart.

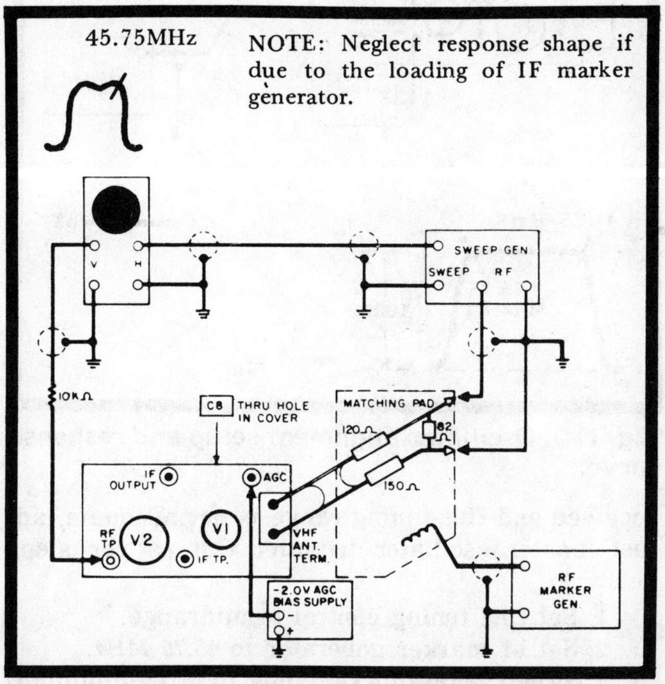

Fig. 11-2. Equipment setup, response curve, and adjustment locations.

frequency chart for frequencies of specific channels. Should picture marker fall below 50 percent or sound carrier below 20 percent on any channel, the tuner must be realigned.

2. If proper response curve cannot be obtained, adjust C8 (Fig. 11-2) as follows. Increase bias to —15V. Increase generator output until visible on scope. Adjust C8 for minimum output. Decrease bias to —2.0V.

3. Set tuner to Channel 10, sweep generator to 195 MHz, rf marker generator to 193.25 MHz; and adjust C20 until 45.75 MHz marker coincides with 193.25 marker.

4. Set fine tuning to midrange and detune mixer plate coil L14.

VHF OSCILLATOR ALIGNMENT (TUBE-TYPE TUNER)

This tuner employs a preset oscillator adjustment screw for each channel. No other adjustment is necessary other than the normal fine tuning adjustment. If all available stations cannot be

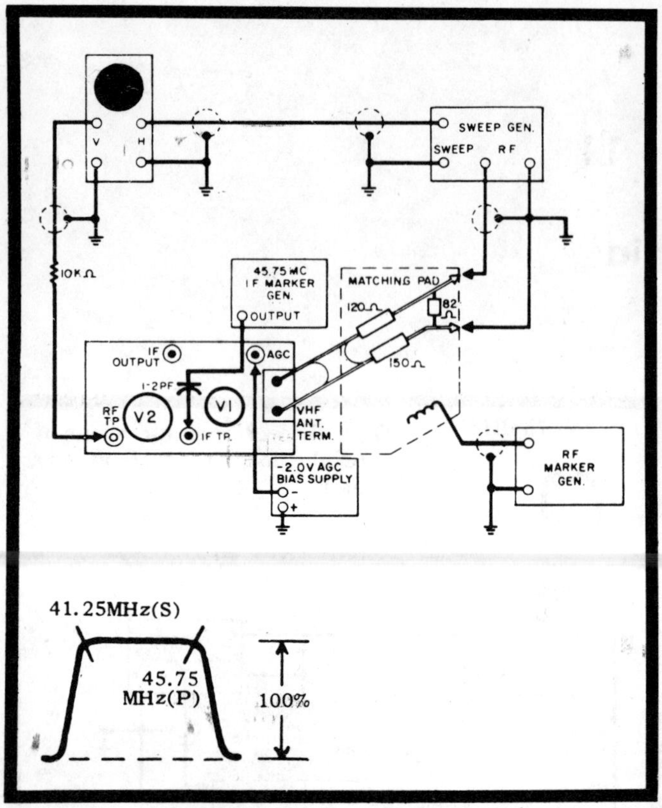

Fig. 11-3. Oscillator alignment setup and response curve.

received and fine-tuning range is not adequate, adjust master oscillator trimmer C20 as per steps below:

1. Set fine tuning control to midrange.
2. Set i-f marker generator to 45.75 MHz.
3. Sweep generator sweeping 10 MHz minimum.
4. Detune mixer place coil L14.
5. Set tuner to Channel 10, sweep generator to 195 MHz, rf marker generator to 193.25 and 197.75 MHz; and adjust C12 and C14 to place markers on shoulders of curve as shown in Fig. 11-3.
6. Recheck all channels for picture and sound marker locations on the i-f response curve. If picture and sound markers fall below 50 percent or 20 percent

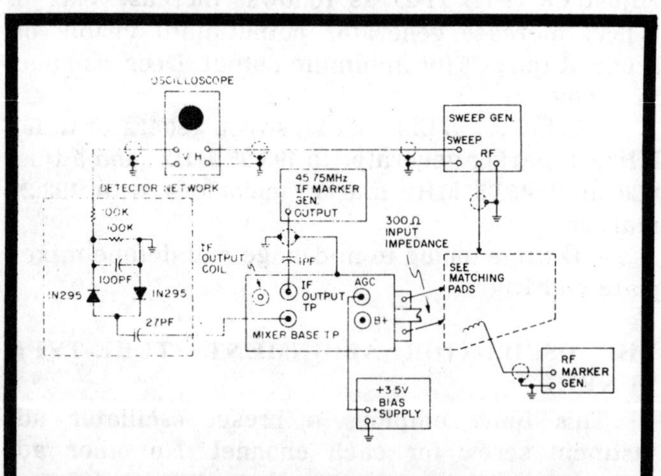

Fig. 11-4. Test equipment interconnections.

respectively, knife or compress L8, L10, and L12 to obtain proper response.

7. Readjust tuner i-f output coil for proper response curve before operating tuner with main chassis assembly.

TRANSISTOR TUNER ALIGNMENT

The preliminary instructions apply to these tuners as well as those employing vacuum tubes. Be sure oscillator is in correct alignment before attempting rf alignment procedure.

Oscillator Alignment

Set VHF tuner to Channel 13, and connect test equipment as shown in Fig. 11-4. The sweep generator is to be connected to the antenna terminal through the appropriate matching pad. (Figure 11-5 shows matching pads for 50, 75, and 100-ohm coaxial cables.)

Set sweep generator to 21 MHz, marker generator (rf) to 211.25 MHz, and i-f marker to 45.75 MHz. Be sure to adjust i-f and rf markers so they are readily visible on the i-f curve oscilloscope presentation.

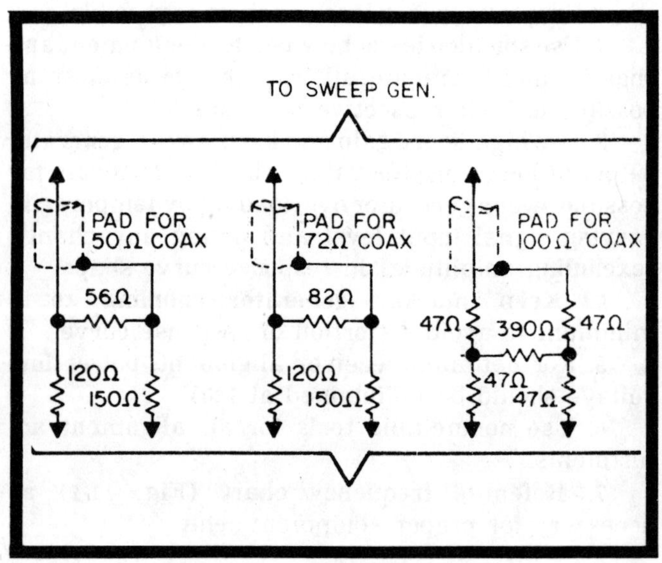

Fig. 11-5. Matching pads.

Adjust appropriate Channel 13 coil by rotating preset fine-tuning control as shown on frequency chart of Fig. 11-1 for coincidence of rf marker and i-f marker frequencies.

After performing the procedure for Channel 1, repeat for all remaining VHF channels.

RF Alignment

1. Remove tuner cover and apply +3.5V bias to agc test point on tuner, as shown in Fig. 11-4. (The i-f marker generator is the only test-equipment item shown that will not be used in this procedure.)
2. Refer to frequency chart (Fig. 11-1) and set marker generator for appropriate picture and sound frequencies for each channel to be aligned.

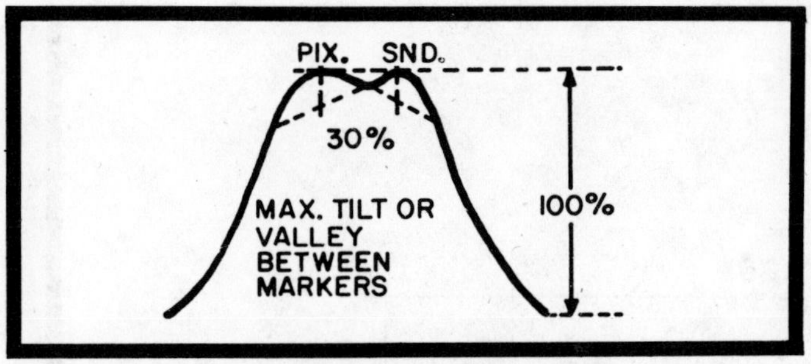

Fig. 11-6. Response curve.

3. Check each channel for response curve as shown in Fig. 11-6. If necessary, squeeze or spread the following coils (applicable only to high-band channels):

LA13—for response tilt
LP13—for correct pix marker position
LG13—for correct sound marker position
L101, L102—for optimizing bandwidth.

4. Switch tuner channel selector to low-band channels (2 through 6) and repeat steps 2 and 3 above. Coils to be squeezed or spread are:

LA2 through LA6 (for channels 2 through 6) on switch S5R.
LP2 through LP6 on switch S3F.
LG2 through LG6 on switch S2R.

NOTE: LA coils govern response tilt; LG coils govern marker positions, and LP, LG affect bandwidth.

TRANSISTOR TUNER ALIGNMENT

The oscillator alignment is to be performed **before** rf alignment.

GRAPHIC DATA

The balance of this chapter consists of schematics and pictorials of various Sylvania VHF and UHF models. The sequence is as follows:

Fig. 11-7. Sylvania 54-15967-4 tuner schematic.

Fig. 11-8. Sylvania 54-27848-3 tuner schematic.

Fig. 11-9. Sylvania 54-23858-4, -13 tuner schematic.

Fig. 11-10. Sylvania 54-23857-2, -3 tuner schematic.

Fig. 11-11. Sylvania 54-29331-3, -5 tuner schematic.

Fig. 11-12. Sylvania 54-29331-7, -8 tuner schematic.

Fig. 11-13. Sylvania 54-29331-4, -9 tuner schematic.

Fig. 11-14. Sylvania 54-29332-2 tuner schematic.

Fig. 11-15. Sylvania 54-29332-4, -5 tuner schematic.

Fig. 11-16. Sylvania 54-33911-1 through -4 tuner schematic.

Fig. 11-17. Sylvania 54-33915-1 tuner schematic.

Fig. 11-18. Sylvania 54-35163-1 tuner schematic.

Fig. 11-19. Sylvania 54-35576-1 tuner schematic.

Fig. 11-20. Sylvania 2420611H tuner schematic.

Fig. 11-21. Sylvania 54-17965-3 tuner schematic.

Fig. 11-22. Sylvania 54-27396-1 tuner schematic.

Fig. 11-23. Sylvania 54-27843-1 tuner schematic.

Fig. 11-24. Sylvania 54-29184-1 tuner schematic.

Fig. 11-25. Sylvania 54-27843-3 tuner schematic.

Fig. 11-26. Sylvania 54-29859-2 tuner schematic.

Fig. 11-27. Sylvania 54-33774-1 tuner schematic.

Fig. 11-28. Sylvania 54-33914-1 tuner schematic.

Fig. 11-29. Sylvania 54-15967-4 tuner schematic.

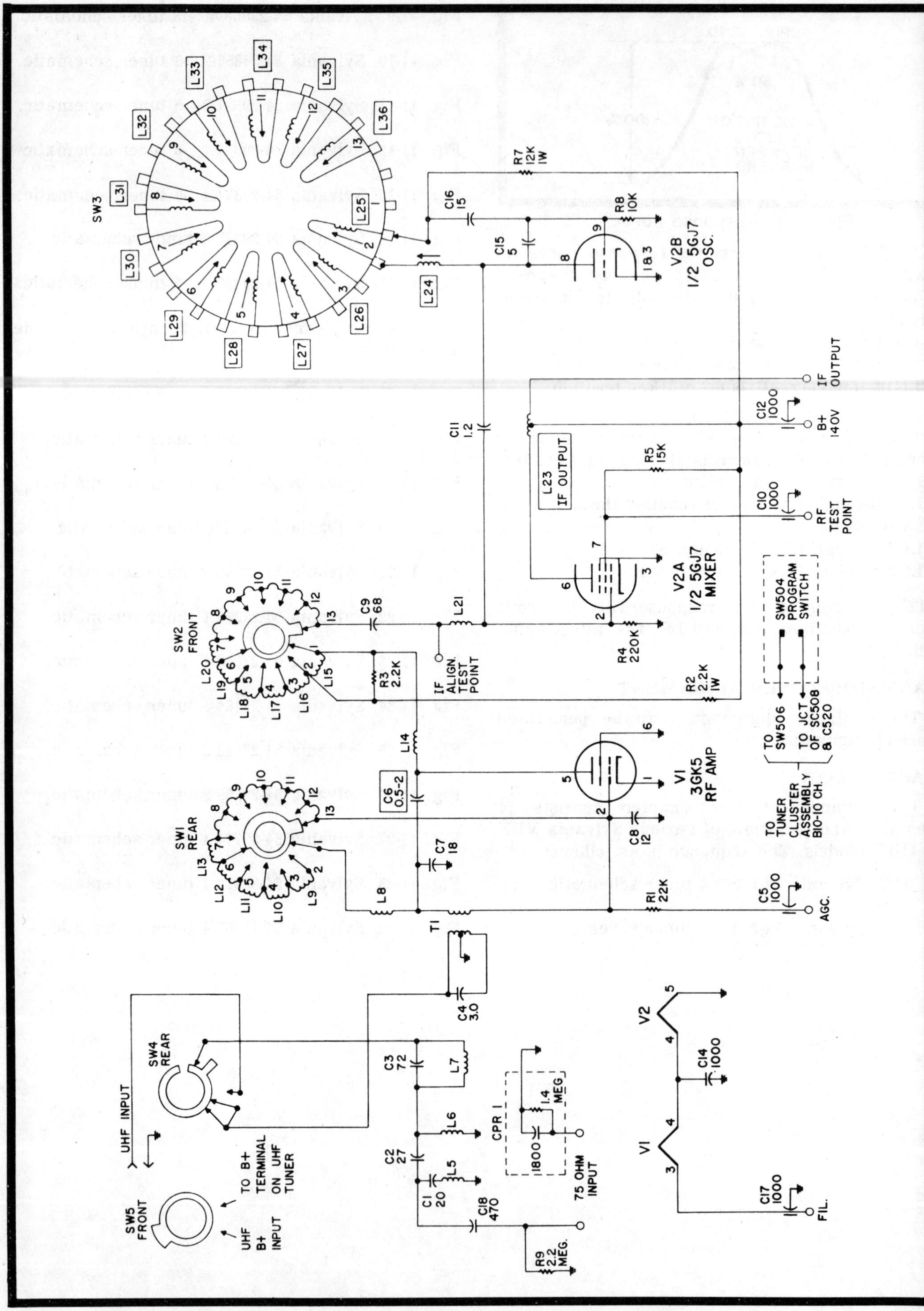

Fig. 11-7. Sylvania 54-15967-4 tuner schematic.

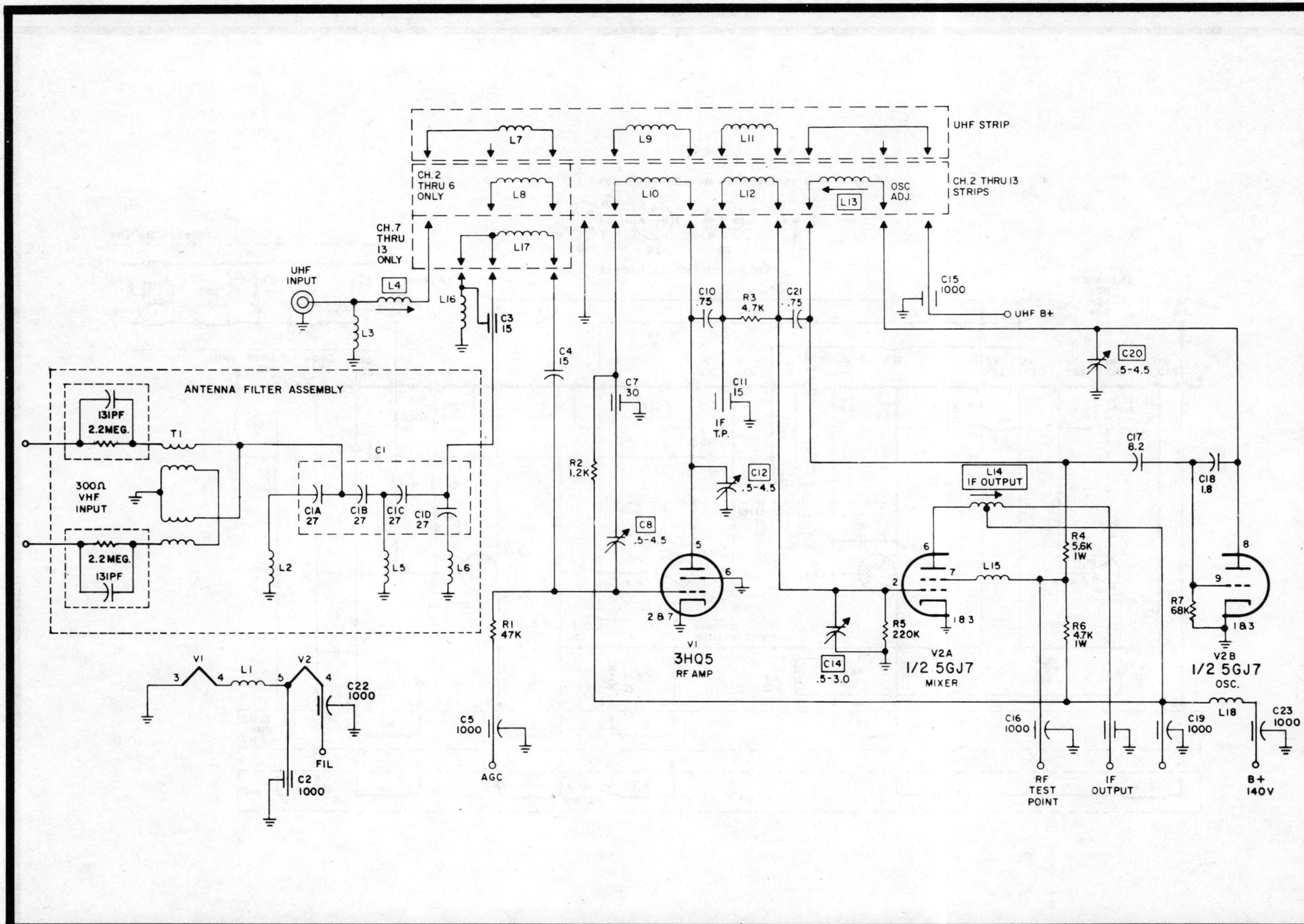

Fig. 11-8. Sylvania 54-27848-3 tuner schematic.

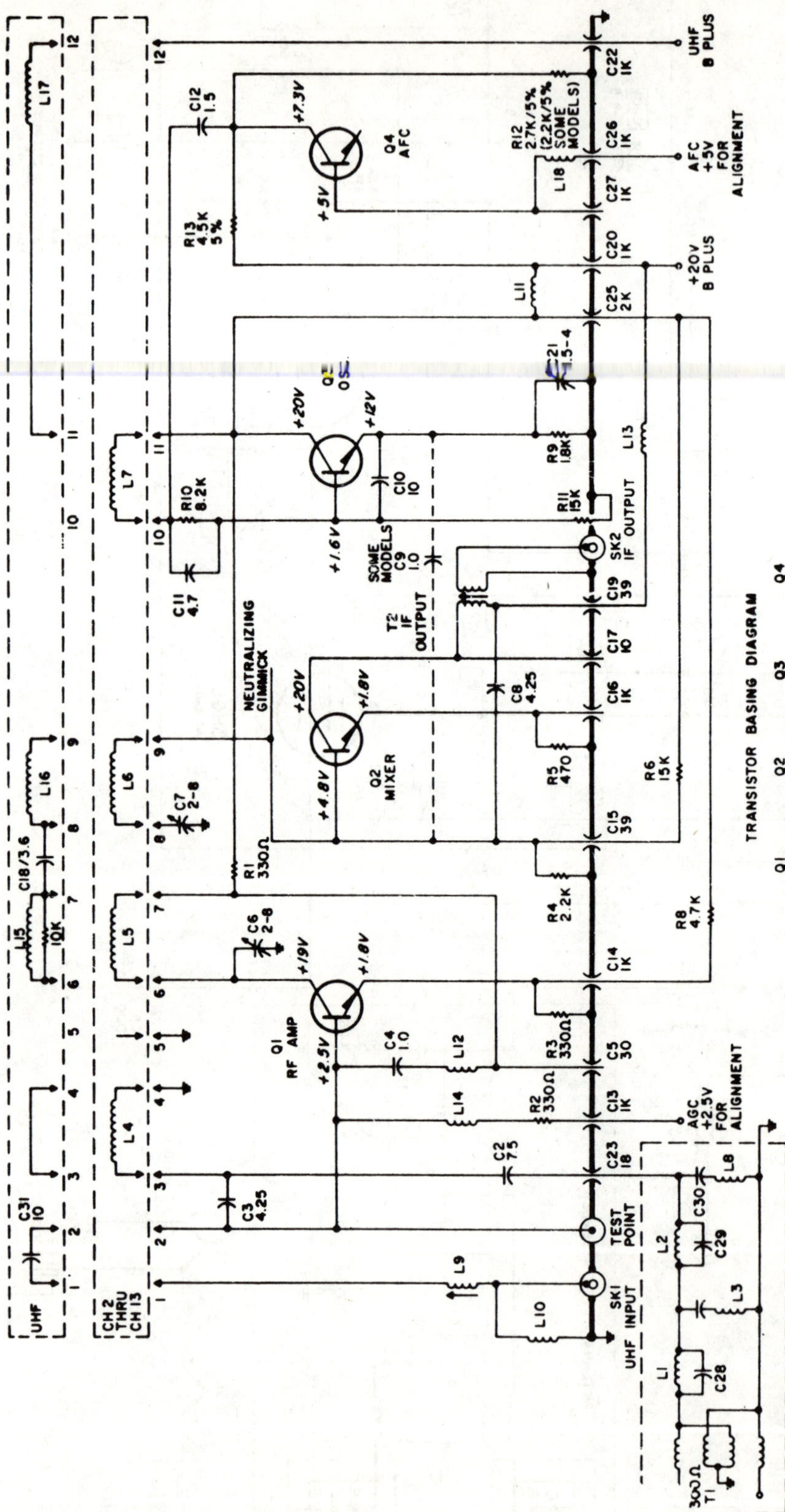

Fig. 11-9. Sylvania 54-23858-4, -13 tuner schematic.

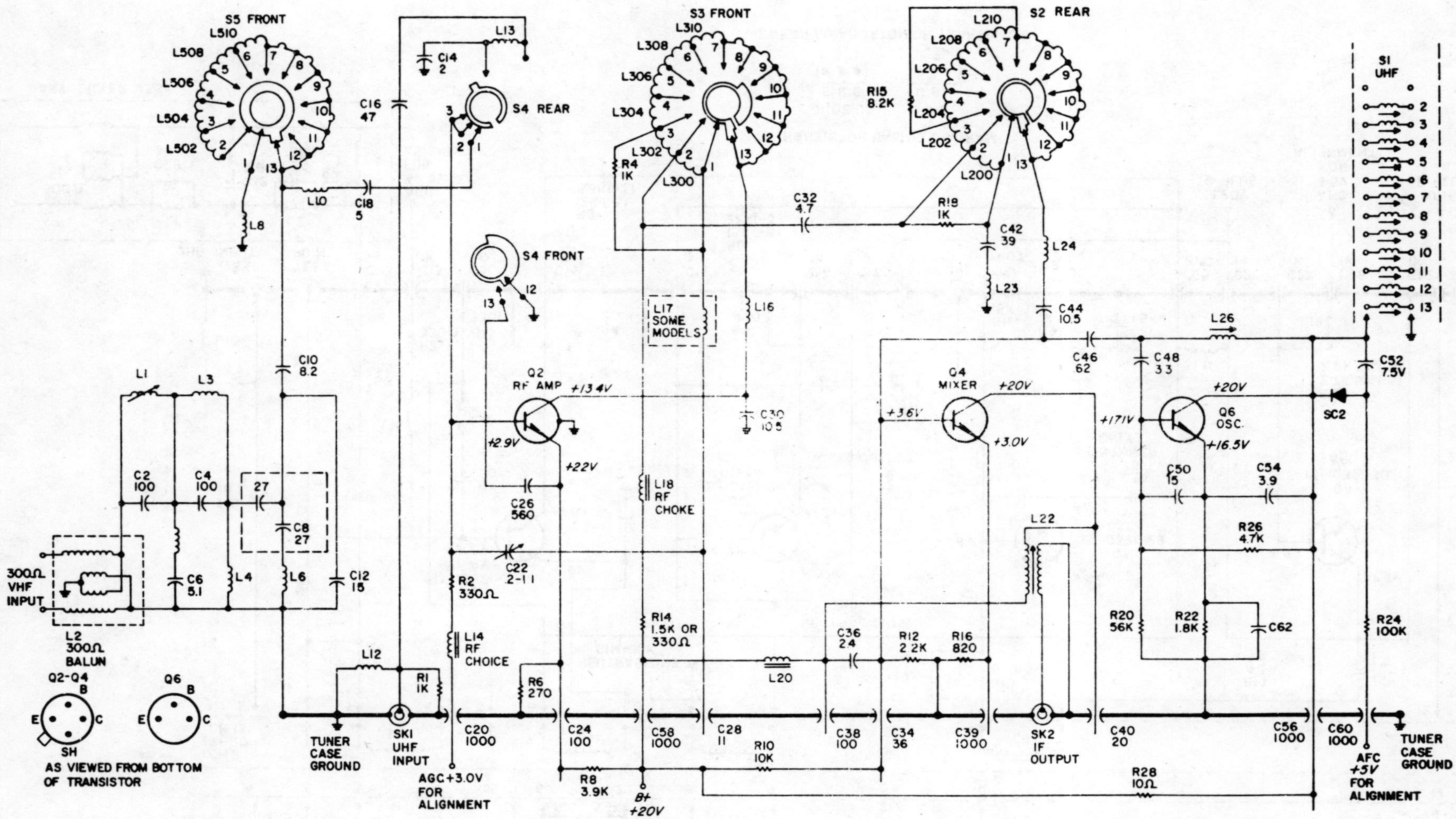

Fig. 11-10. Sylvania 54-23857-2, -3 tuner schematic.

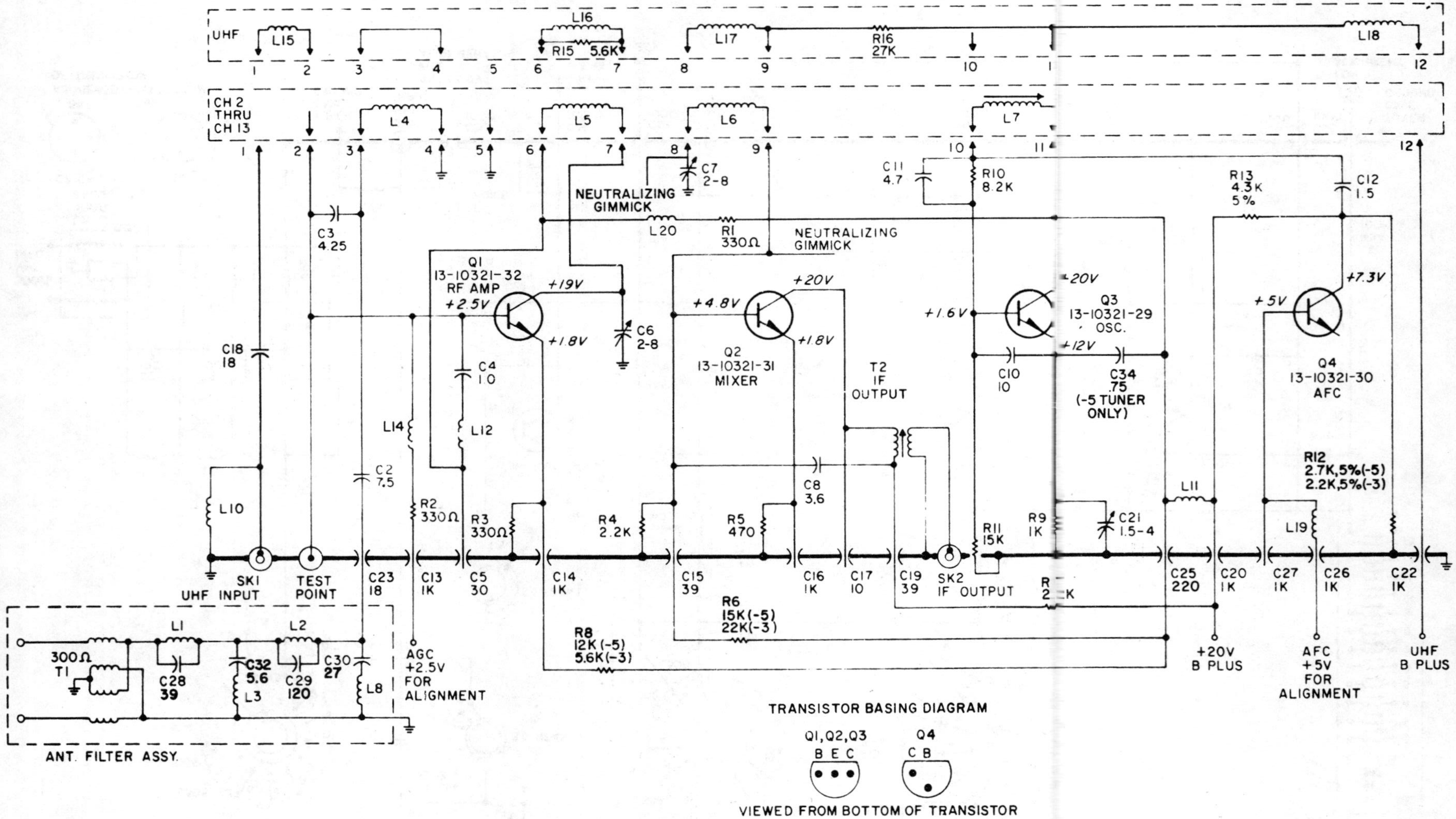

Fig. 11-11. Sylvania 54-29331-3, -5 tuner schematic.

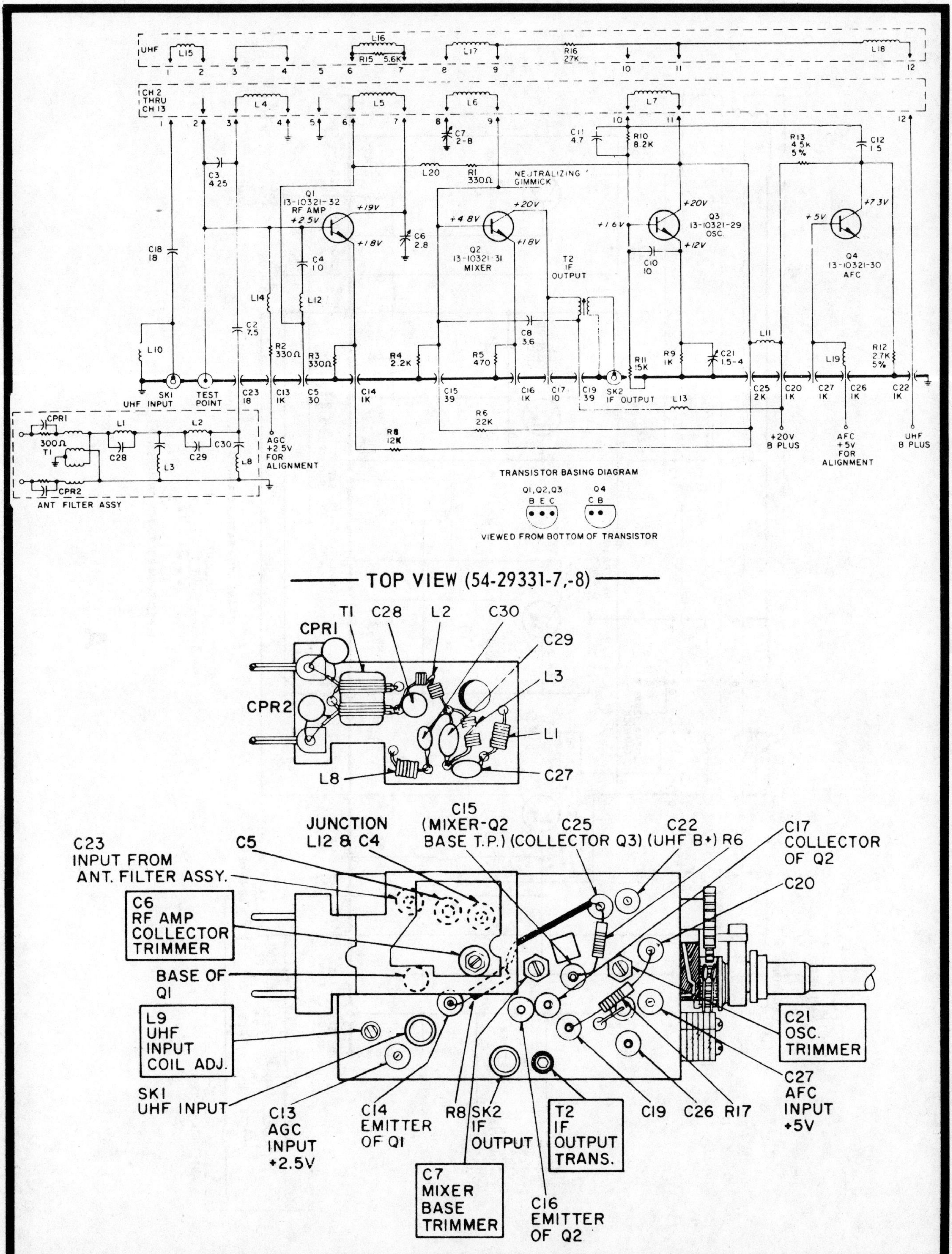

Fig. 11-12. Sylvania 54-29331-7, -8 tuner schematic.

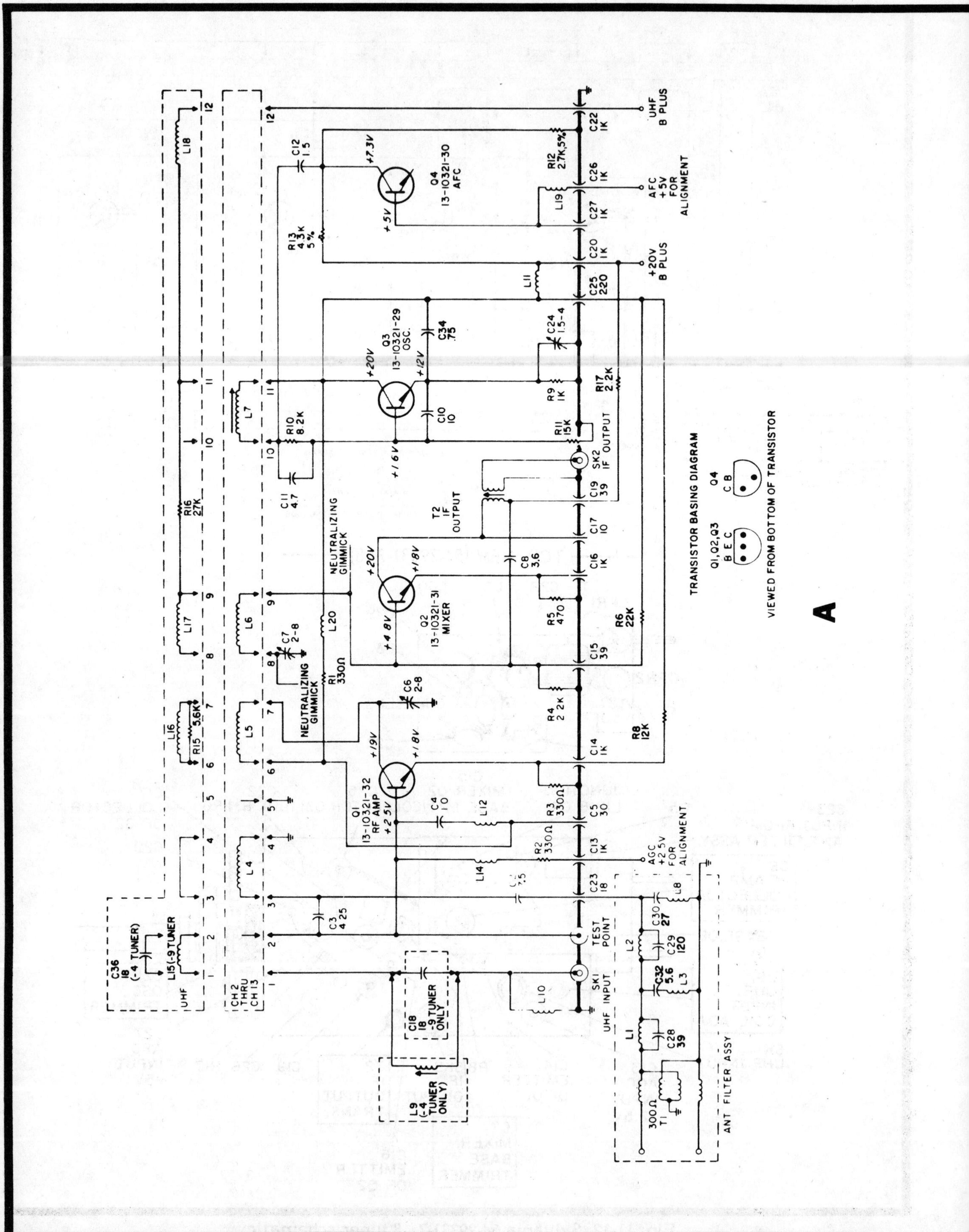
Q1 13-10321-32 RF AMP
Q2 13-10321-31 MIXER
Q3 13-10321-29 OSC.
Q4 13-10321-30 AFC
NEUTRALIZING GIMMICK
NEUTRALIZING GIMMICK
T2 IF OUTPUT
SK1 UHF INPUT
TEST POINT
SK2 IF OUTPUT
AGC +2.5V FOR ALIGNMENT
+20V B PLUS
AFC +5V FOR ALIGNMENT
UHF B PLUS
ANT FILTER ASSY
CH2 THRU CH13
UHF
C36 18 (-4 TUNER)
L15 (-9 TUNER)
C18 18 -9 TUNER ONLY
L9 (-4 TUNER ONLY)
TRANSISTOR BASING DIAGRAM
Q1,Q2,Q3 B E C
Q4 C B
VIEWED FROM BOTTOM OF TRANSISTOR
A

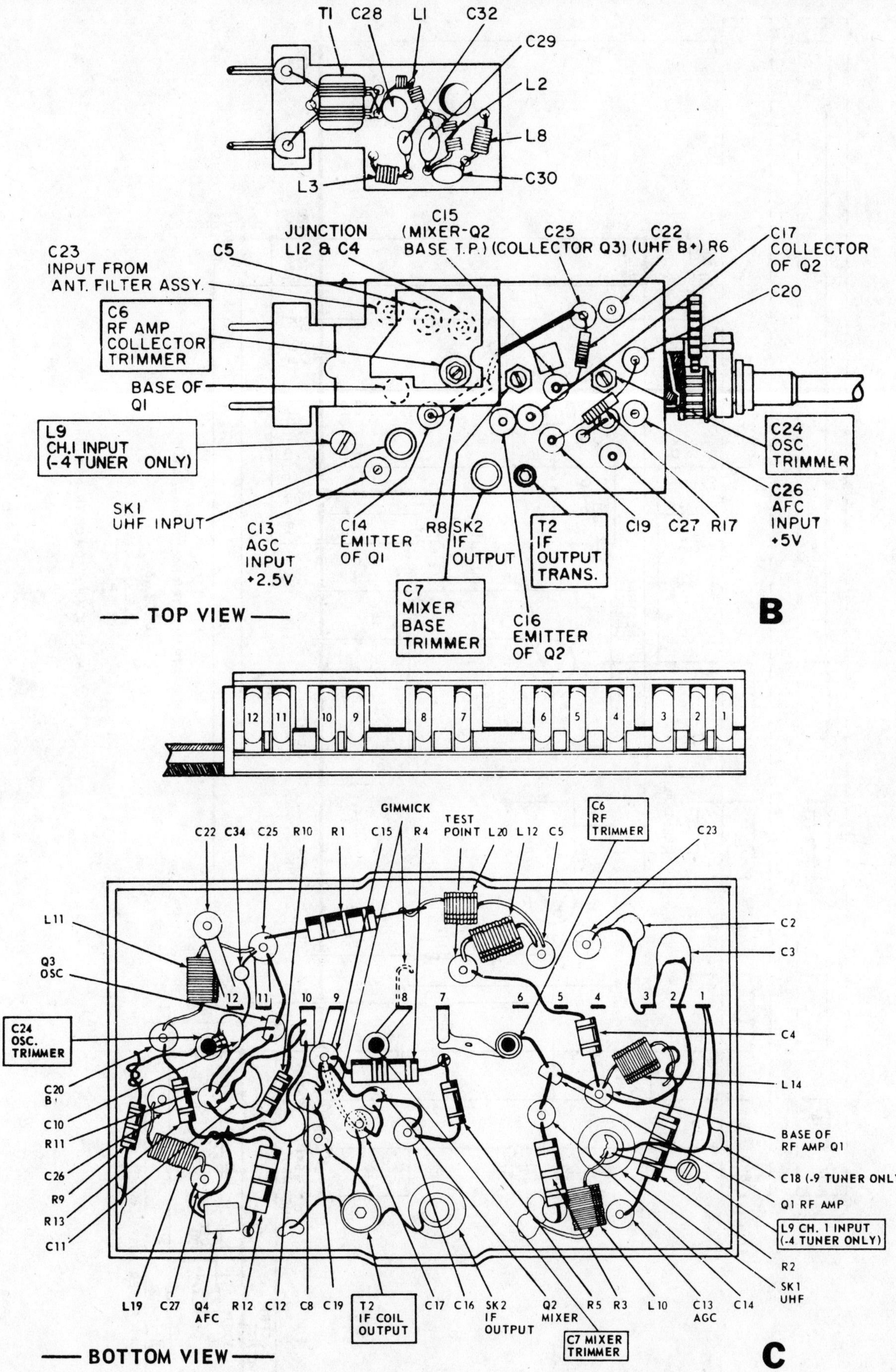

Fig. 11-13. Sylvania 54-29331-4, -9 tuner schematic.

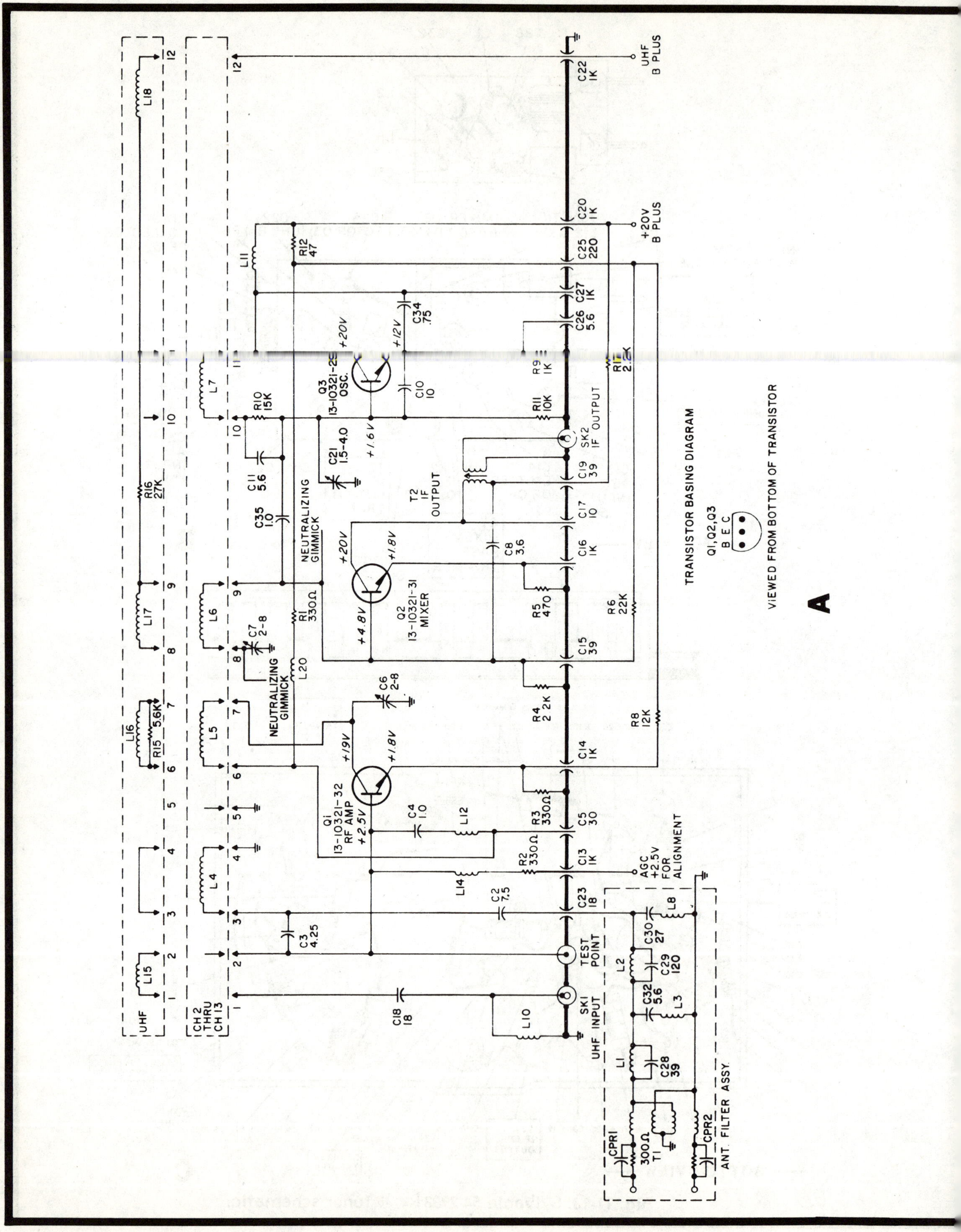
UHF
CH2
THRU
CH 13
NEUTRALIZING
GIMMICK
Q1
13-10321-32
RF AMP
Q2
13-10321-31
MIXER
Q3
13-10321-25
OSC.
T2
IF
OUTPUT
SK1
UHF INPUT
TEST
POINT
SK2
IF OUTPUT
AGC
+2.5V
FOR
ALIGNMENT
ANT. FILTER ASSY.
+20V
B PLUS
UHF
B PLUS
TRANSISTOR BASING DIAGRAM
Q1, Q2, Q3
B E C
VIEWED FROM BOTTOM OF TRANSISTOR
A

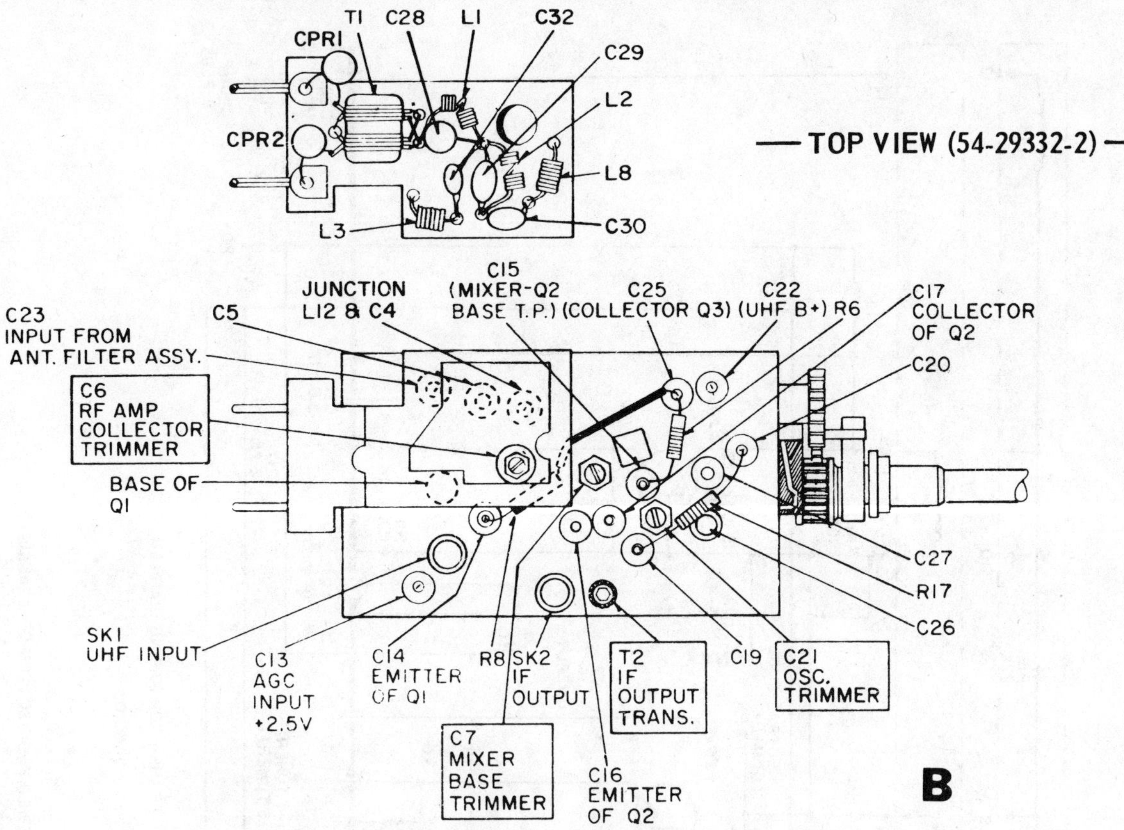

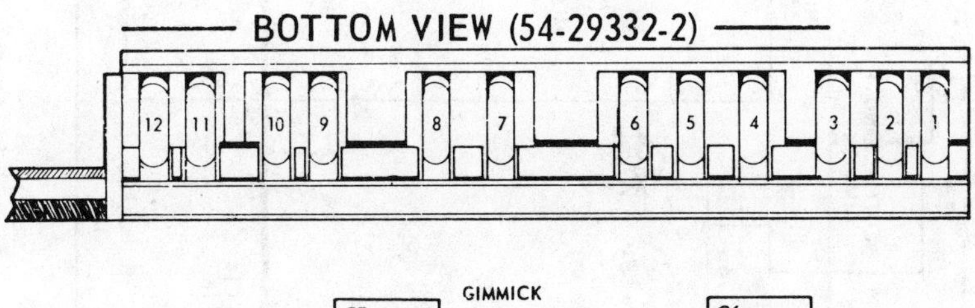

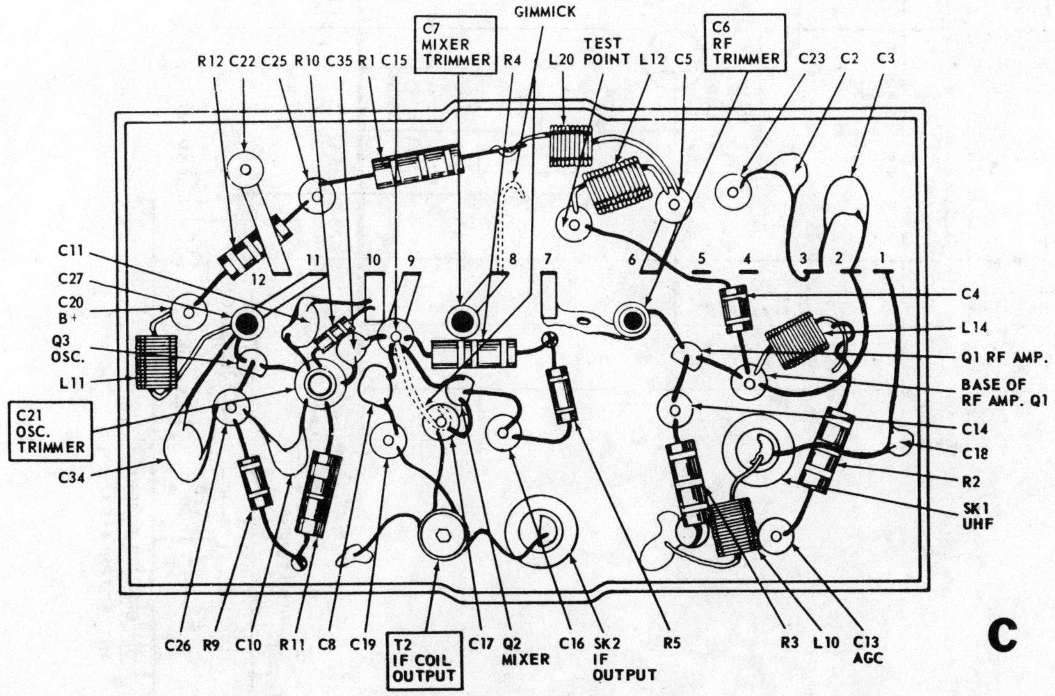

Fig. 11-14. Sylvania 54-29332-2 tuner schematic.

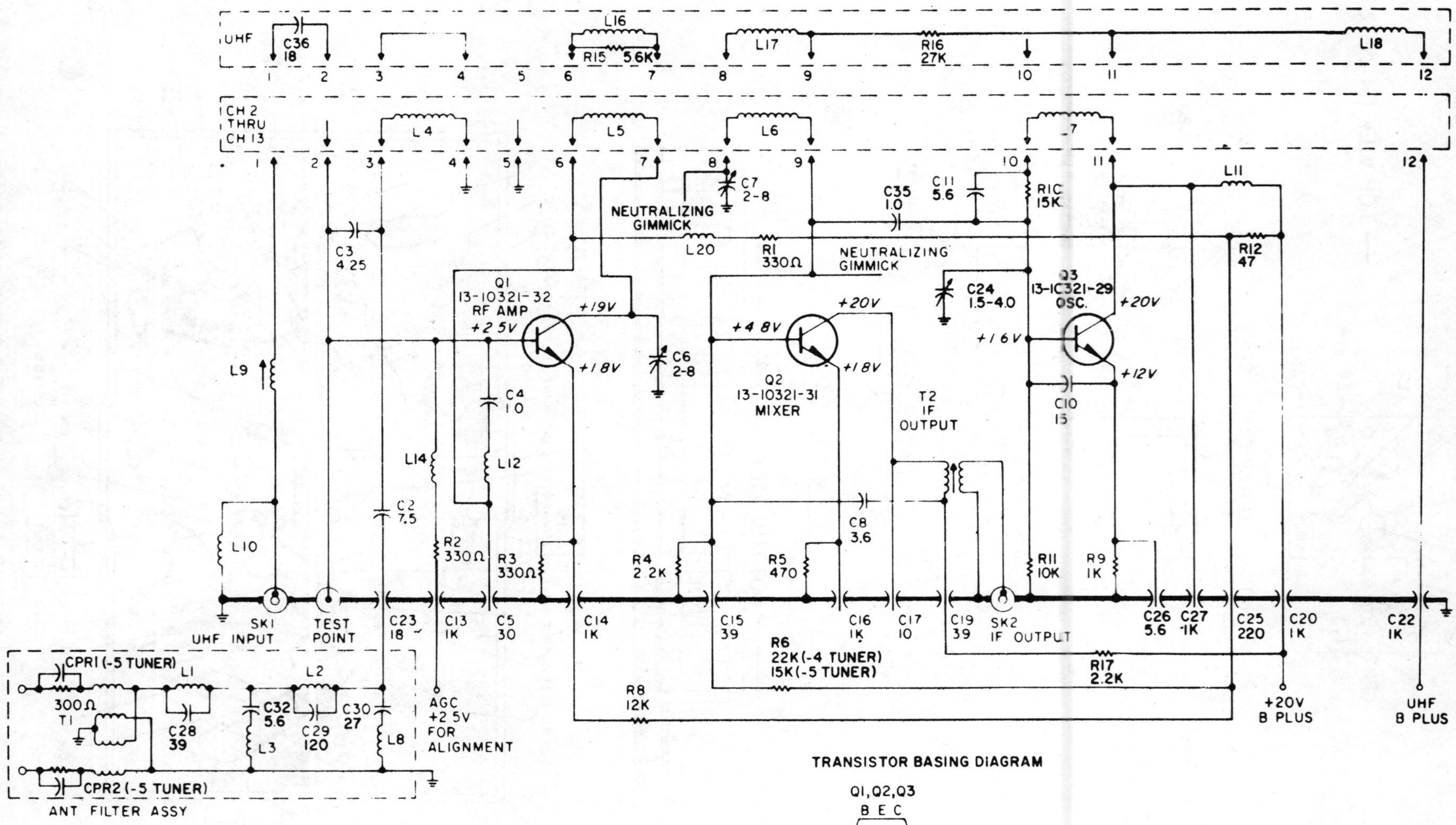

Fig. 11-15. Sylvania 54-29332-4, -5 tuner schematic.

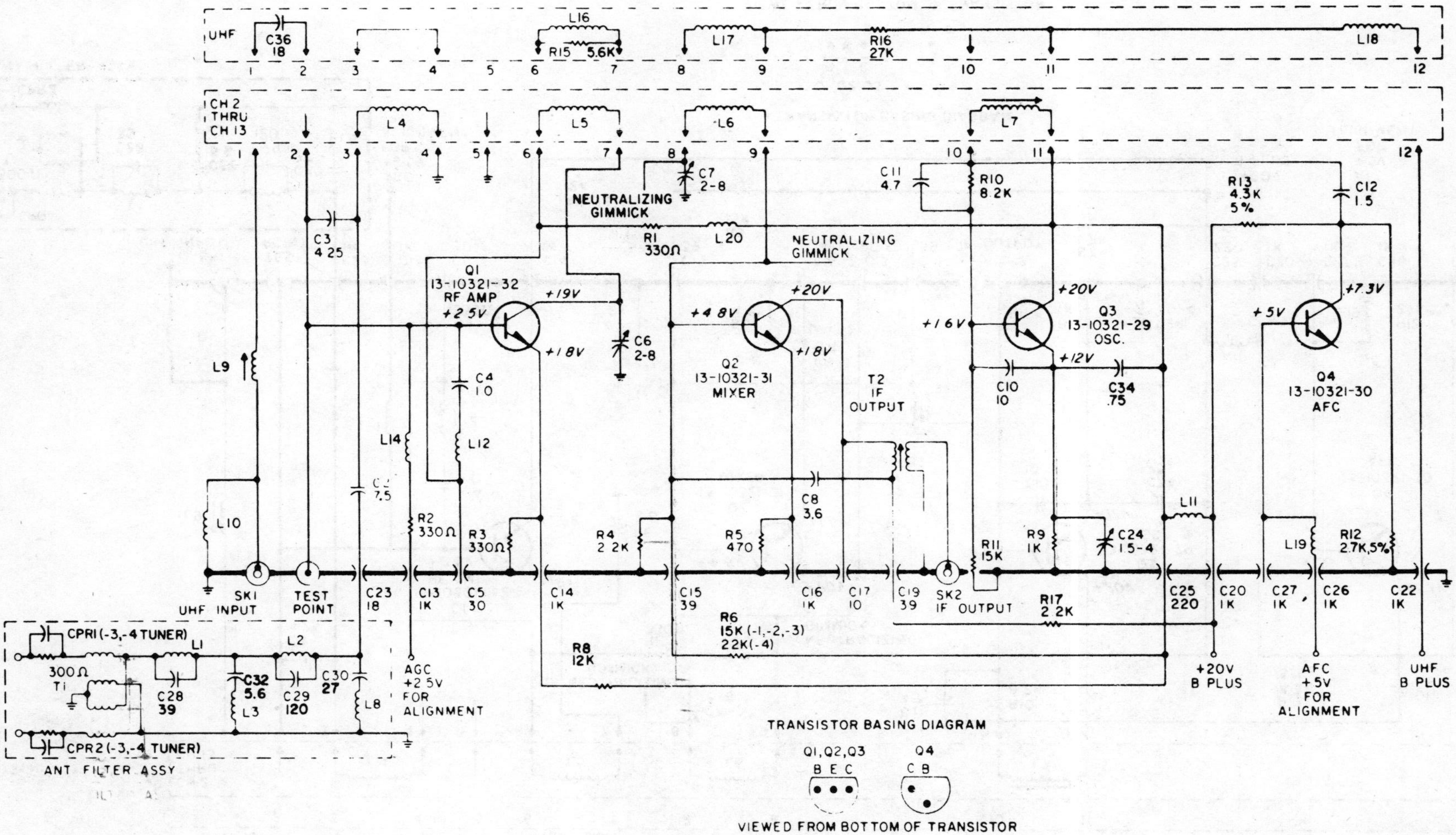

Fig. 11-16. Sylvania 54-33911-1 thru -4 tuner schematic.

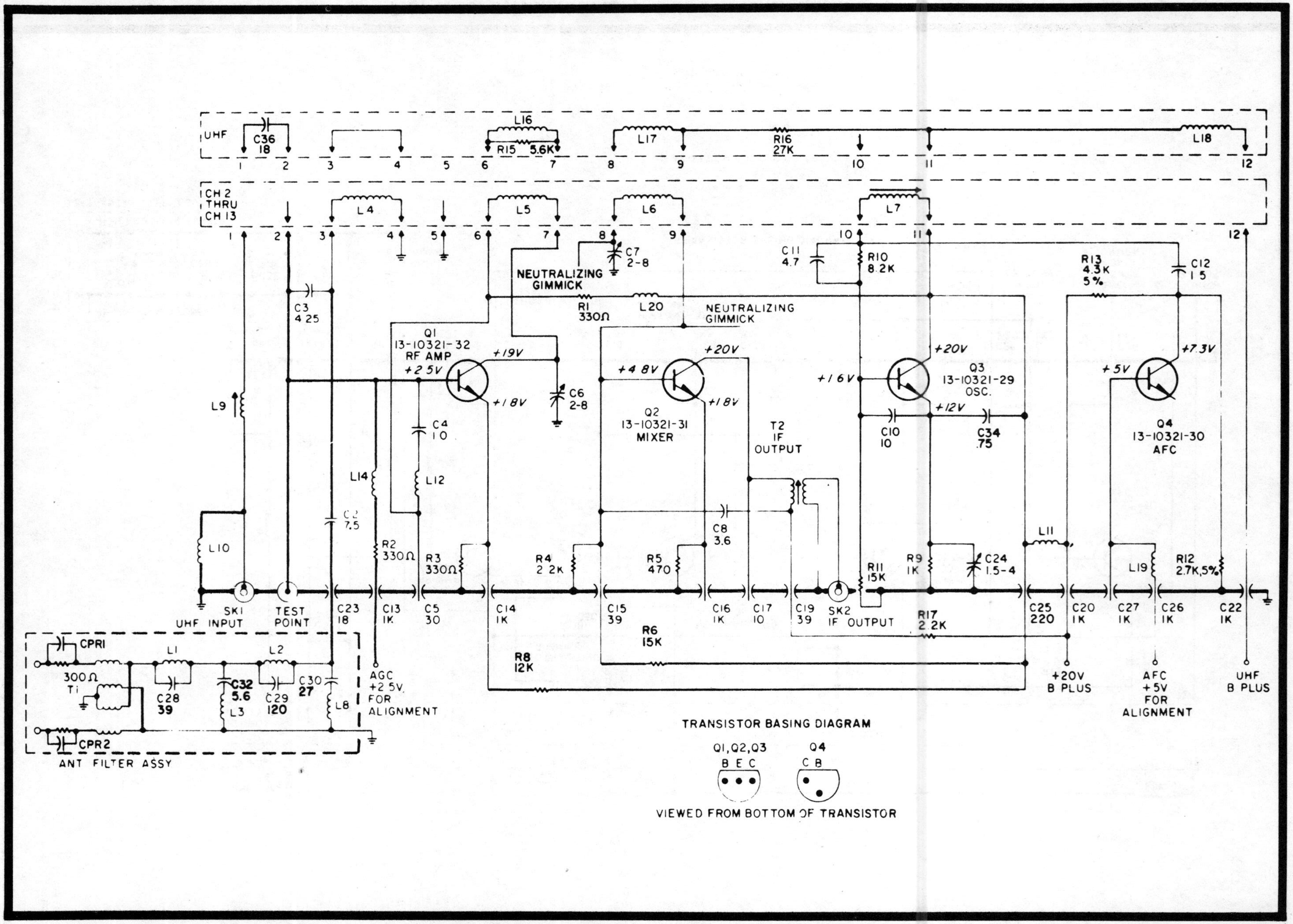

Fig. 11-17. Sylvania 54-33915-1 tuner schematic.

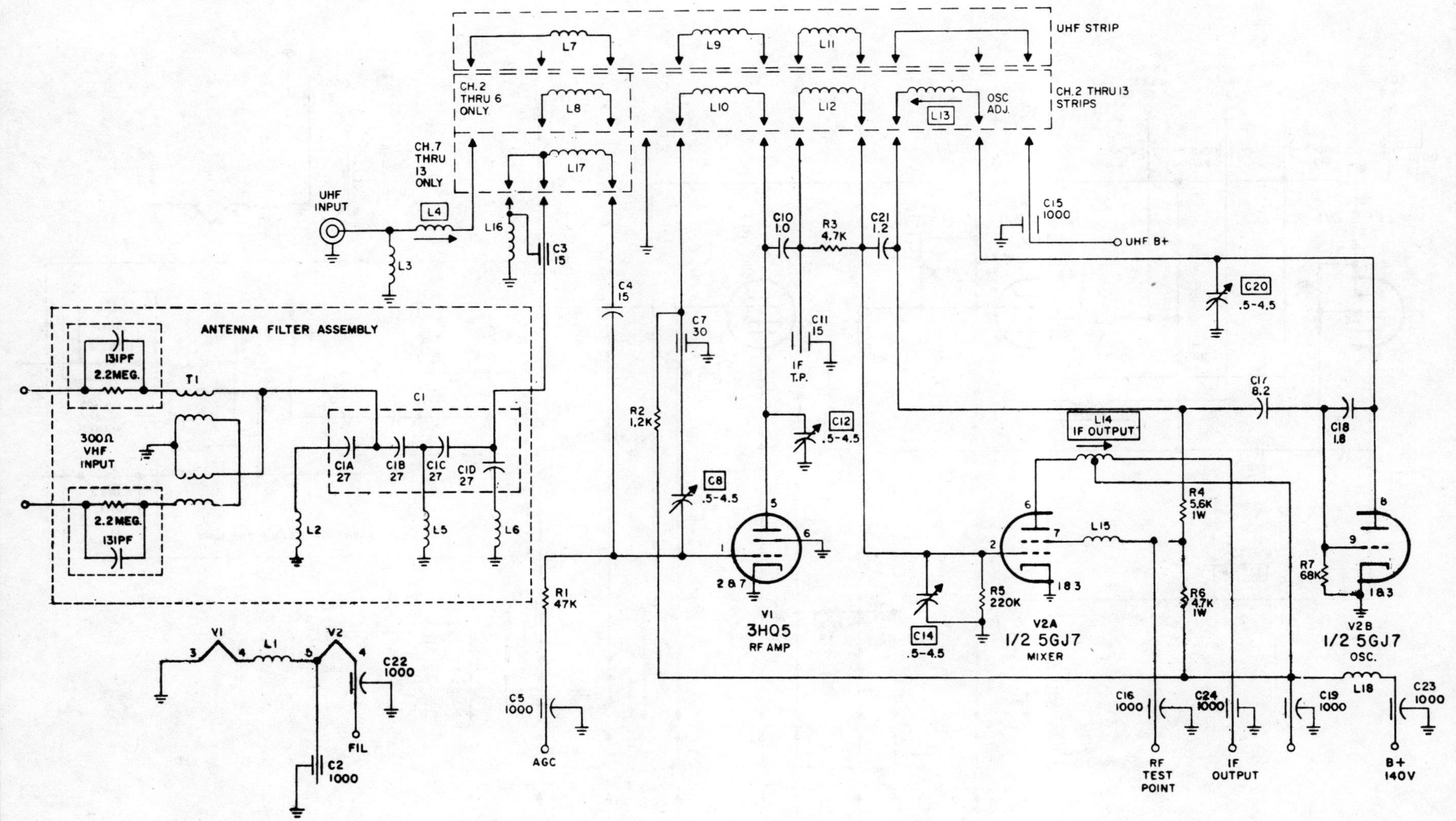

Fig. 11-18. Sylvania 54-35163-1 tuner schematic.

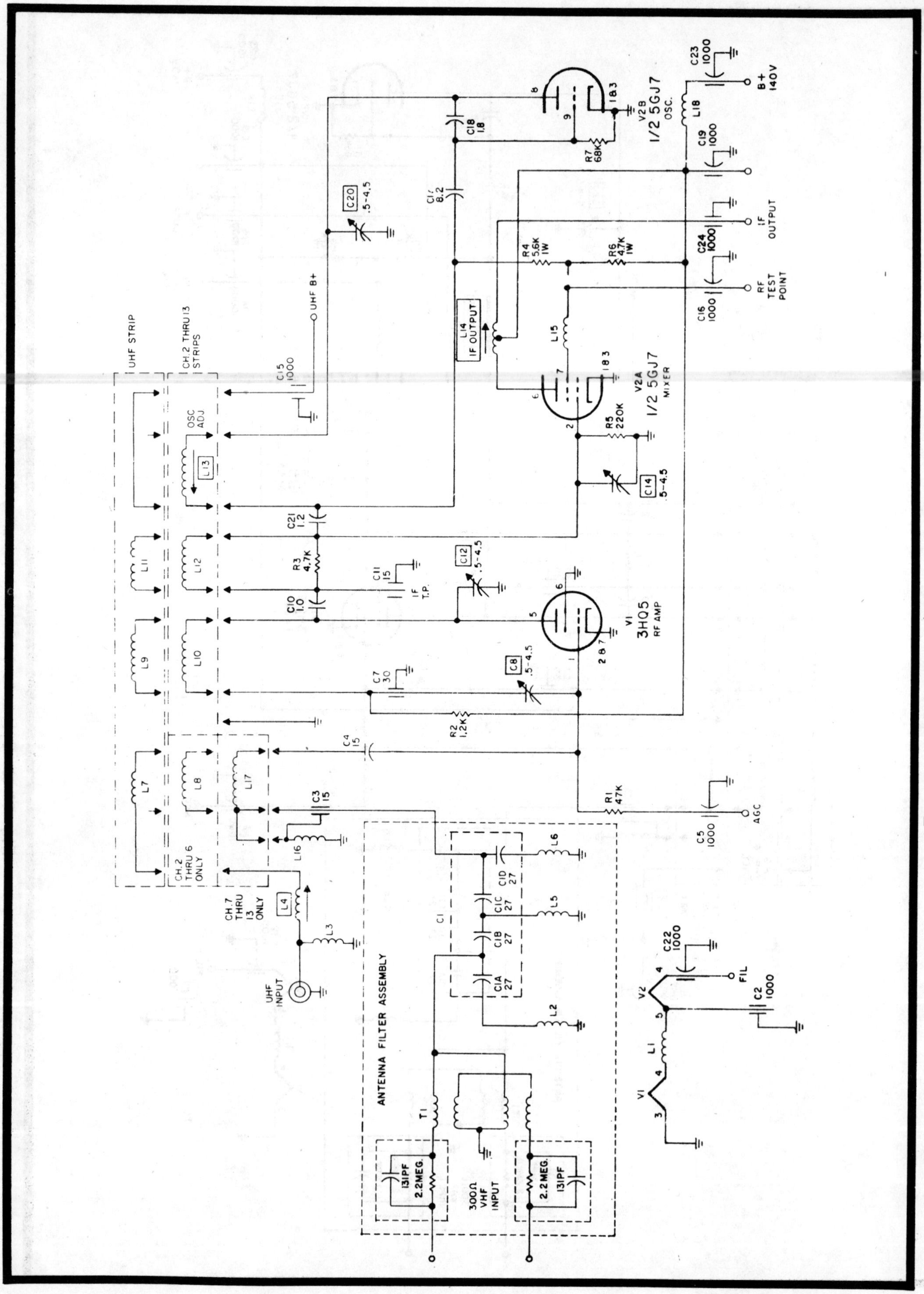

Fig. 11-19. Sylvania 54-35576-1 tuner schematic.

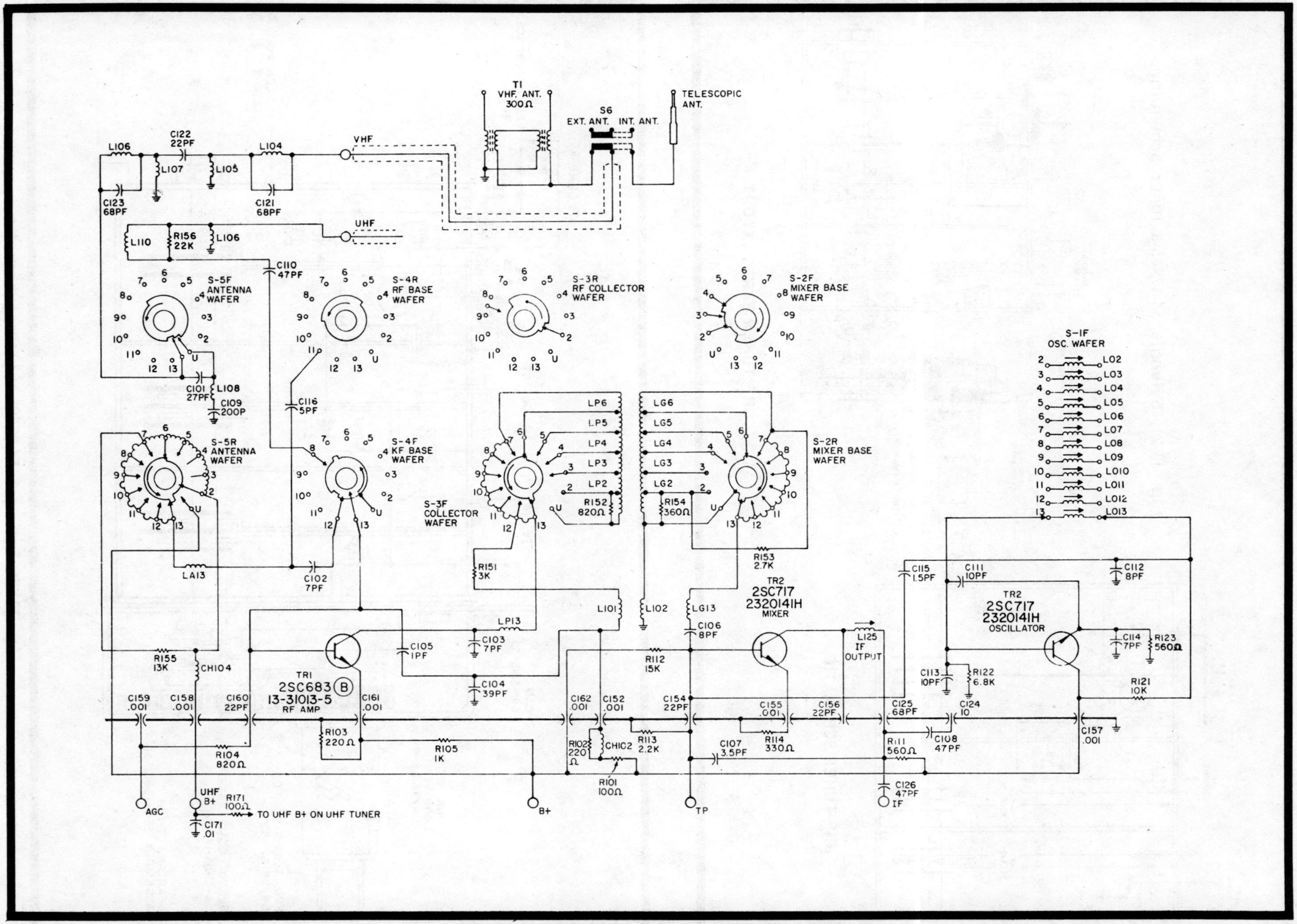

Fig. 11-20. Sylvania 2420611H tuner schematic.

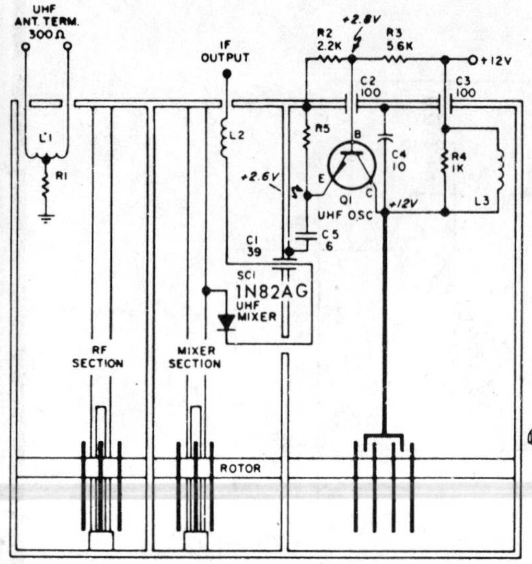

UHF TUNER SCHEMATIC

Fig. 11-21. Sylvania 54-17965-3 tuner schematic.

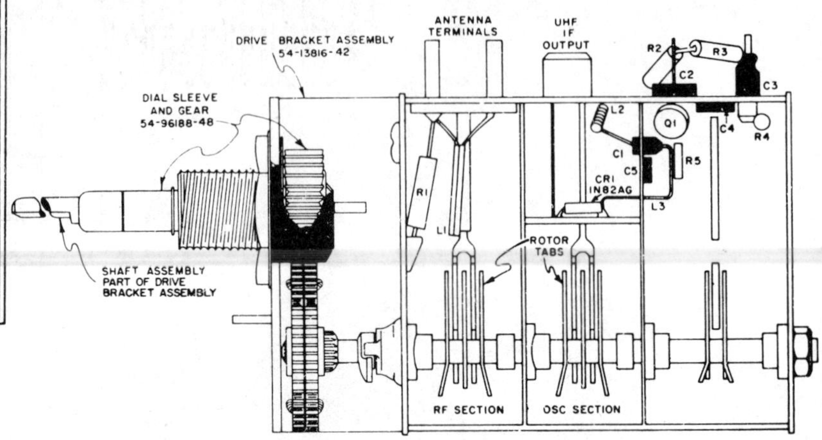

UHF TUNER PARTS LAYOUT

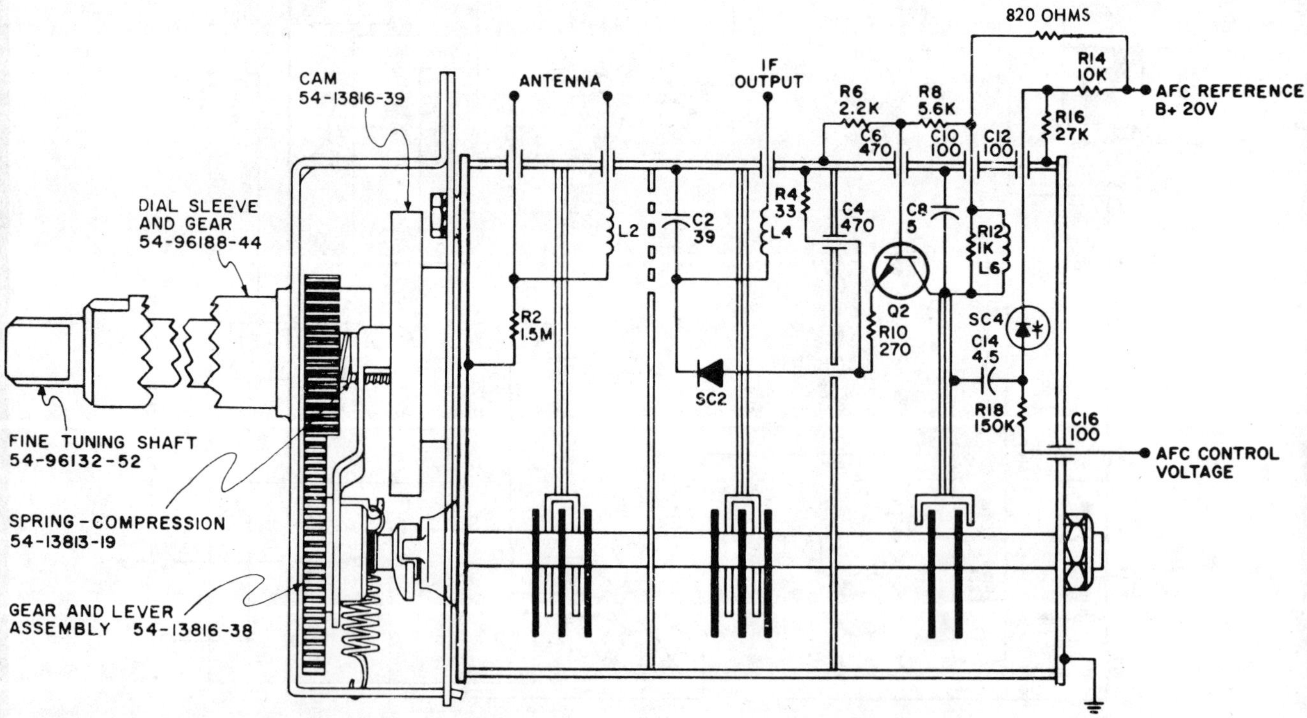

Fig. 11-22. Sylvania 54-27396-1 tuner schematic.

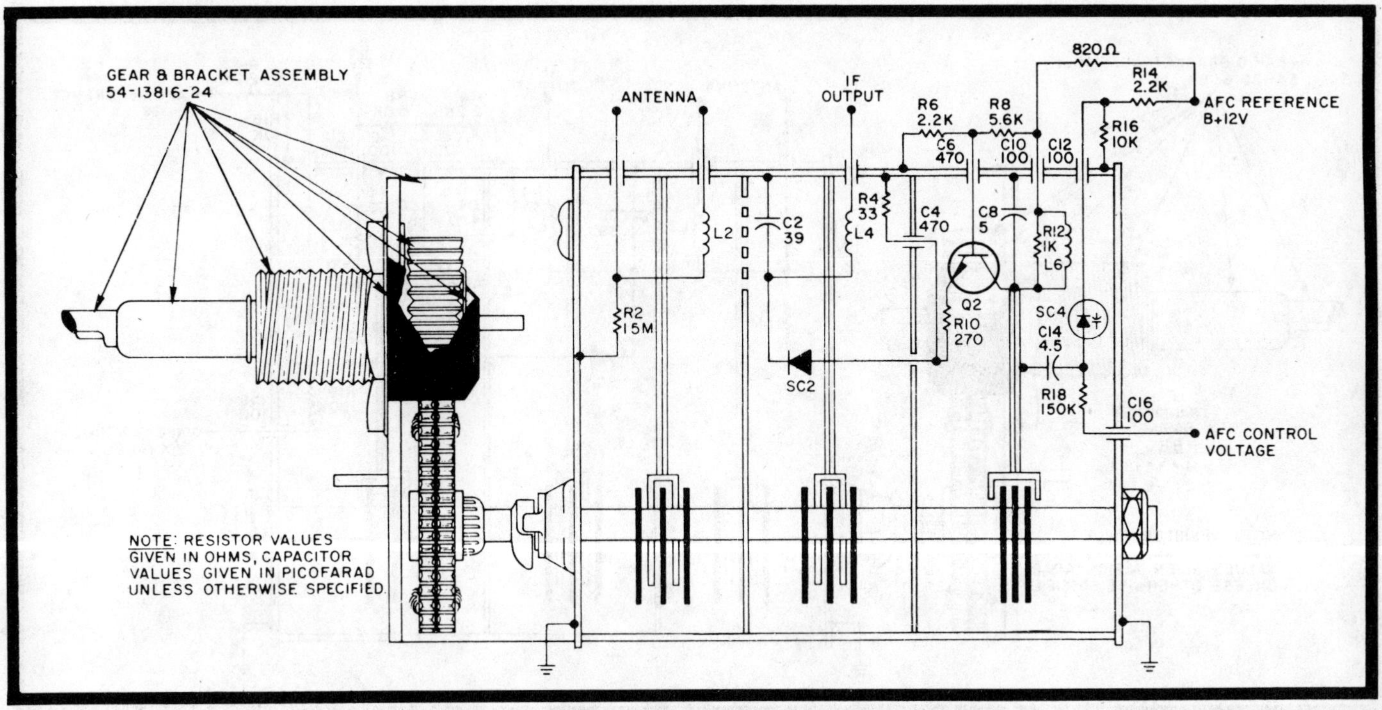

Fig. 11-23. Sylvania 54-27843-1 tuner schematic.

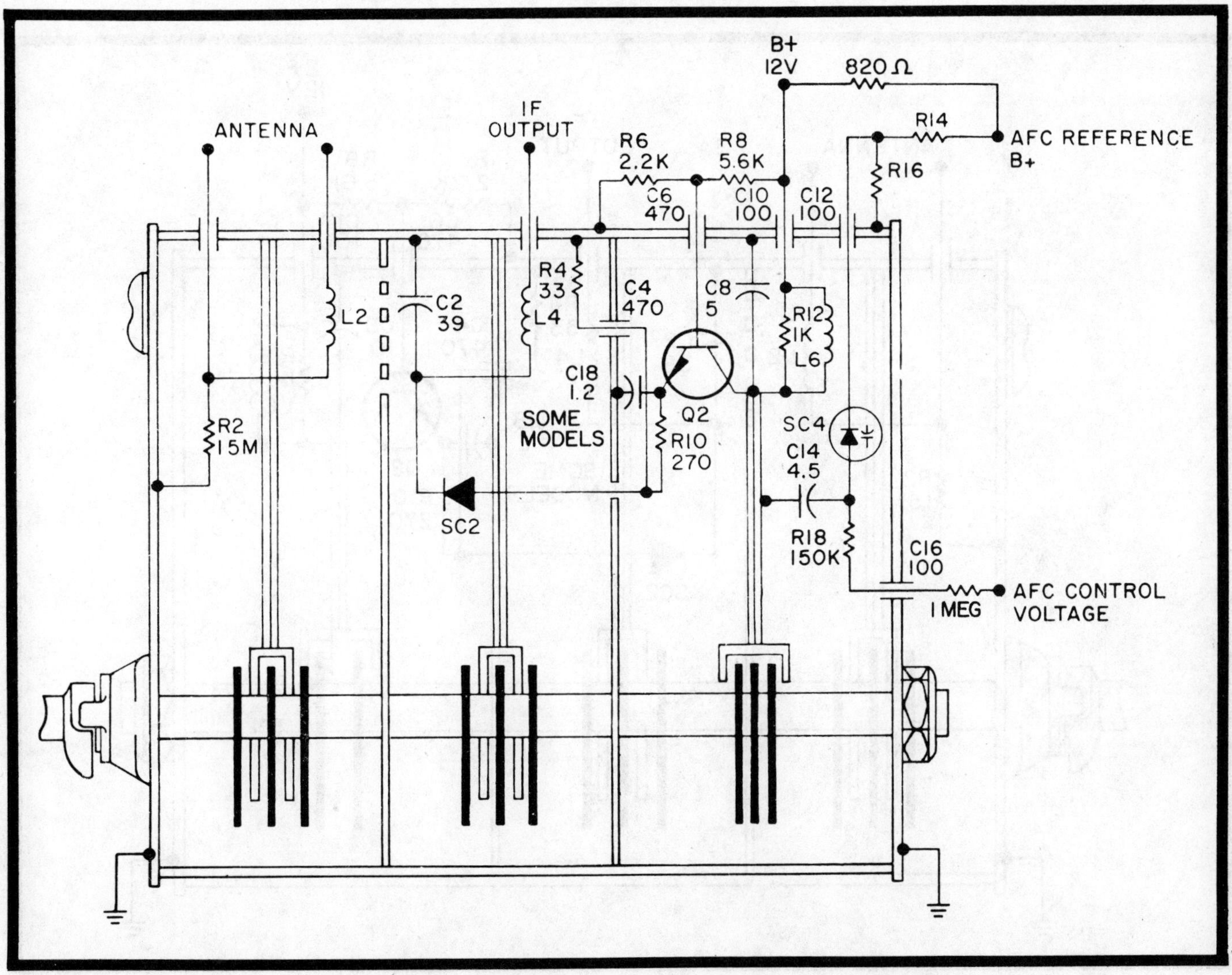

Fig. 11-24. Sylvania 54-29184-1 tuner schematic.

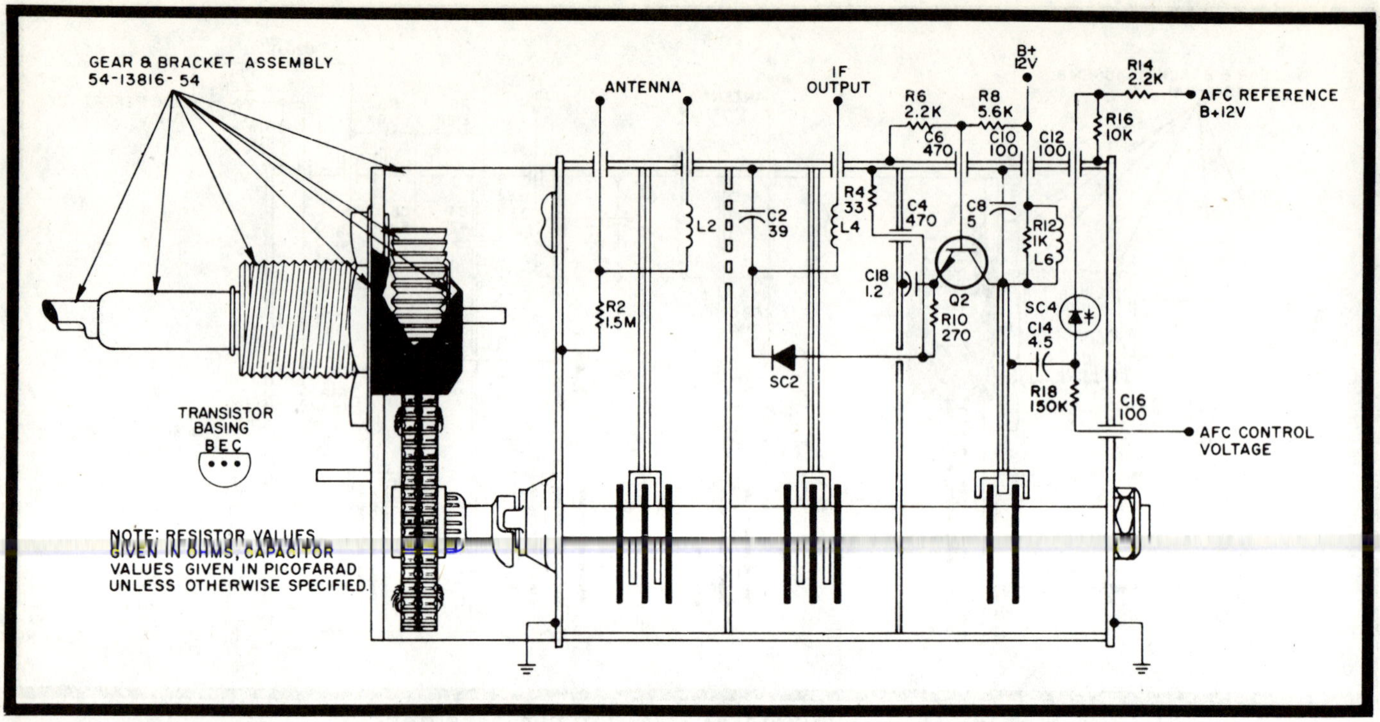

Fig. 11-25. Sylvania 54-27843-3 tuner schematic.

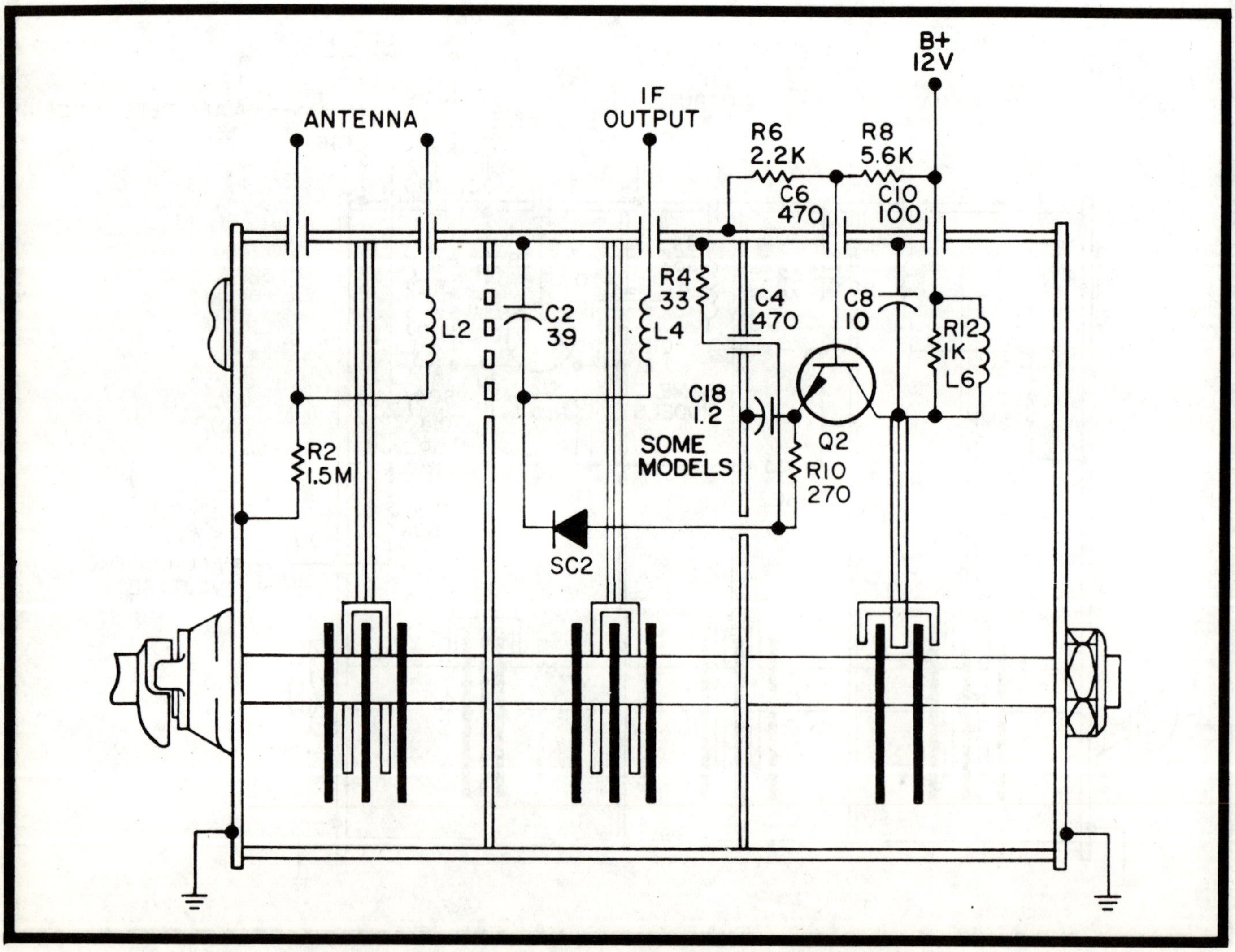

Fig. 11-26. Sylvania 54-29859-2 tuner schematic.

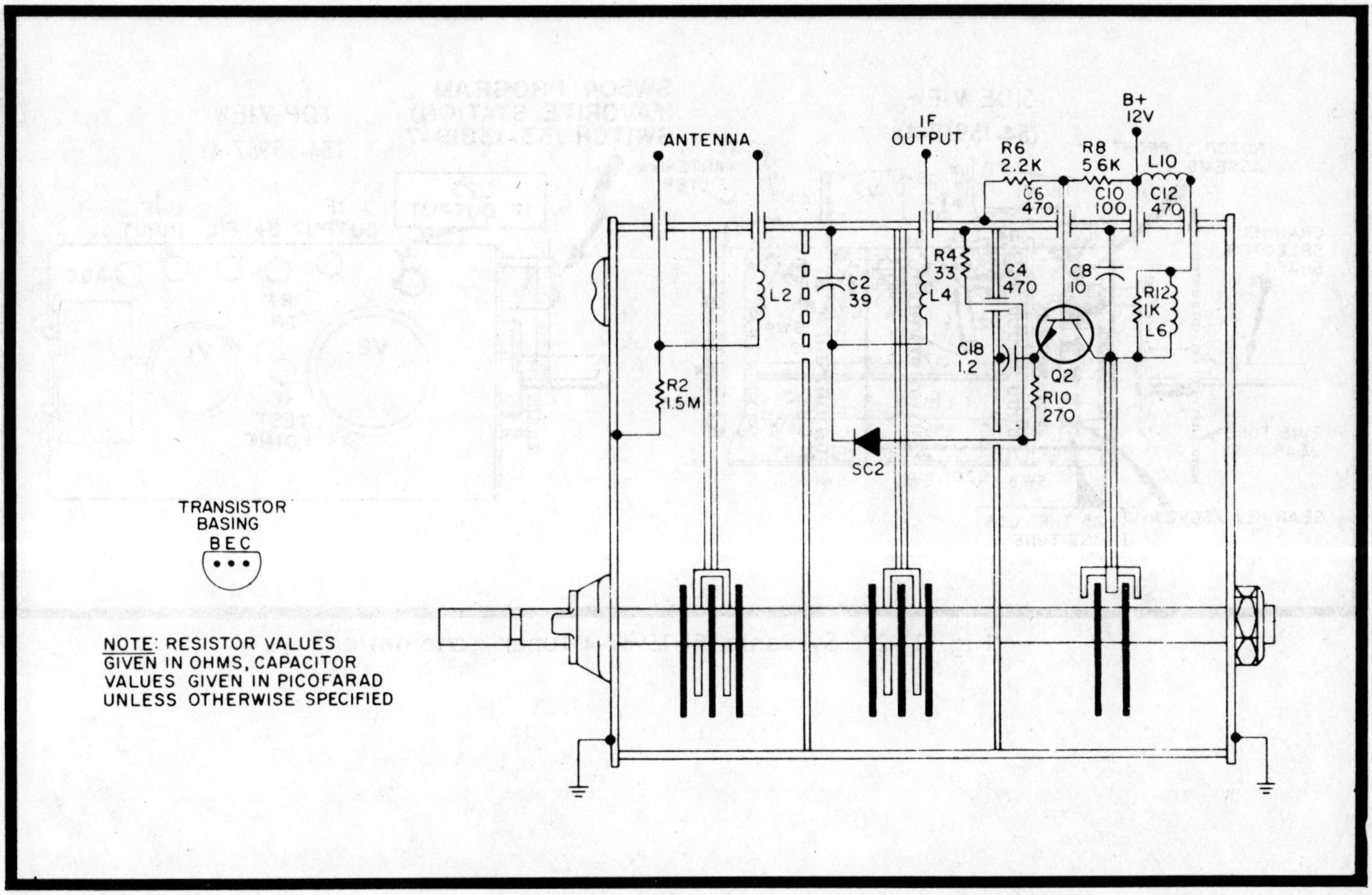

Fig. 11-27. Sylvania 54-33774-1 tuner schematic.

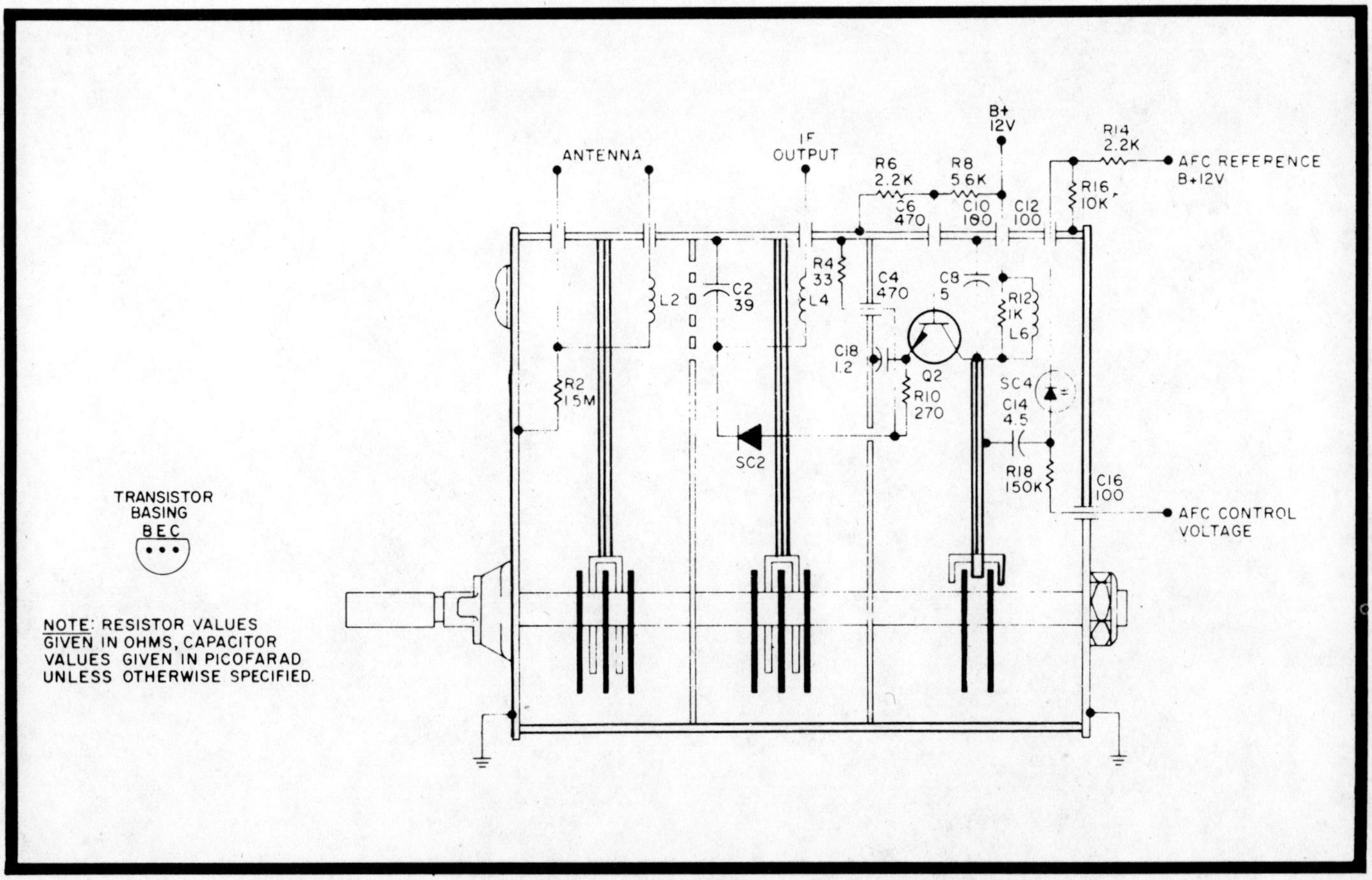

Fig. 11-28. Sylvania 54-33914-1 tuner schematic.

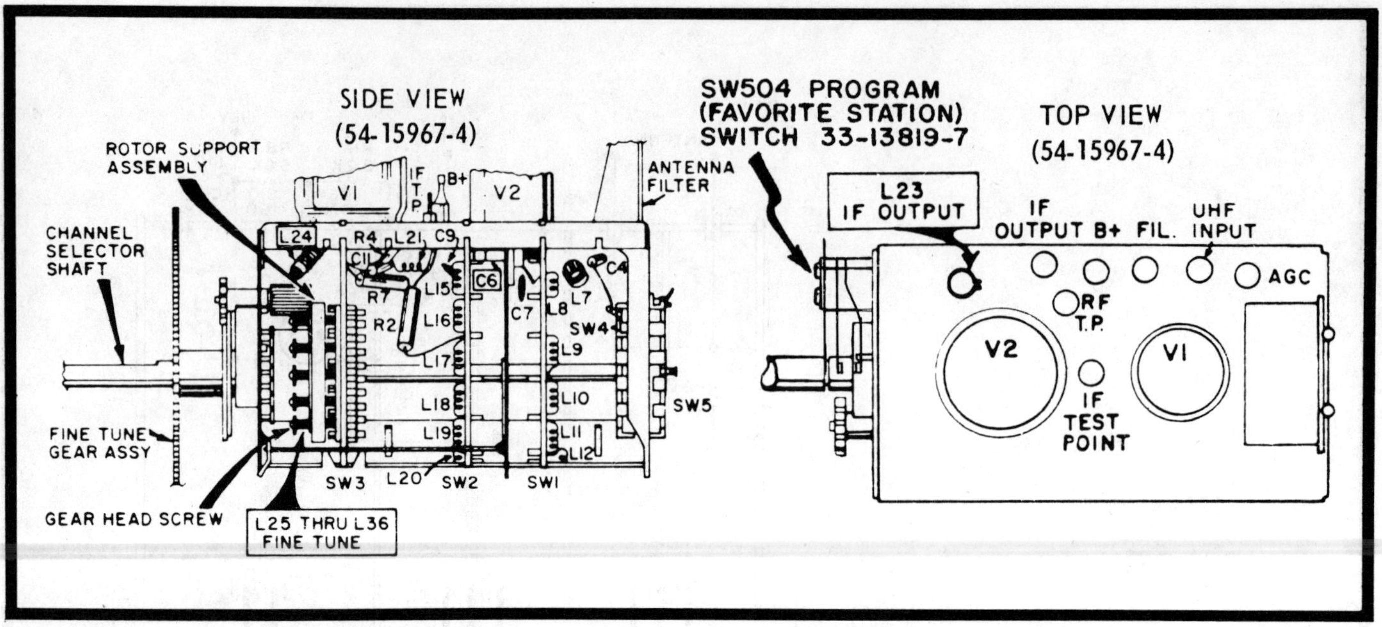

Fig. 11-29. Sylvania 54-15967-4 tuner schematic.

Chapter 12

Zenith Tuner Alignment

For aligning Zenith tuners in this chapter, a suitable VHF and UHF sweep generator with accurate marker should be used (or a postmarker generator alignment system). It is very important to terminate the output cable and check to see if the attenuator is reactive. If it is reactive or if the output cable is improperly terminated, correct alignment cannot be made because the response curve will be distorted. The attenuator should only vary the amplitude and not the shape of the response curve.

TURRET TUNER ALIGNMENT, 175-600 SERIES (ROUND-NOSE PASSBAND)

Round-nose passband tuners are used with chassis that have solid-state i-f modules.

1. Connect the negative lead of 0-20V bias supply to agc terminal; positive lead to chassis. (On AFC tuners add +3V dc to AFC terminal.)
2. Connect scope through a 10K isolation resistor to converter screen grid feedthrough capacitor.
3. Feed sweep generator to antenna terminals. (Use matching transformer shown in Fig. 12-1.)
4. Switch receiver and sweep generator to Channel 10. Set bias for —1.5V.
5. Adjust antenna coil and rf plate and mixer grid trimmers for pattern of Fig. 12-2. Turn fine-tuning oscillator screw of Channel 10 strip to halfway position. Set master oscillator trimmer screw for correct marker positions on waveforms. Use extreme care in spreading or compressing the channel strip coil windings to prevent damage.

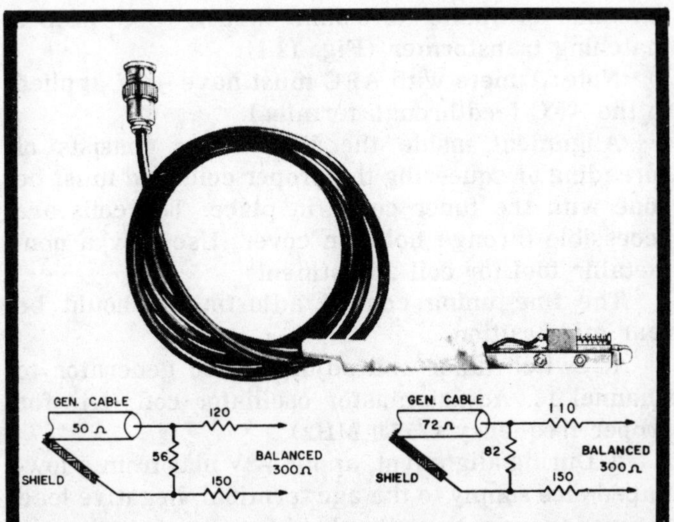

Fig. 12-1. I-f / rf alignment fixtures (Zenith).

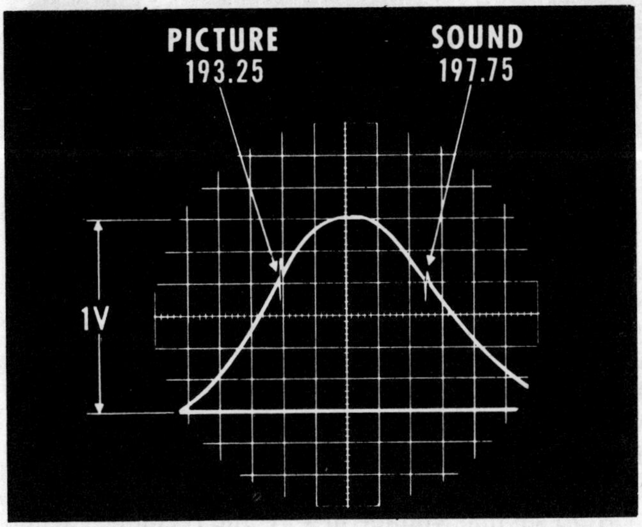

Fig. 12-2. Channel 10 rf response (round nose) (Zenith).

6. To neutralize the rf amplifier, increase the bias to —20V. Without changing the bias, adjust the neutralizing trimmer for minimum response at i-f output on scope.

There is some interaction between the plate and neutralizing trimmers and it may be necessary to repeat steps 4, 5, and 6 until correct results are obtained.

Note: When inserting or removing a strip always turn the fine-tuning oscillator screw fully inward first for proper clearance.

TURRET TUNER ALIGNMENT, 175-500 AND 1200 SERIES (ROUND-NOSE PASSBAND)

1. Connect a negative bias voltage to the tuner agc terminal.
2. Connect scope through a 10K isolation resistor to test point B.
3. Feed sweep generator signal into antenna terminals with proper matching transformer.
4. Switch receiver and sweep generator to Channel 10. Set bias for —3V. To prevent overload do not exceed one-volt peak output during alignment.

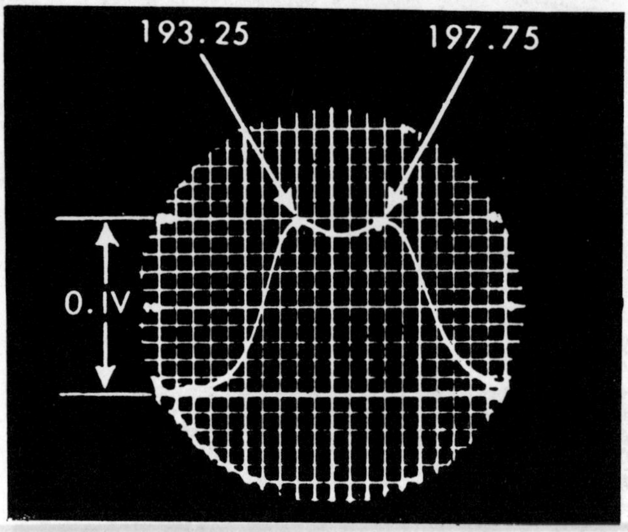

Fig. 12-3. Channel 10 rf response (Zenith).

5. Adjust the rf plate and converter grid trimmers for Fig. 12-2 pattern.

6. For neutralization, connect scope through a 10K isolation resistor to terminal C (detector) on TV chassis and connect ground lead to chassis. Switch sweep generator and TV tuner to Channel 10. Adjust bias to —20V and tune neutralizing capacitor C8 for minimum output as indicated on scope.

7. Reset bias voltage to —3V and reconnect scope to test point B on tuner.

Note: For easier alignment it may be desirable to insert a tuning wand into the coil to determine if adjustment is necessary. An increase in amplitude with brass indicates too much inductance and coil turns must be spread. An increase in amplitude with iron indicates too little inductance and the coil must be squeezed. At resonance, a reduction in amplitude will be noted with both iron and brass.

8. Switch tuner to other channels and observe bandpass. If necessary, adjust interstage (rf plate and mixer grid) coils for proper bandwidth and coupling.

9. The converter plate coil is adjusted as part of the i-f alignment. The antenna filter is factory-adjusted for minimum response at 43.5 MHz. (For tuners with AFC, always apply +3V to the AFC terminal to maintain correct frequency correlation.)

BANDSWITCH TUNER ALIGNMENT, 175-1150, 1160 AND 1170 SERIES (ROUND-NOSE PASSBAND)

1. Connect scope (appox. 1V per inch) to the converter screen grid capacitor (test point B). Use 10K isolation resistor. Set agc to —2V. Connect sweep and marker generator through balun to antenna terminals. (On AFC tuners connect +3V to AFC terminal.)

2. Set tuner to Channel 13 and roughly align channel 13 bandpass (Fig. 12-3) by "knifing" plate coil with insulated tool.

3. Turn fine-tuning control clockwise until Channel 13 oscillator screw is all the way in (point is reached when oscillator marker stops movement on waveform).

4. Adjust master oscillator inductance coil L6 until the oscillator frequency marker is about 4.5 MHz high (near sound marker).

5. Turn fine-tuning knob until oscillator marker is set on correct frequency (near sound marker).

6. Align i-f output coil L4 to peak at 44 MHz.

7. Repeat step 5 on Channels 12 through 2.

8. Set tuner to Channel 6 and roughly align using the Channel 6 rf plate, converter grid, and antenna coils.

9. Set agc to about —14V. Raise sweep generator to full output and adjust neutralizing trimmer C3 for minimum output and equal peaks on scope pattern.

10. Set tuner and sweep generator to Channel 13 and adjust neutralizing coil L3 for minimum output. (Recheck neutralization on Channel 6.)

11. Set tuner and sweep generator to Channel 13 and readjust agc to —2V. Detune i-f output coil L4 by setting core into coil for maximum inductance.

12. Align rf plate, converter grid, and high-band coupling wires for smooth round-nose response.

13. With an insulated tool, carefully bend channels 7-12 rf plate segments for a proper pattern response (Fig. 12-3).

14. Align channels 6 through 1 for proper response pattern by adjusting rf plate, converter grid, and antenna coils.

VHF TRANSISTOR TUNER ALIGNMENT, 175-878, -879, -880, AND 175-881

Connect the scope to test point B on VHF tuner through an rf bandpass isolation network. The 40 MHz i-f output and input jacks on top of the tuner must be terminated with their respective load circuits or with a 75-ohm carbon resistor to prevent distortion of patterns. Connect sweep generator to the antenna terminals (300-ohm input). Use proper matching transformer (Fig. 12-1).

Note: Tuners with AFC must have +3V applied to the AFC feedthrough terminal.

Alignment inside the VHF tuner consists of spreading or squeezing the proper coils and must be done with the tuner cover in place. The coils are accessible through holes in cover. Use only a non-metallic tool for coil adjustments.

The fine-tuning control adjustment should be near mid-position.

1. Switch tuner and adjust sweep generator to Channel 13. Adjust master oscillator coil (L4) for proper frequency (257.0 MHz).

2. During alignment, apply +2V bias from a low-impedance supply to the agc terminal, negative lead to tuner chassis. Never apply high voltages to the agc terminal. The rf transistor is designed for forward bias only.

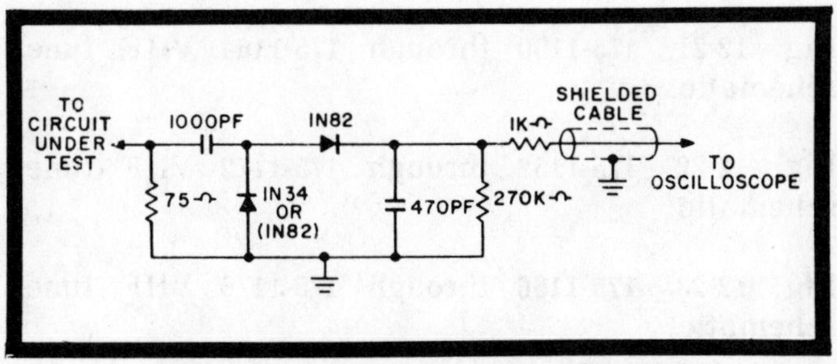

Fig. 12-4. RF detector (Zenith).

3. **High-band alignment**—To obtain proper response, location of markers, and maximum gain on Channel 13, spread or squeeze the rf collector coil located on wafer section next to metal shield and the mixer base coil.

4. **Low-band alignment**—Set tuner and sweep generator to Channel 6. Adjust antenna coil for resonance between picture and sound markers.

5. Adjust all other low-band antenna coils (channels 5, 4, 3, and 2) for proper resonance and center of bandpass. (Interstage coils for channels 5, 4, 3, and 2 are not adjustable.)

6. Adjust Channel 6 interstage coils for proper bandpass. For Channel 1 alignment, apply a 40 MHz signal to Channel 1 UHF input and adjust peaking coil for maximum response.

The antenna filter is an i-f and FM filter. Misadjustment of this filter can cause attenuation and tilt on Channel 2.

VHF TURRET TUNER ALIGNMENT, 175-1014, 175-1015, 175-1016, 175-1019 AND 175-1023 SERIES

1. Connect a +1.85V bias supply voltage to the agc terminal (negative lead to chassis). Connect 3V dc to AFC terminals (on tuners with AFC connection).

2. Connect scope through detector probe to test point B on tuner.

3. Feed sweep generator to the 300-ohm antenna terminals or use a matching network.

4. Switch receiver and set sweep generator to Channel 10.

5. Center oscillator screw on the channel strip and set oscillator to correct frequency by use of master oscillator trimmer C19.

6. Adjust rf collector and converter base trimmers for scope pattern (Fig. 12-3). Spread or squeeze the antenna coil on the channel strip for maximum amplitude of the response curve.

7. Switch tuner to other channels and observe bandpass. If necessary, adjust antenna coils for proper bandpass tilt and adjust interstage (rf collector and mixer base) coils on the strip for proper bandwidth and coupling.

8. The converter collector coil is adjusted as part of the i-f alignment. (The antenna filter is factory adjusted for minimum response at 43.5 MHz; damage to this filter can cause attenuation on Channel 2.)

VHF TUNER ALIGNMENT (1800 SERIES TRANSISTORIZED)

Connect 300-ohm load across the primary of the mixer output coil; then connect the rf detector and scope (Fig. 12-4) to the i-f output jack. Connect sweep generator to antenna terminals of tuner with antenna termination pad. On the 175-1803 tuner only, apply 24V through a 2.2 megohm resistor to the agc terminal. This voltage sets the two clamping diodes (transistors used as diodes) for proper operation.

All the 1800 series Gold Video Guard VHF tuners are self-biased for maximum gain and do not require an external voltage to be applied to the agc terminal for alignment purposes.

Apply +3V to AFC terminal.

Alignment inside the VHF tuner consists of spreading or squeezing the proper coils and adjusting rf output and mixer input trimmers. Use only a nonmetallic tool for coil adjustment.

1. Adjust oscillator frequencies on all channels to within ±300 kHz of nominal.

2. Set receiver and tune sweep generator to Channel 13. To prevent overload do not exceed 100 mV peak output during alignment.

3. Adjust rf output and mixer input trimmers, and spread or squeeze antenna coil on channel strip for a pattern with maximum amplitude. Check carrier marker locations.

4. For alignment of other channels, spread or squeeze antenna, rf output, and mixer input coils for same pattern and maximum amplitude of response as shown in Fig. 12-5.

5. The antenna filter is factory adjusted for minimum response at 45.5 MHz and also at FM broadcast frequencies. Damage to this filter can cause attenuation on Channel 2 or 6, and also cause rf interference. Tuner alignment connections are illustrated in Fig. 12-6.

The balance of this chapter is composed of schematic diagrams for the following Zenith tuners.

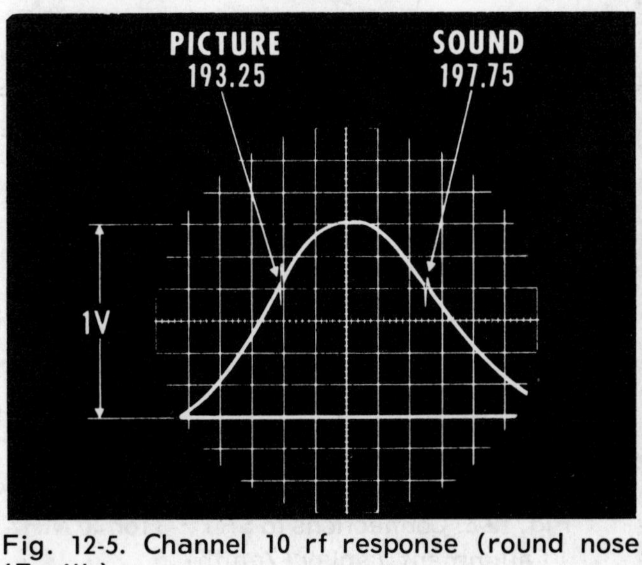

Fig. 12-5. Channel 10 rf response (round nose) (Zenith).

Fig. 12-7. 175-501 through 175-589 VHF tuner schematic.

Fig. 12-8. 175-552 VHF tuner schematic.

Fig. 12-9. 175-540 through 175-556 VHF tuner schematic.

Fig. 12-10. 175-651 series VHF tuner schematic.

Fig. 12-11. 175-663 VHF turret tuner schematic.

Fig. 12-12. 175-707 through 175-716 VHF tuner schematic.

Fig. 12-13. 175-719 switch-type VHF tuner schematic.

Fig. 12-14. 175-743 through 175-749 VHF tuner schematic.

Fig. 12-15. 175-856A and 175-857A VHF tuner schematic.

Fig. 12-16. 175-878 and 175-880 VHF tuner schematic.

Fig. 12-17. 175-879 (AFC) VHF tuner schematic.

Fig. 12-18. 175-1006A and 175-1007A VHF tuner schematic.

Fig. 12-19. 175-1015 series turret tuner schematic.

Fig. 12-20. 175-1101 through 175-1122 VHF tuner schematic.

Fig. 12-21. 175-1150 through 175-1164 VHF tuner schematic.

Fig. 12-22. 175-1152 through 175-1172 VHF tuner schematic.

Fig. 12-23. 175-1166 through 175-1179 VHF tuner schematic.

Fig. 12-24. 175-1201 Gold Video Guard tuner schematic.

Fig. 12-25. 175-1402 VHF tuner schematic.

Fig. 12-26. 175-1401 and 175-1405 VHF tuner schematic.

Fig. 12-27. 175-1502 varactor tuner schematic.

Fig. 12-28. 175-1800 series (AFC) VHF tuner schematic.

Fig. 12-29. 175-82 through 175-89 UHF tuner schematic.

Fig. 12-30. 175-84 and 175-85 UHF tuner schematic.

Fig. 12-31. 175-910 UHF tuner schematic.

Fig. 12-32. 175-1606 UHF varactor tuner schematic.

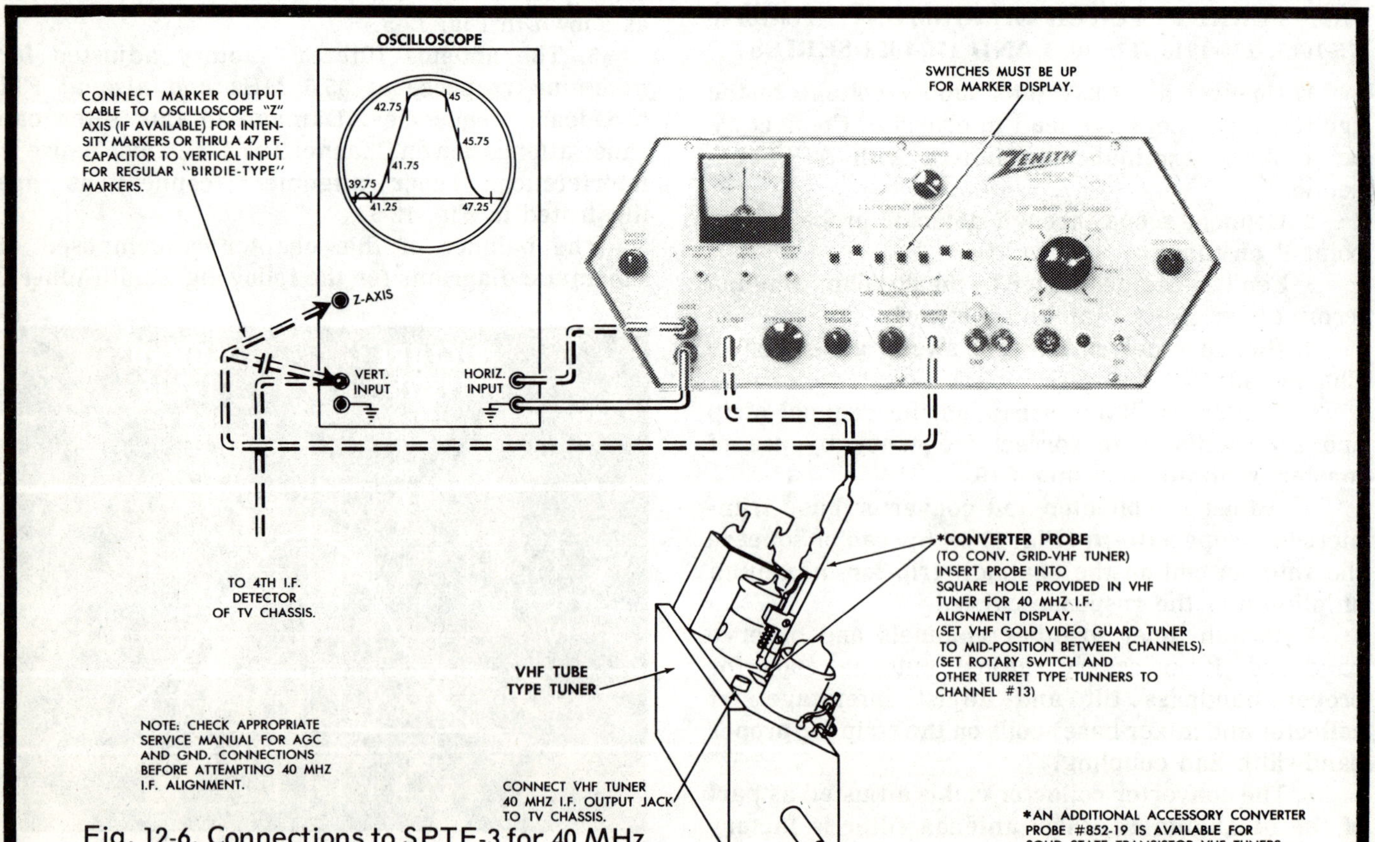

Fig. 12-6. Connections to SPTE-3 for 40 MHz i-f alignment display (Zenith).

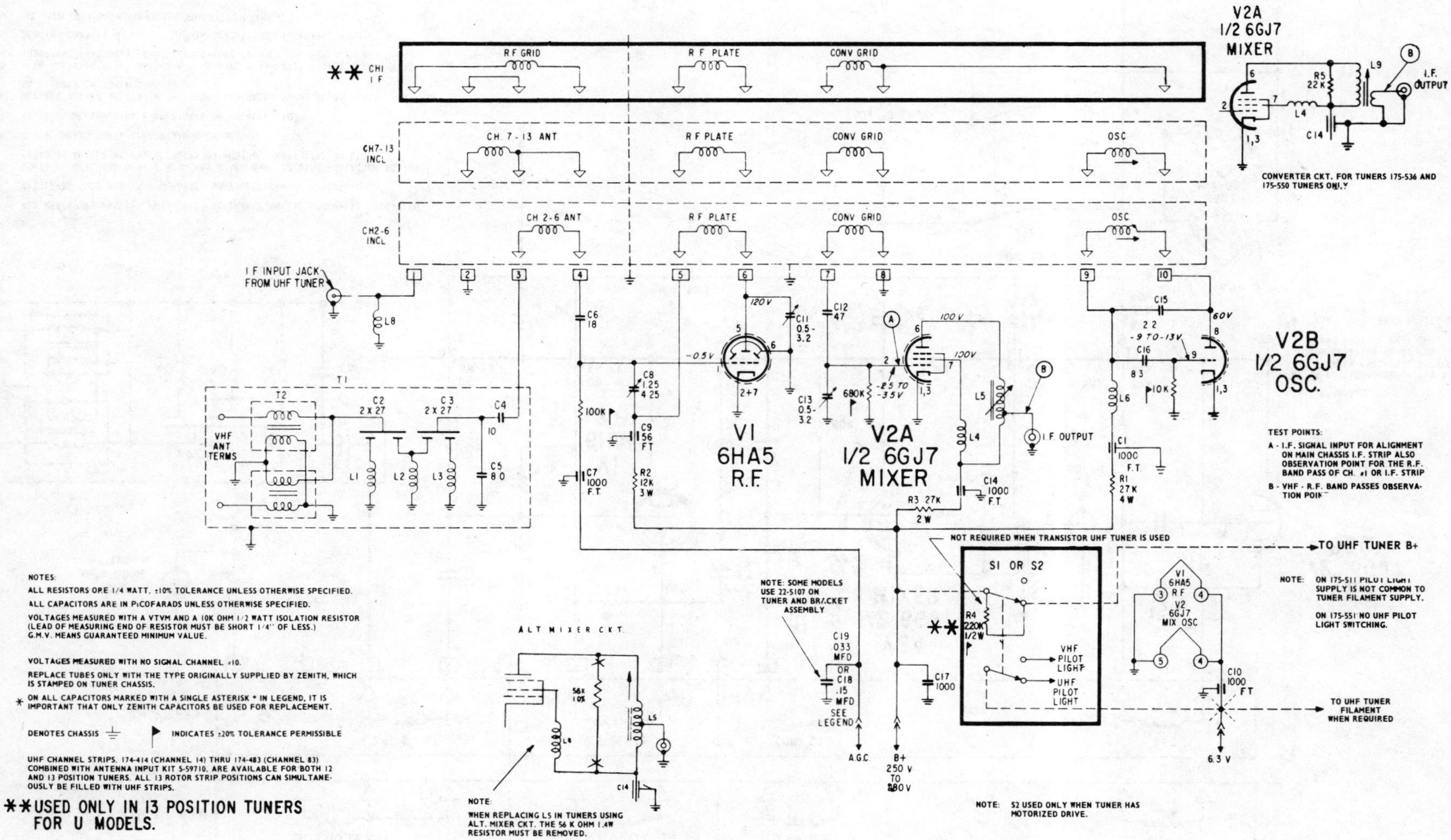

Fig. 12-7. 175-501 through 175-589 VHF tuner schematic.

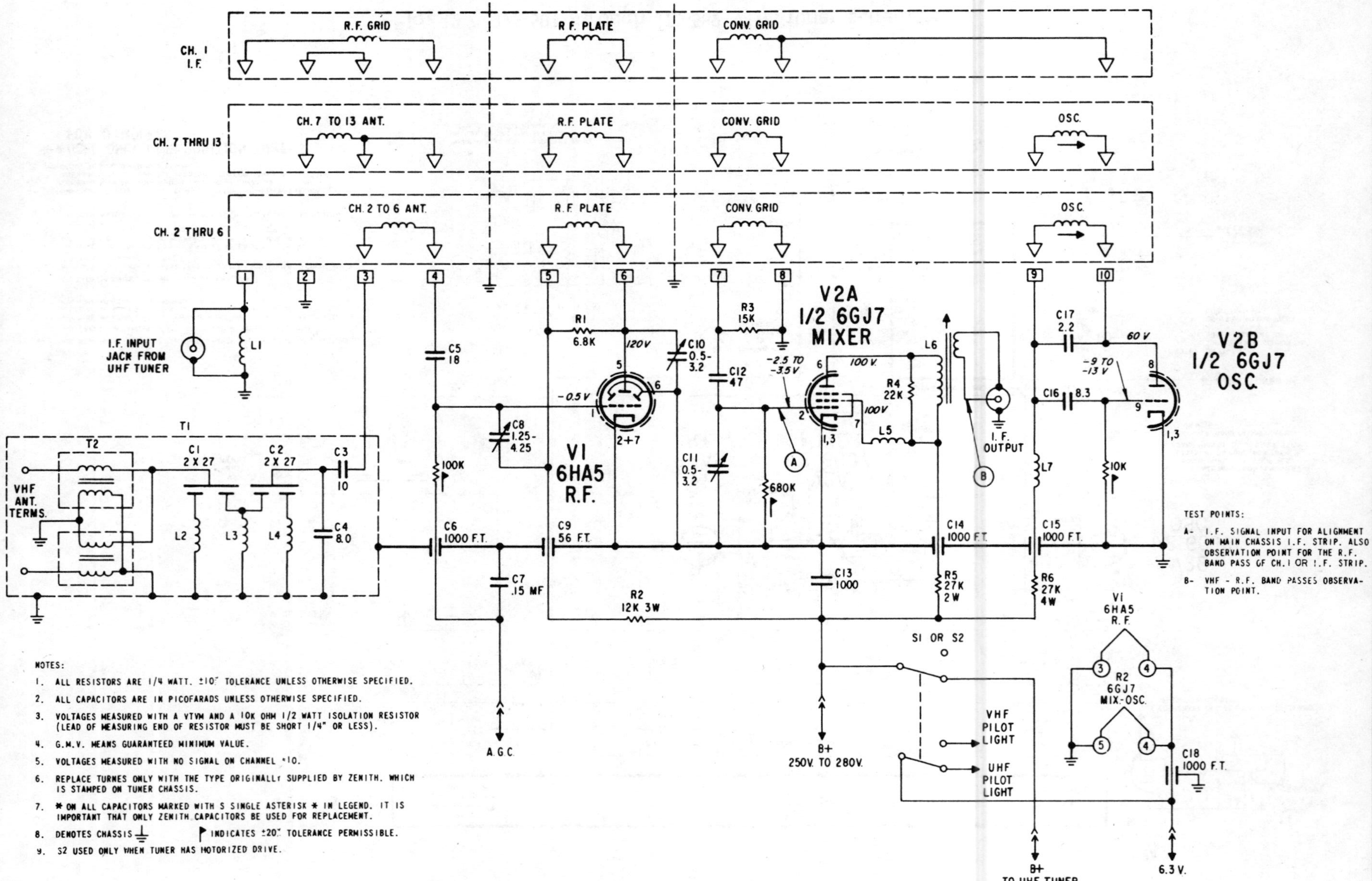

Fig. 12-8. 175-552 VHF tuner schematic.

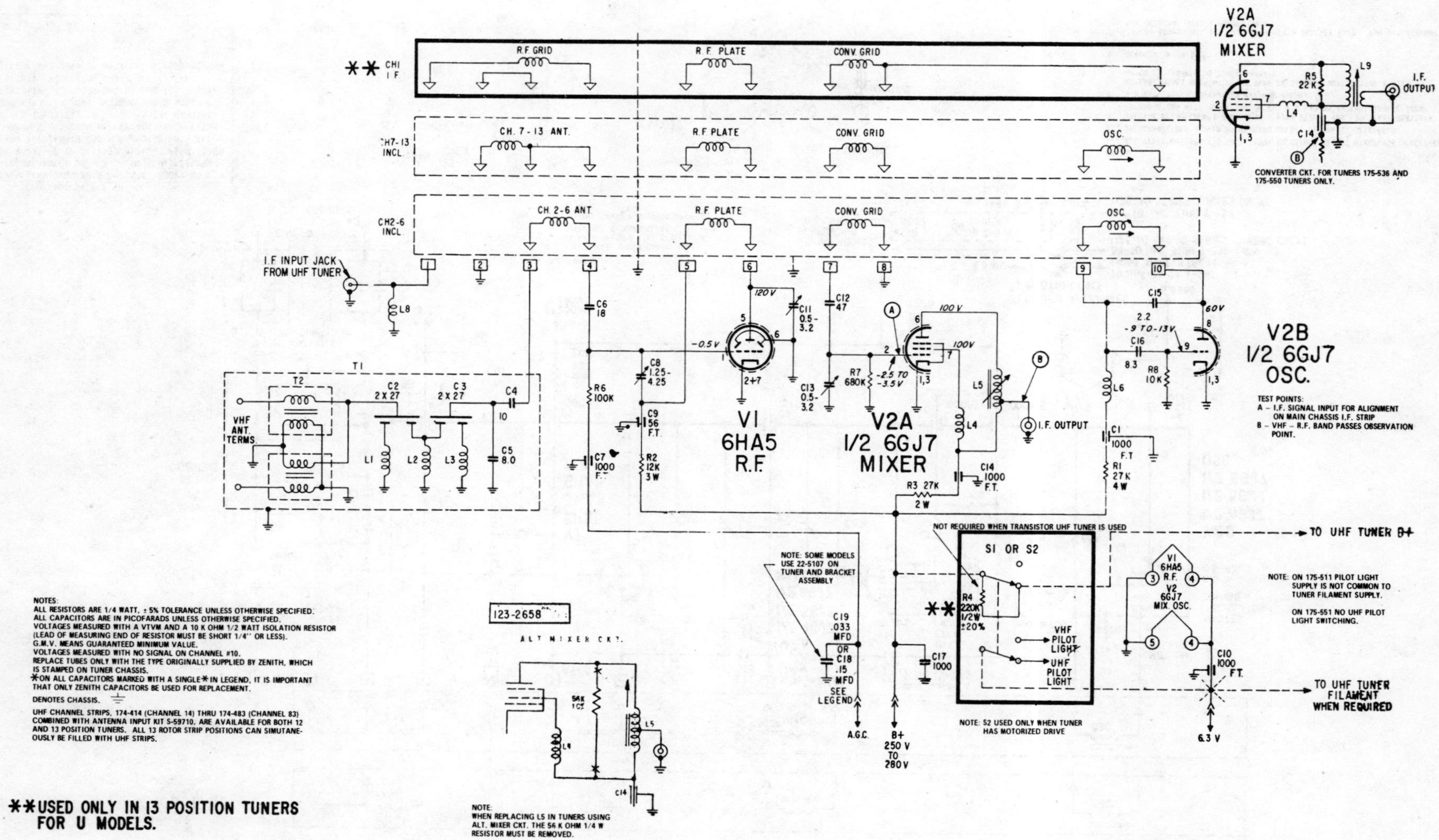

Fig. 12-9. 175-540 through 175-556 VHF tuner schematic.

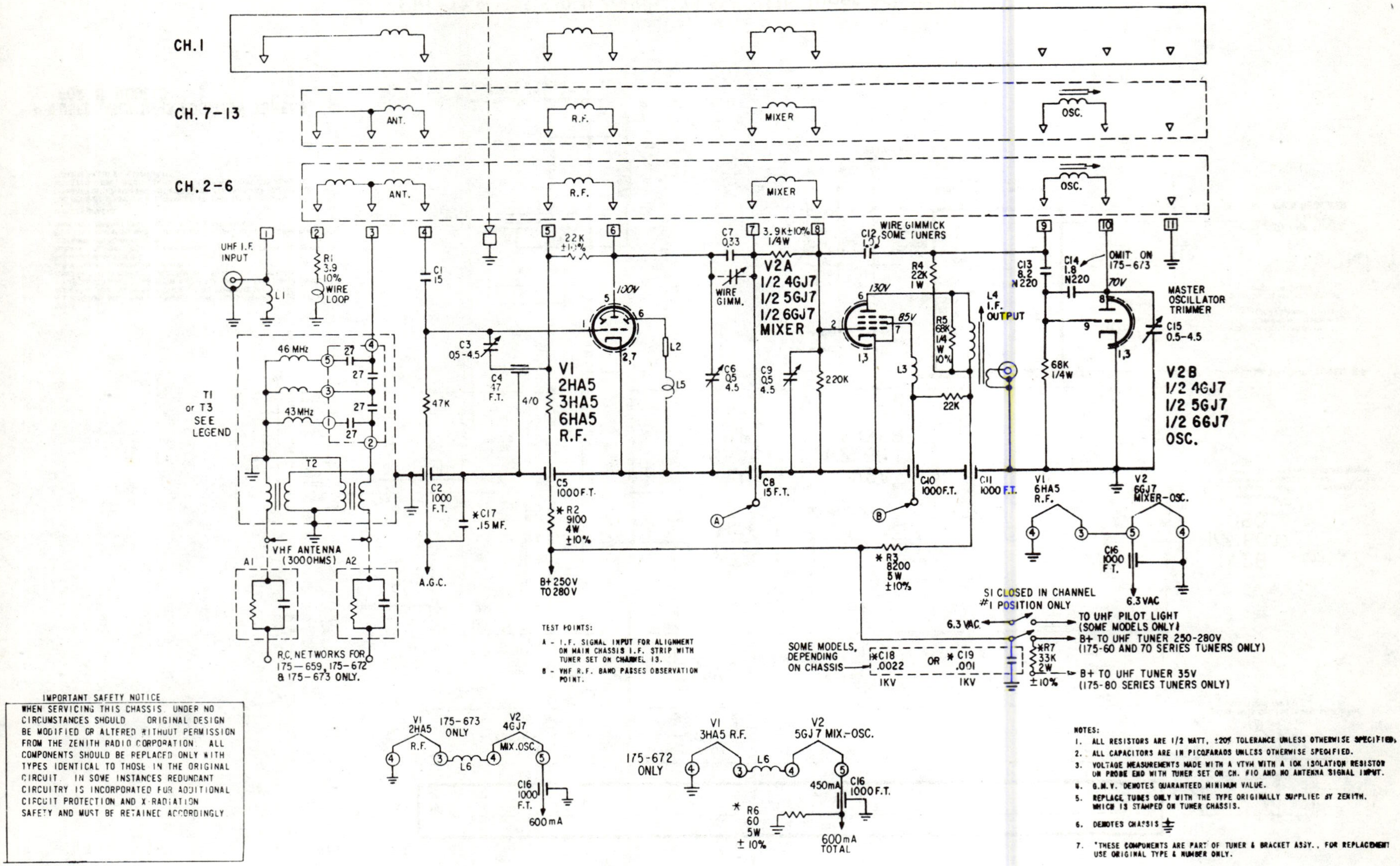

Fig. 12-10. 175-651 series VHF tuner schematic.

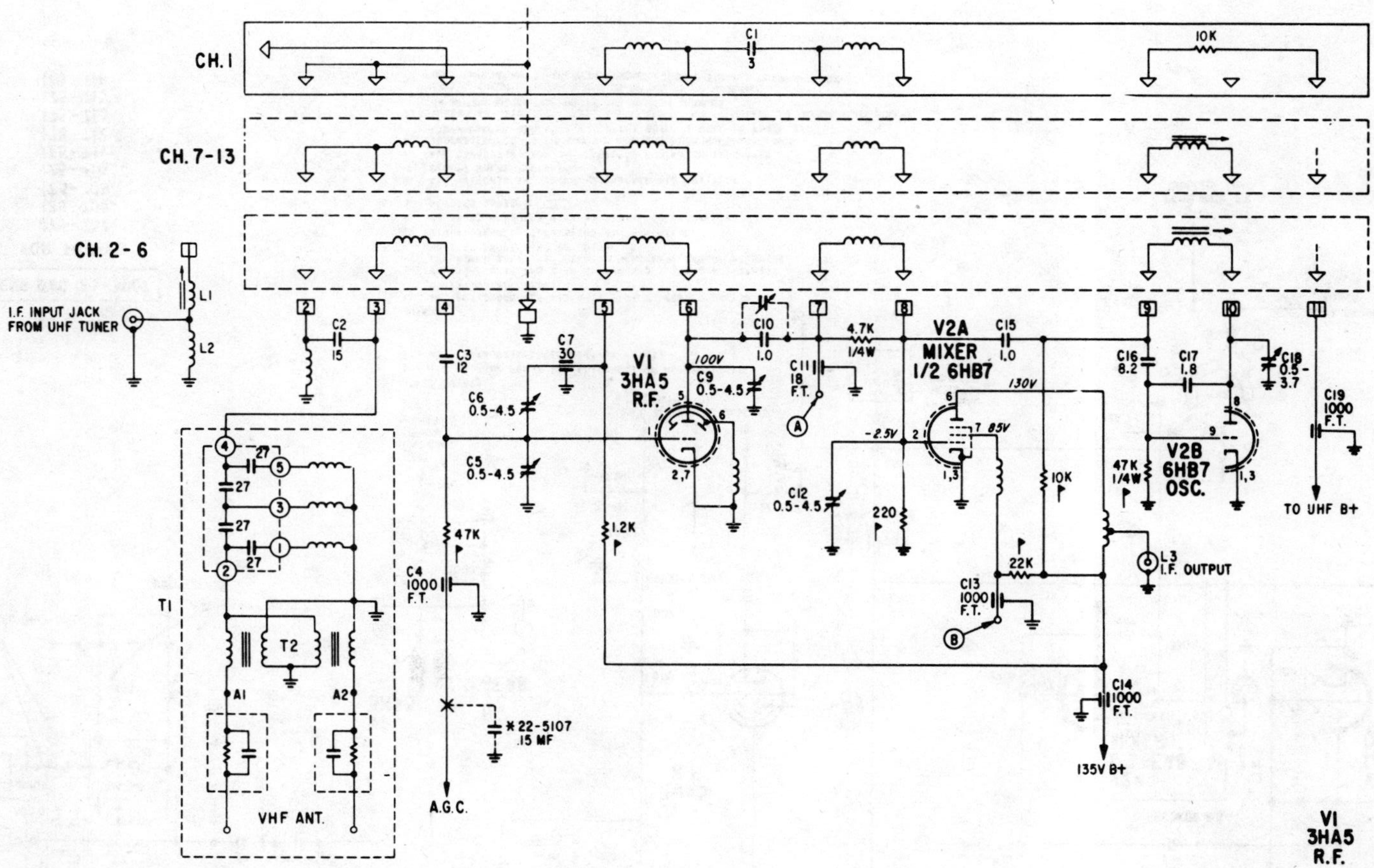

NOTES:-
1. ALL RESISTORS ARE 1/2W, ± 10% TOLERANCE UNLESS OTHERWISE SPECIFIED.
2. ALL CAPACITORS ARE IN PICOFARADS UNLESS OTHERWISE SPECIFIED.
3. VOLTAGE MEASUREMENTS MADE WITH A VTVM, WITH A 10K ISOLATION RESISTOR ON PROBE END WITH TUNER SET ON CH.#10 AND NO ANTENNA SIGNAL INPUT.
4. G.M.V. DENOTES GUARANTEED MINIMUM VALUE.
5. REPLACE TUBES ONLY WITH THE TYPE ORIGINALLY SUPPLIED BY ZENITH, WHICH IS STAMPED ON TUNER CHASSIS.
6. DENOTES CHASSIS ⏚
7. P INDICATES 20% MAY BE USED.

TEST POINTS:-
A - I.F. SIGNAL INPUT FOR ALIGNMENT ON MAIN CHASSIS I.F. STRIP WITH TUNER SET ON CHANNEL 13
B - VHF R.F. BAND PASSES OBSERVATION POINT.

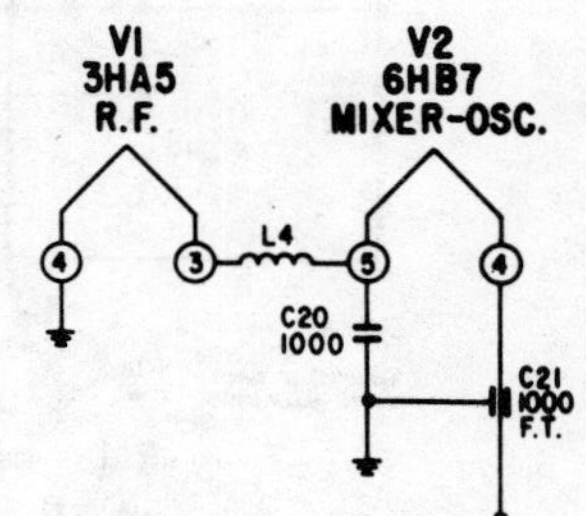

* THESE COMPONENTS ARE ON TUNER AND BRACKET ASSEMBLY.

Fig. 12-11. 175-663 VHF turret tuner schematic.

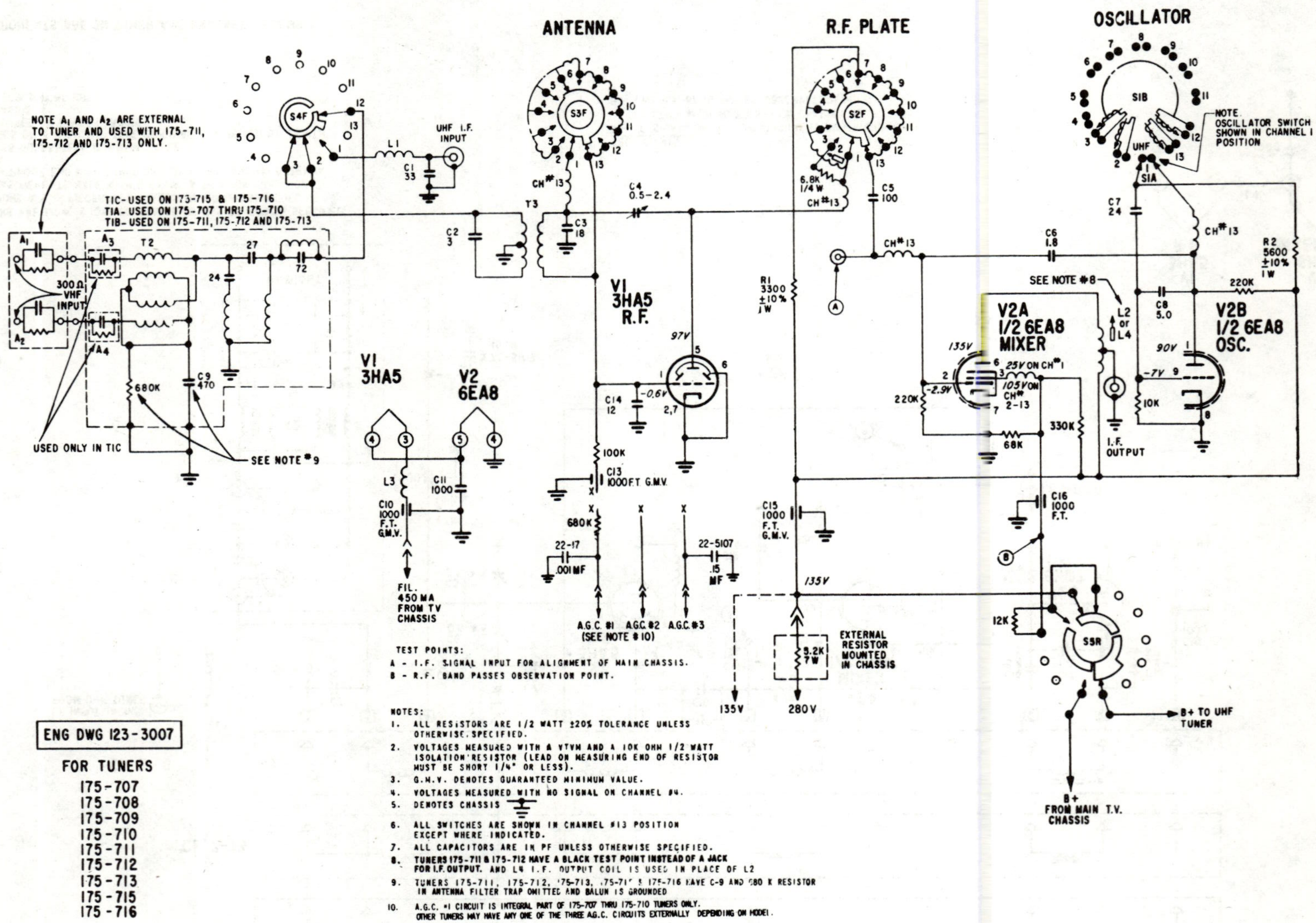

Fig. 12-12. 175-707 through 175-716 VHF tuner schematic.

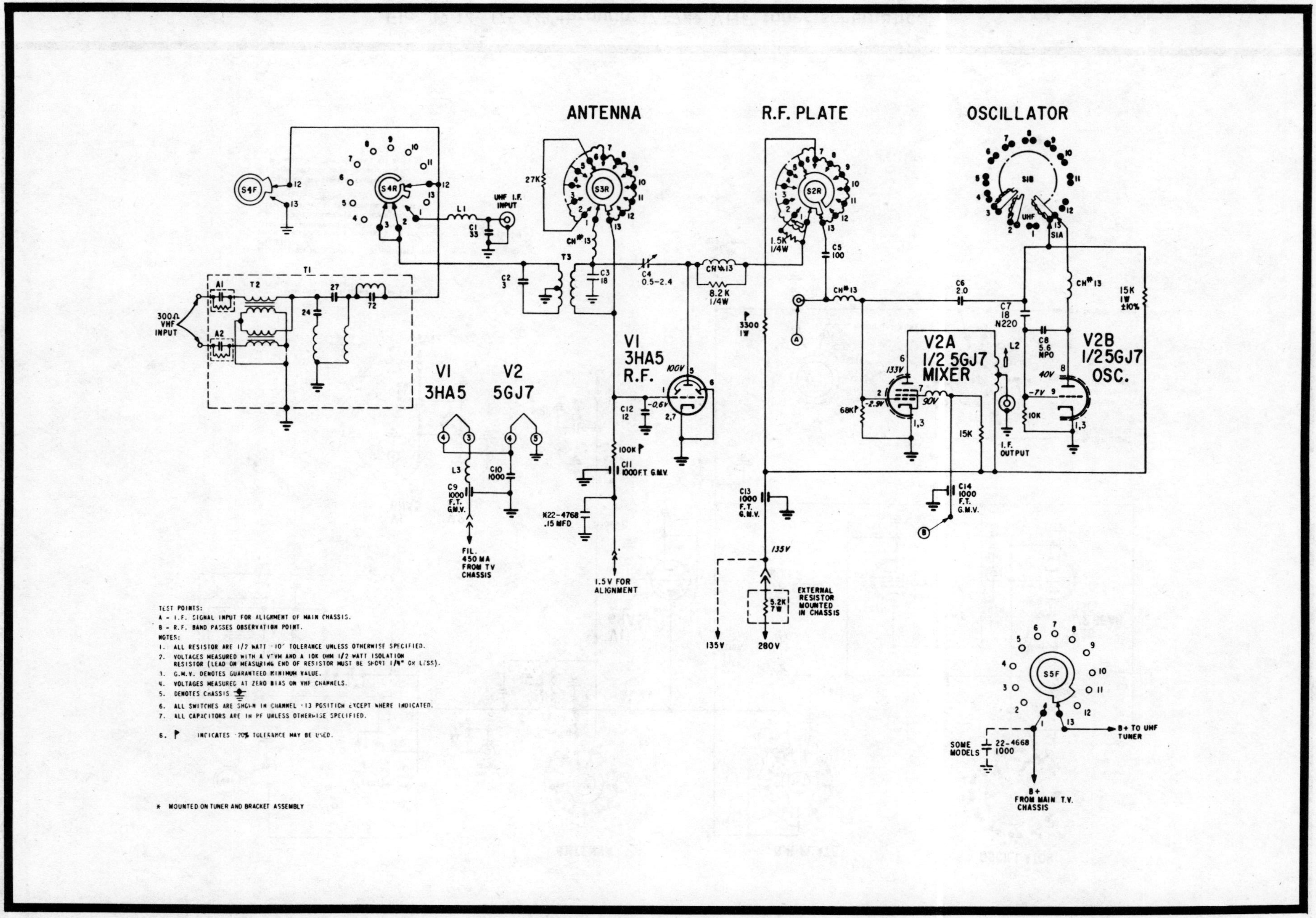

Fig. 12-13. 175-719 switch-type VHF tuner schematic.

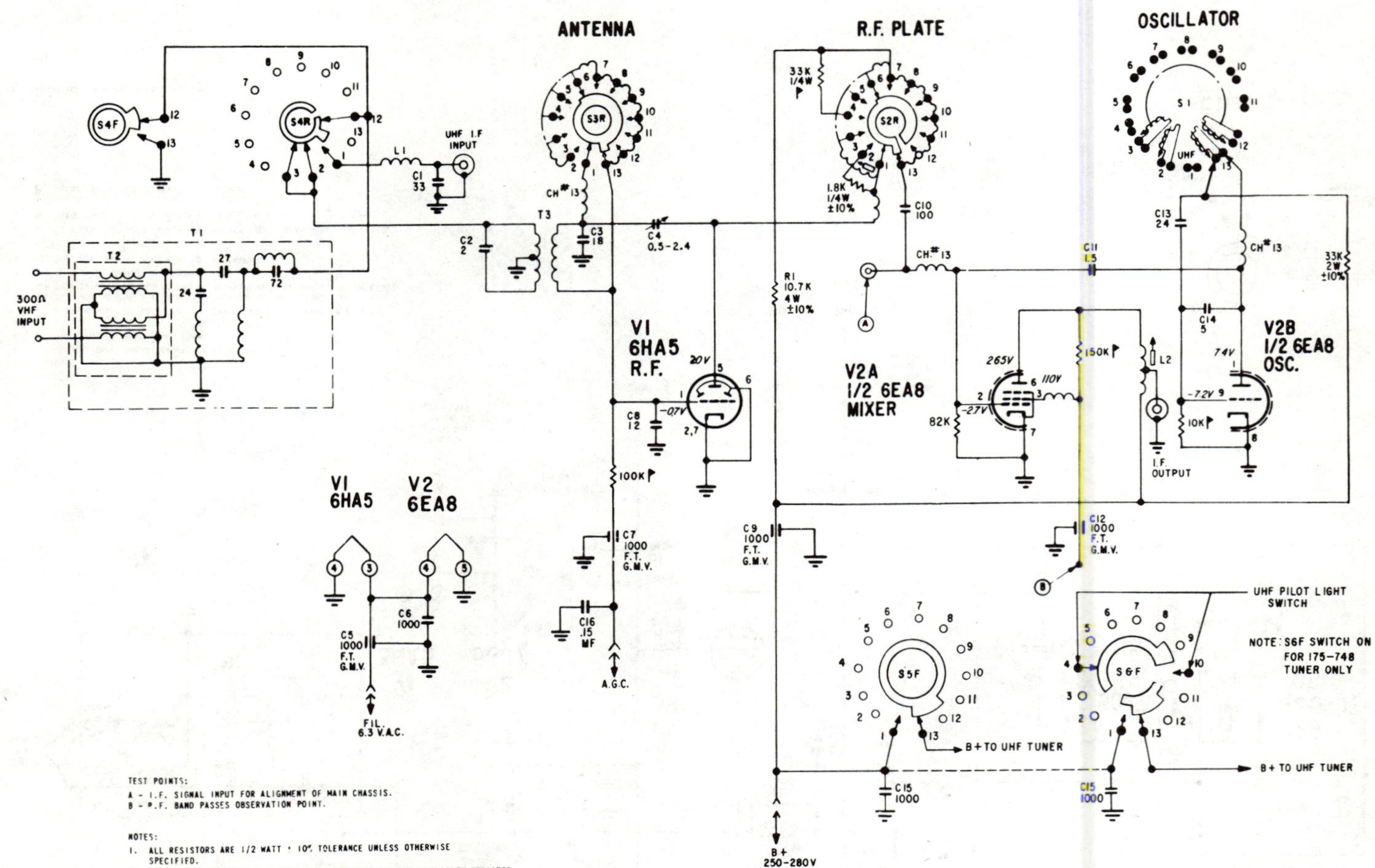

TEST POINTS:

A – I.F. SIGNAL INPUT FOR ALIGNMENT OF MAIN CHASSIS.
B – R.F. BAND PASSES OBSERVATION POINT.

NOTES:

1. ALL RESISTORS ARE 1/2 WATT ± 10% TOLERANCE UNLESS OTHERWISE SPECIFIED.
2. VOLTAGES MEASURED WITH A VTVM AND A 10K OHM 1/2 WATT ISOLATION RESISTOR (LEAD ON MEASURING END OF RESISTOR MUST BE SHORT 1/4" OR LESS).
3. G.M.V. DENOTES GUARANTEED MINIMUM VALUE.
4. VOLTAGES MEASURED AT ZERO BIAS ON VHF CHANNELS.
5. DENOTES CHASSIS
6. ALL SWITCHES ARE SHOWN IN CHANNEL •13 POSITION.
7. ALL CAPACITORS ARE IN PF UNLESS OTHERWISE SPECIFIED.
8. INDICATES ± 20% TOLERANCE MAY BE USED.

Fig. 12-14. 175-743 through 175-749 VHF tuner schematic.

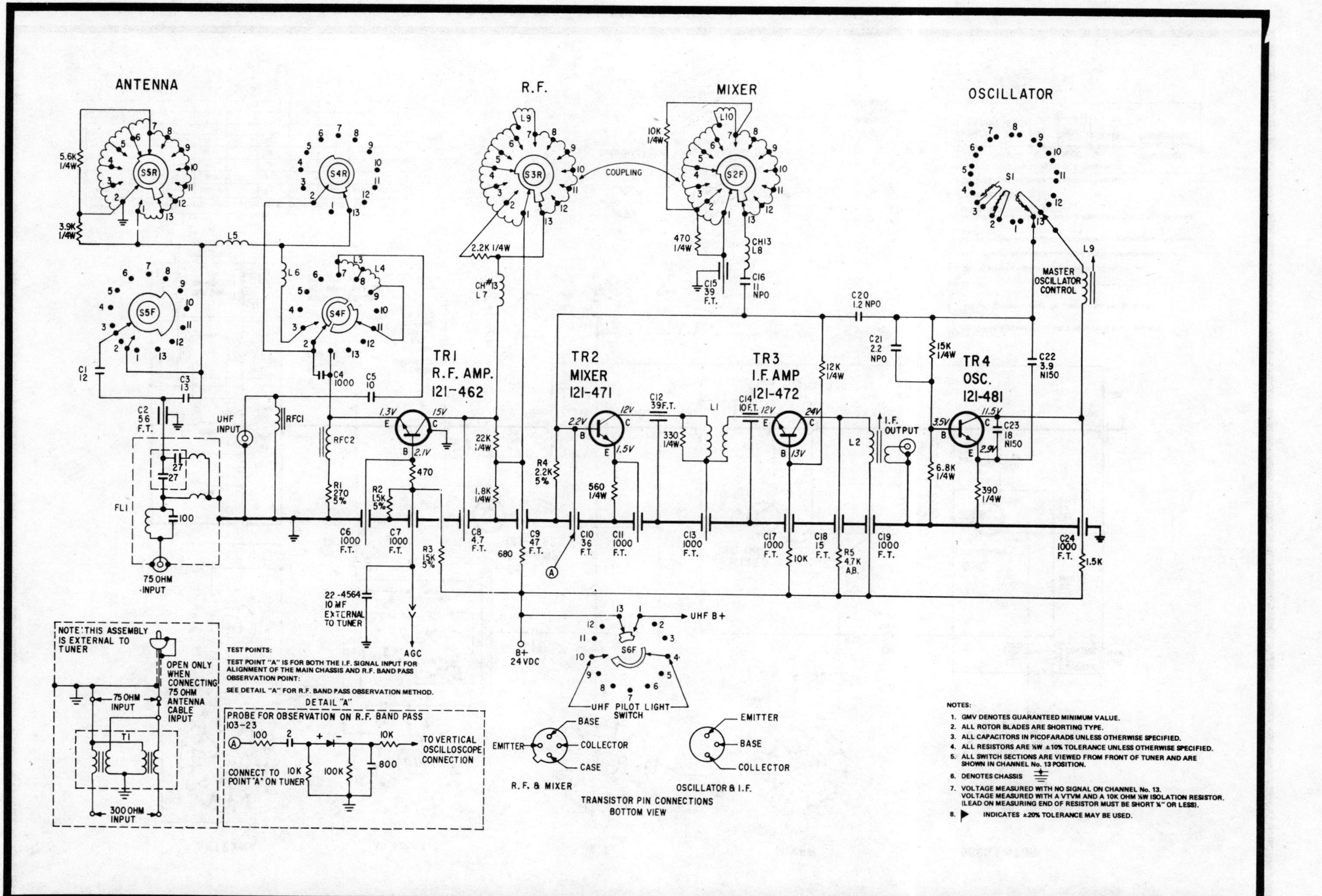

Fig. 12-15. 175-856A and 175-857A VHF tuner schematic.

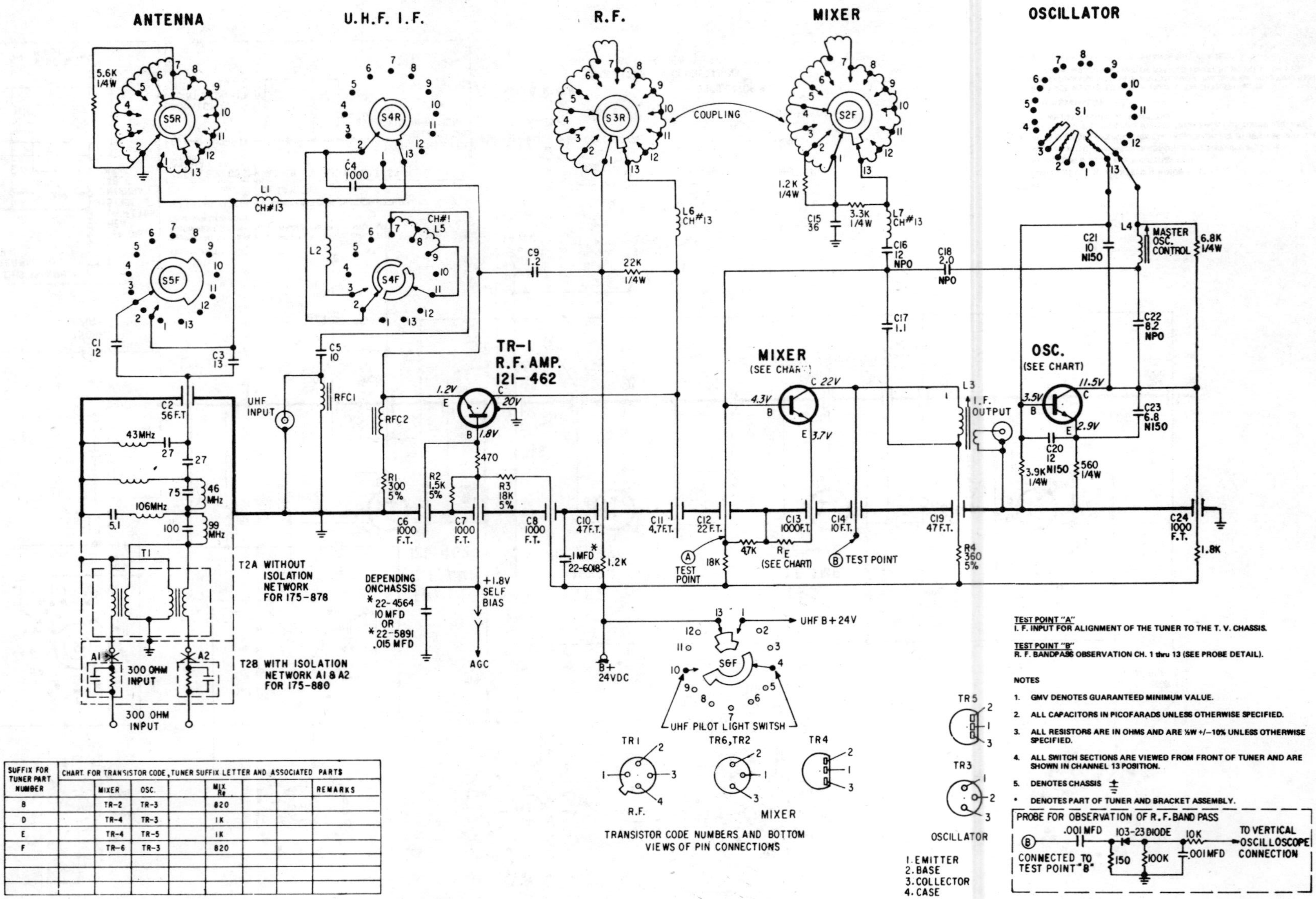

SUFFIX FOR TUNER PART NUMBER	CHART FOR TRANSISTOR CODE, TUNER SUFFIX LETTER AND ASSOCIATED PARTS							
		MIXER	OSC.		MIX R_e			REMARKS
B		TR-2	TR-3		820			
D		TR-4	TR-3		1K			
E		TR-4	TR-5		1K			
F		TR-6	TR-3		820			

TEST POINT "A"
I. F. INPUT FOR ALIGNMENT OF THE TUNER TO THE T. V. CHASSIS.

TEST POINT "B"
R. F. BANDPASS OBSERVATION CH. 1 thru 13 (SEE PROBE DETAIL).

NOTES

1. GMV DENOTES GUARANTEED MINIMUM VALUE.
2. ALL CAPACITORS IN PICOFARADS UNLESS OTHERWISE SPECIFIED.
3. ALL RESISTORS ARE IN OHMS AND ARE ½W +/−10% UNLESS OTHERWISE SPECIFIED.
4. ALL SWITCH SECTIONS ARE VIEWED FROM FRONT OF TUNER AND ARE SHOWN IN CHANNEL 13 POSITION.
5. DENOTES CHASSIS ⏚

• DENOTES PART OF TUNER AND BRACKET ASSEMBLY.

PROBE FOR OBSERVATION OF R.F. BAND PASS
.001 MFD
103-23 DIODE
10K
TO VERTICAL OSCILLOSCOPE CONNECTION
CONNECTED TO TEST POINT "B"
150
100K
.001 MFD

Fig. 12-16. 175-878 and 175-880 VHF tuner schematic.

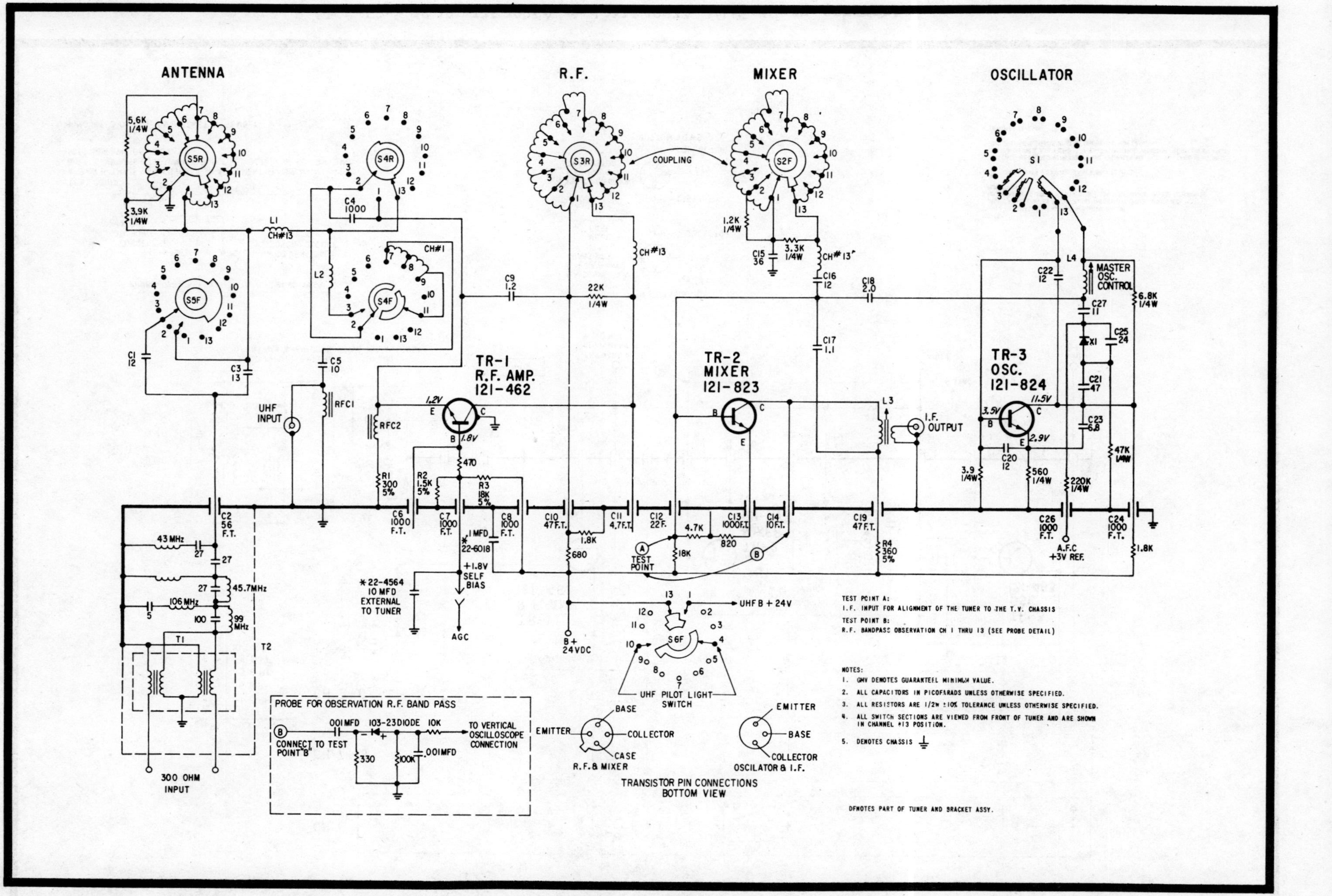

Fig. 12-17. 175-879 (AFC) VHF tuner schematic.

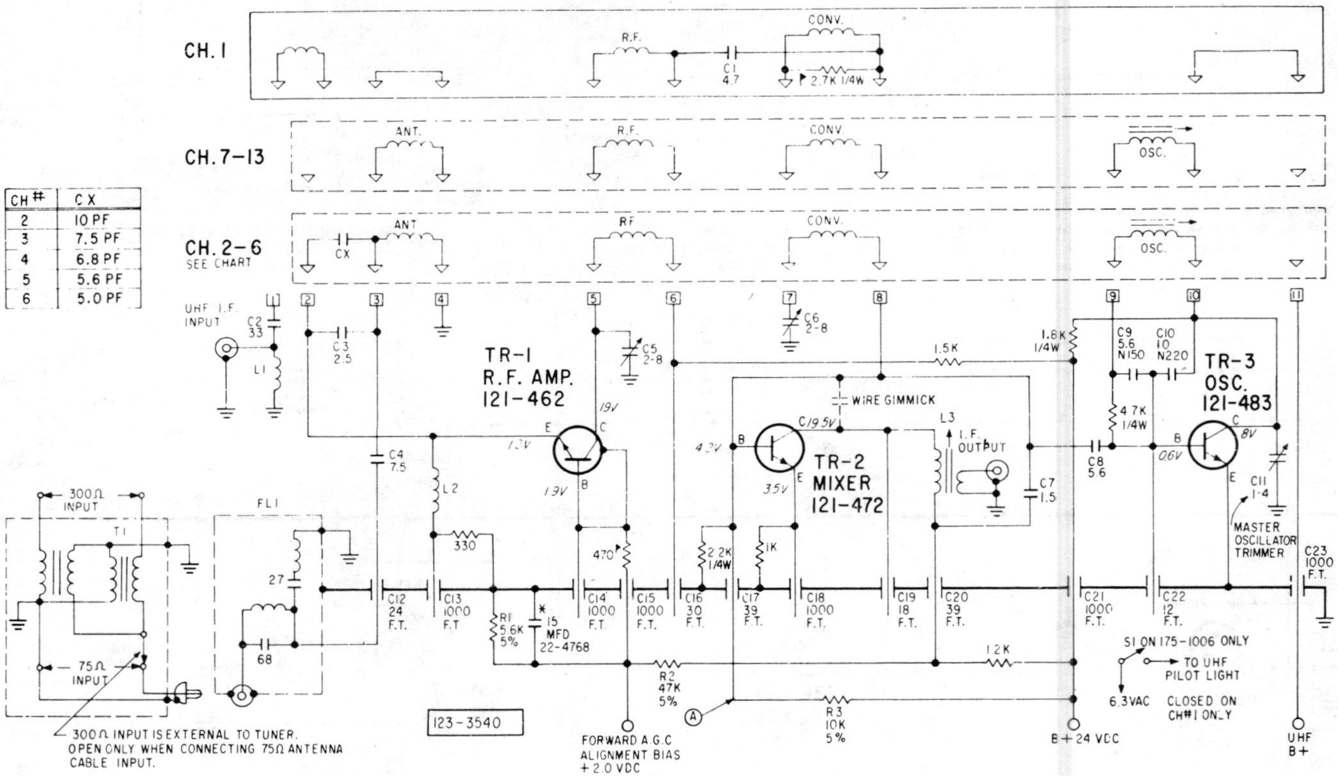

CH #	CX
2	10 PF
3	7.5 PF
4	6.8 PF
5	5.6 PF
6	5.0 PF

TEST POINT:
A—FOR BOTH I.F. SIGNAL INPUT FOR ALIGNMENT OF THE MAIN CHASSIS & R.F. BAND PASS OBSERVATION POINT. SEE DETAIL "A" FOR R.F. BAND PASS OBSERVATION POINT.

* MOUNTED ON TUNER & BRACKET ASSEMBLY

NOTES:
1. GMV DENOTES GUARANTEED MINIMUM VALUE.
2. ALL CAPACITORS IN PICOFARADS UNLESS OTHERWISE SPECIFIED.
3. ALL RESISTORS ARE 1/2 WATT AND ± 10% TOLERANCE UNLESS OTHERWISE SPECIFIED.
4. ▶ INDICATES ± 20% TOLERANCE MAY BE USED.
5. VOLTAGE MEASURED WITH NO SIGNAL ON CHANNEL #13. VOLTAGE MEASURED WITH A VTVM & A 10 K OHM, 1/2 WATT ISOLATION RESISTOR (LEAD ON MEASURING END OF RESISTOR MUST BE SHORT 1/4" OR LESS).
6. DENOTES CHASSIS

Fig. 12-18. 175-1006A and 175-1007A VHF tuner schematic.

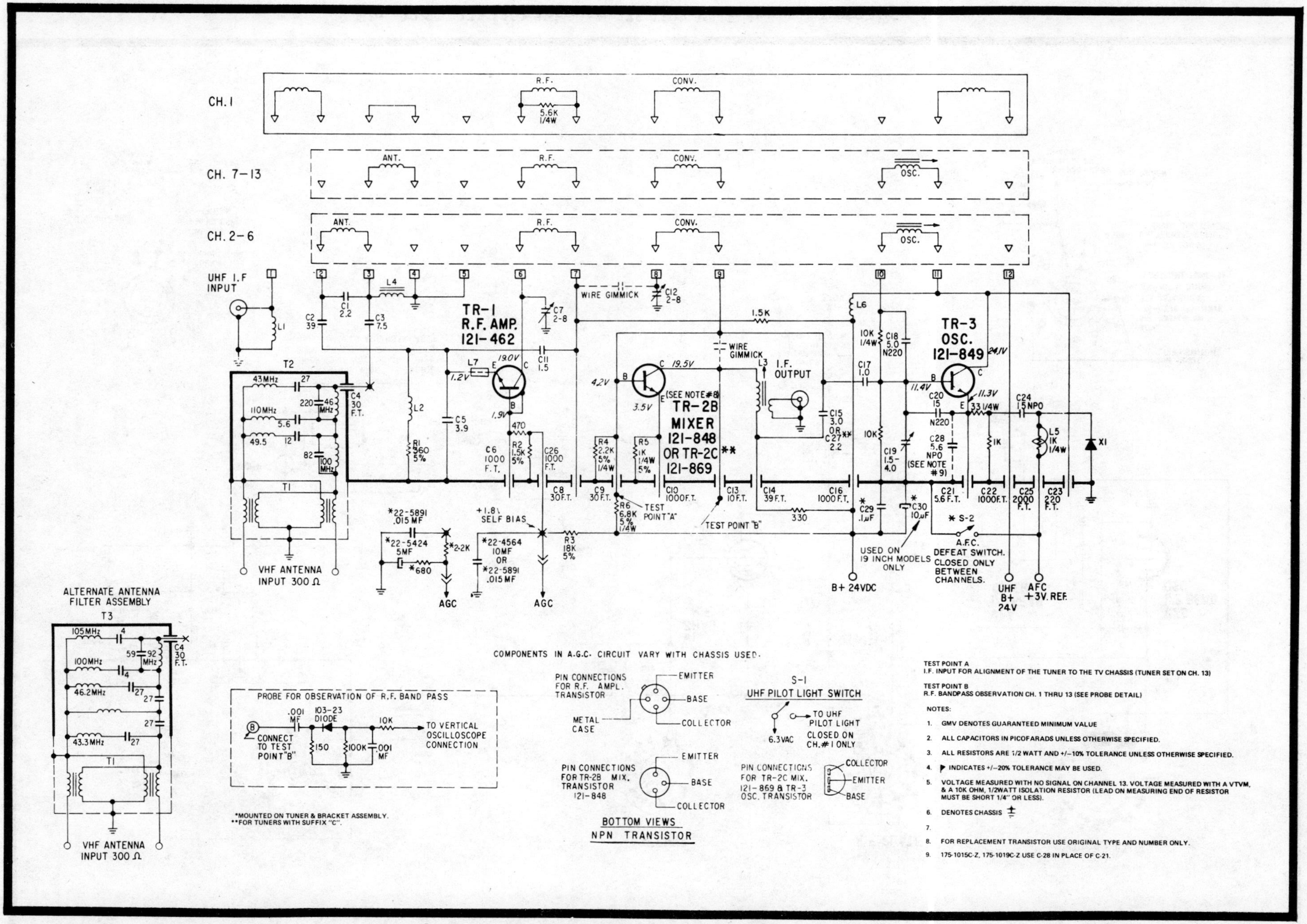

Fig. 12-19. 175-1015 series turret tuner schematic.

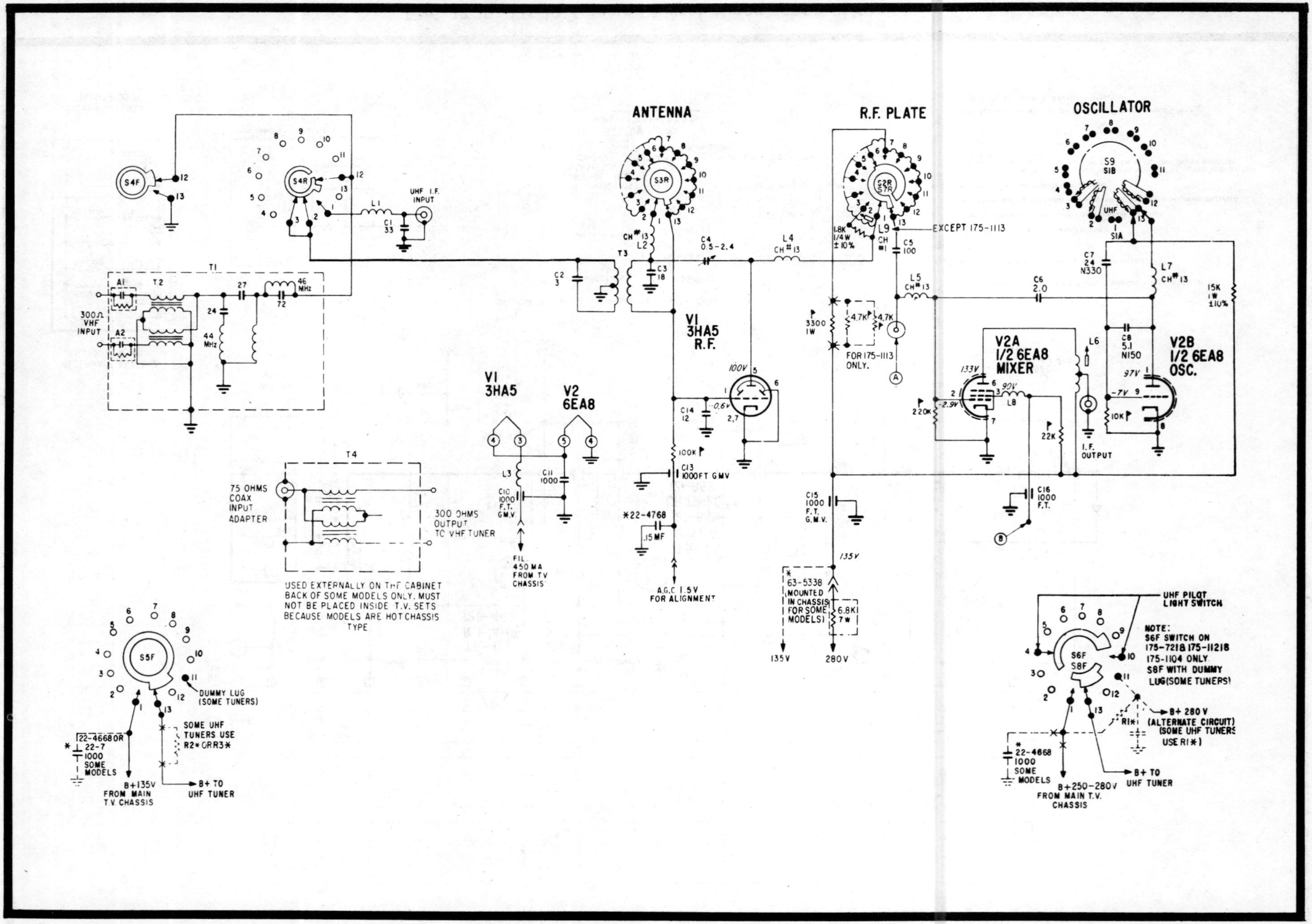

Fig. 12-20. 175-1101 through 175-1122 VHF tuner schematic.

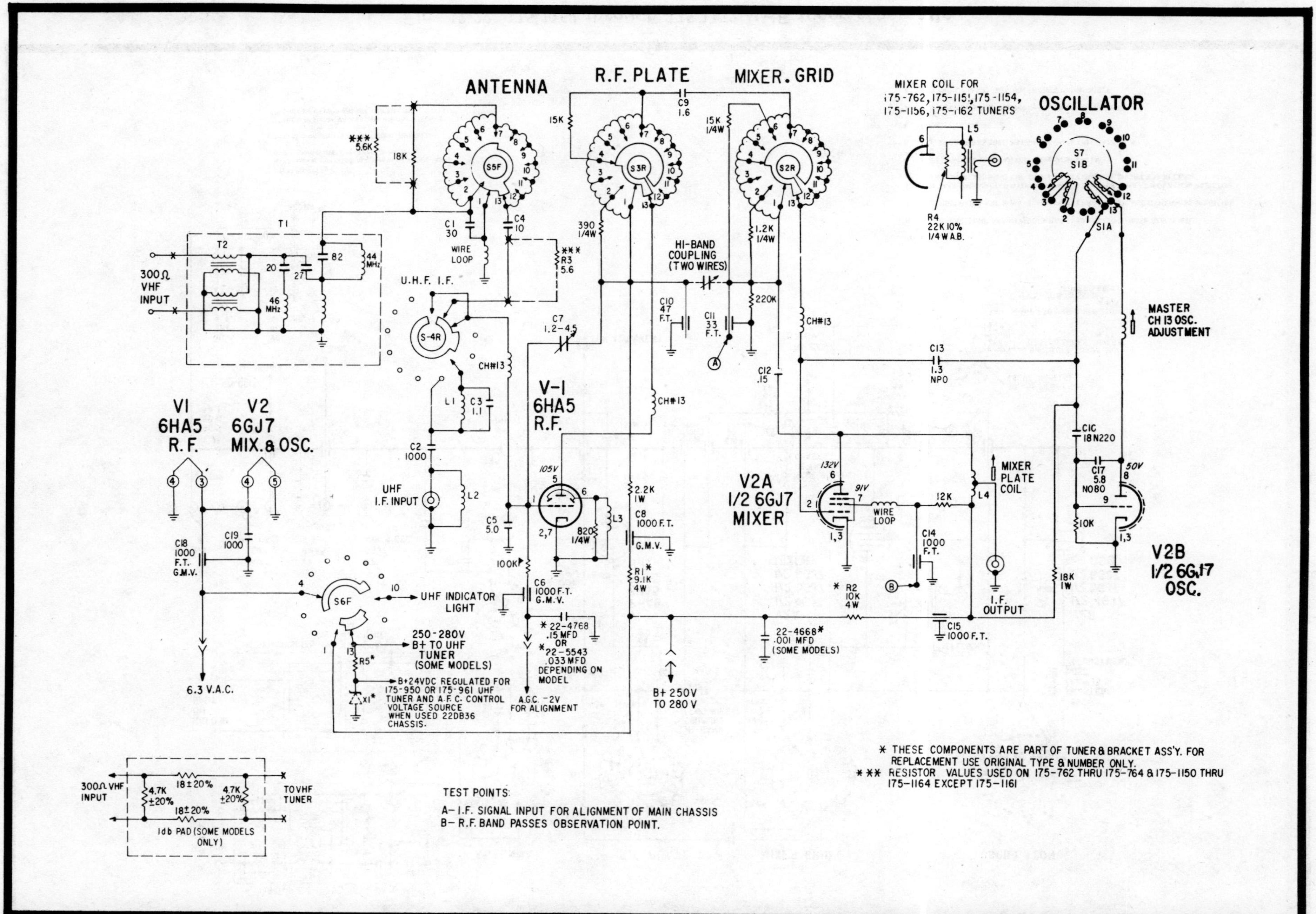

Fig. 12-21. 175-1150 through 175-1164 VHF tuner schematic.

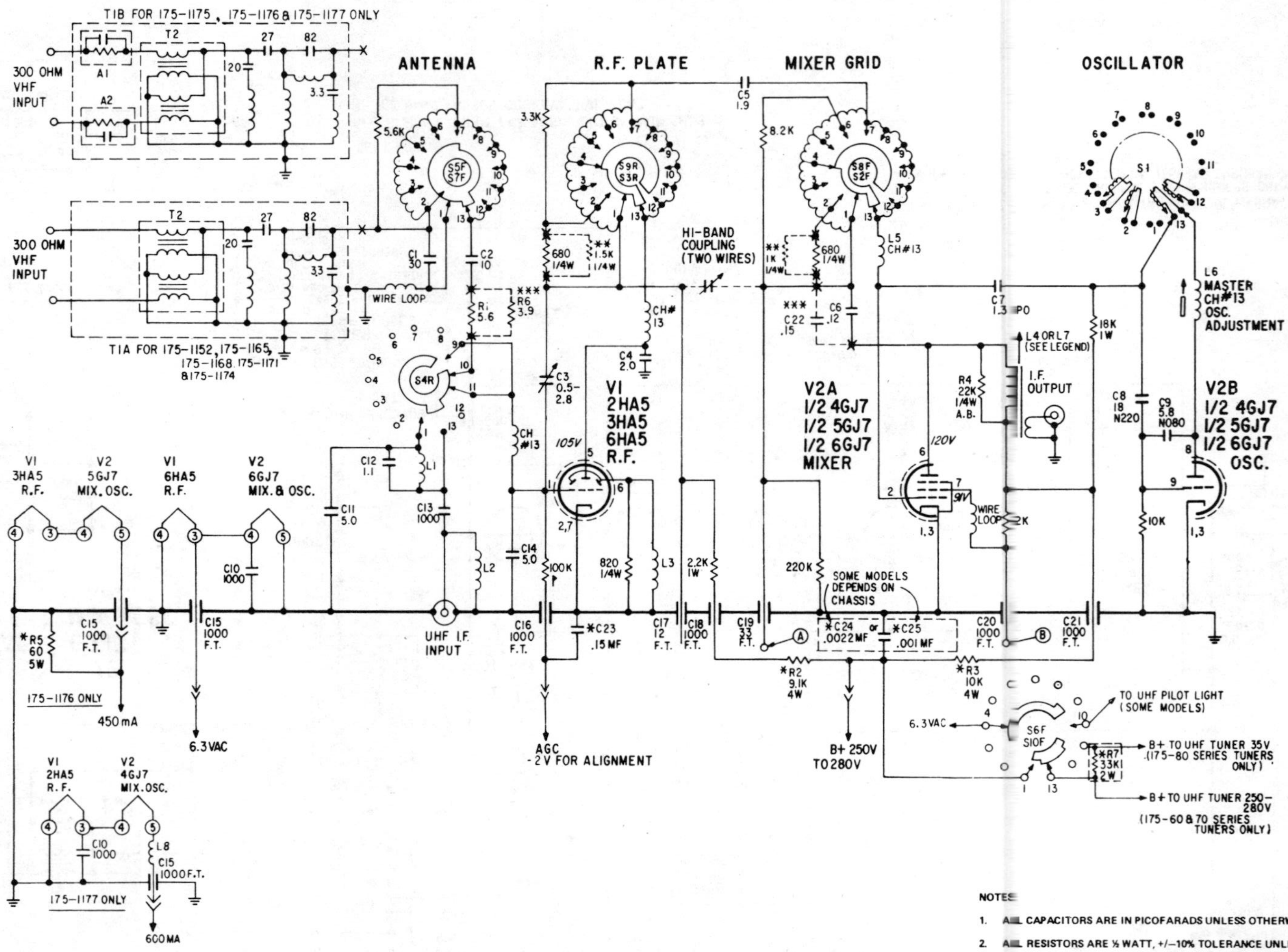

TEST POINTS:

A—I.F. SIGNAL INPUT FOR ALIGNMENT OF MAIN CHASSIS

B—R.F. BAND PASSES OBSERVATION POINT

* THESE COMPONENTS ARE EXTERNAL TO TUNER

** THESE RESISTOR VALUES USED ON 175–1175 ONLY.

*** THE COMPONENTS ARE USED ON 175–1152 & 175–1177

NOTES

1. ALL CAPACITORS ARE IN PICOFARADS UNLESS OTHERWISE SPECIFIED.
2. ALL RESISTORS ARE ½ WATT, +/–10% TOLERANCE UNLESS OTHERWISE SPECIFIED.
3. VOLTAGES MEASURED WITH A VTVM & A 10K OHM ½ WATT ISOLATION RESISTOR, (LEAD ON MEASURING END OF RESISTOR MUST BE SHORT ¼" OR LESS).
4. GMV DENOTES GUARANTEED MINIMUM VALUE.
5. VOLTAGES MEASURED WITH NO SIGNAL ON CHANNEL 4.
6. DENOTES CHASSIS ⏚
7. ALL SWITCHES ARE SHOWN IN CHANNEL 13 POSITION.
8. ◤ INDICATES +/–20% TOLERANCE PERMISSIBLE.

Fig. 12-22. 175-1152 through 175-1172 VHF tuner schematic.

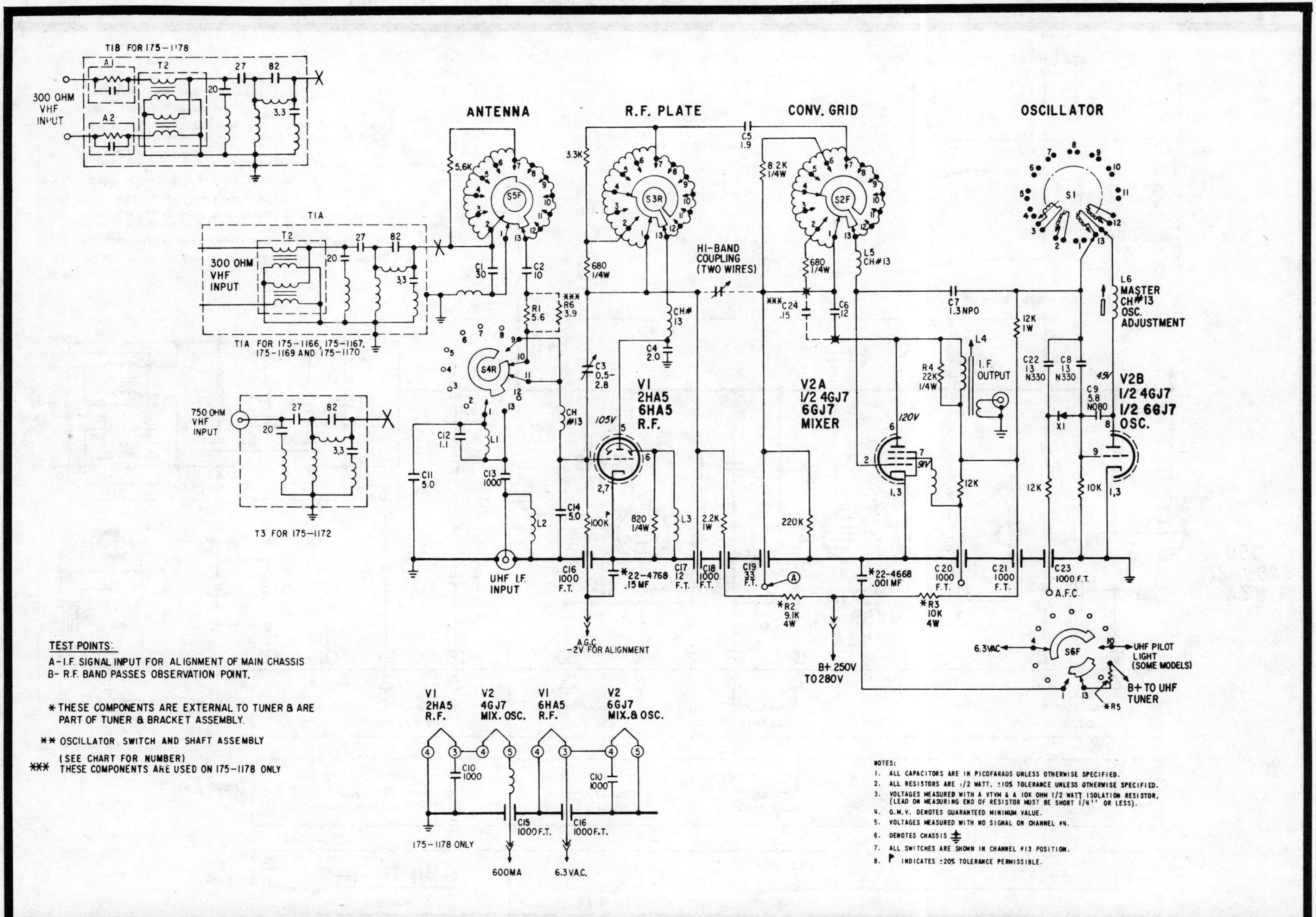

Fig. 12-23. 175-1166 through 175-1179 VHF tuner schematic.

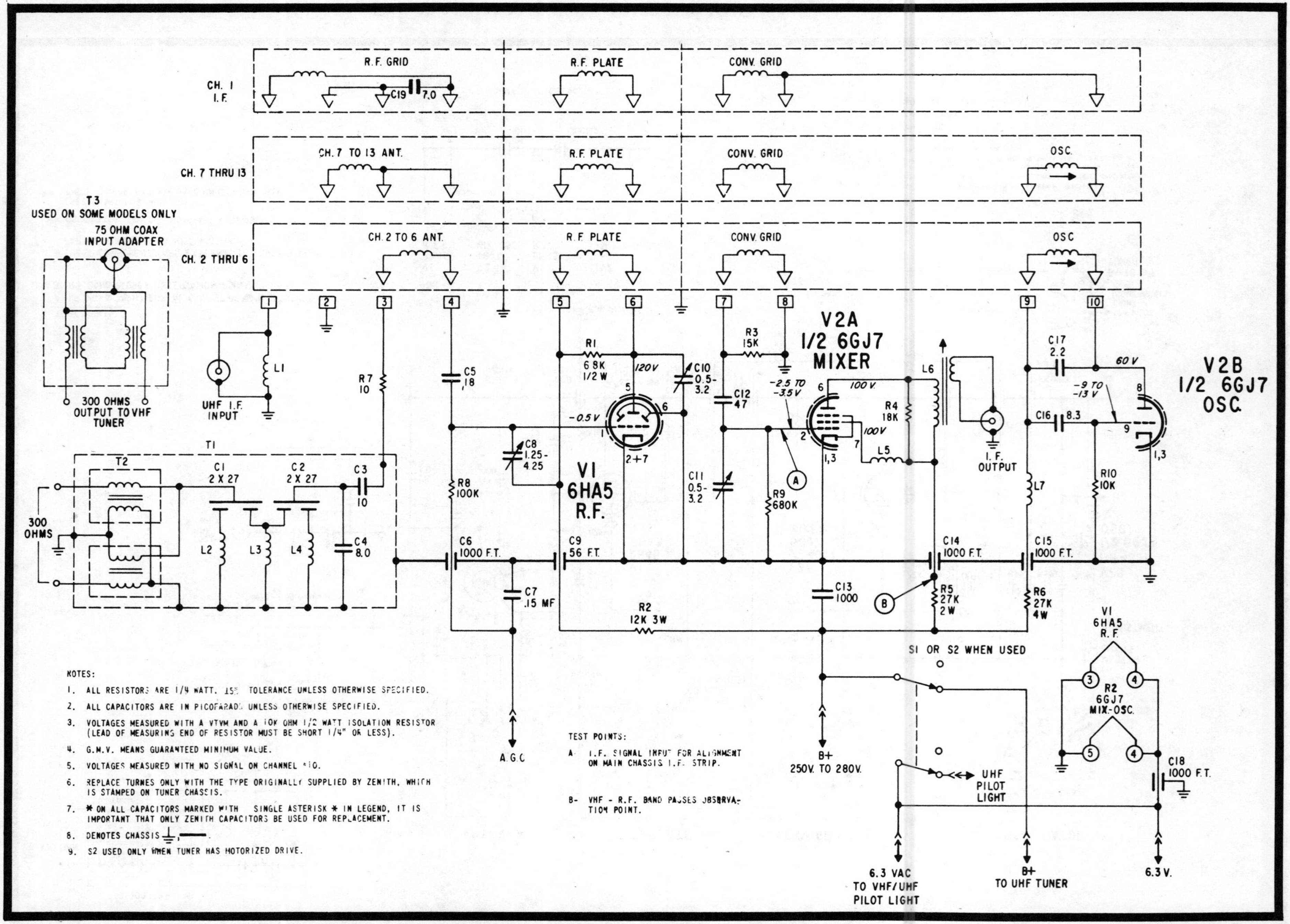

Fig. 12-24. 175-1201 Gold Video Guard tuner schematic.

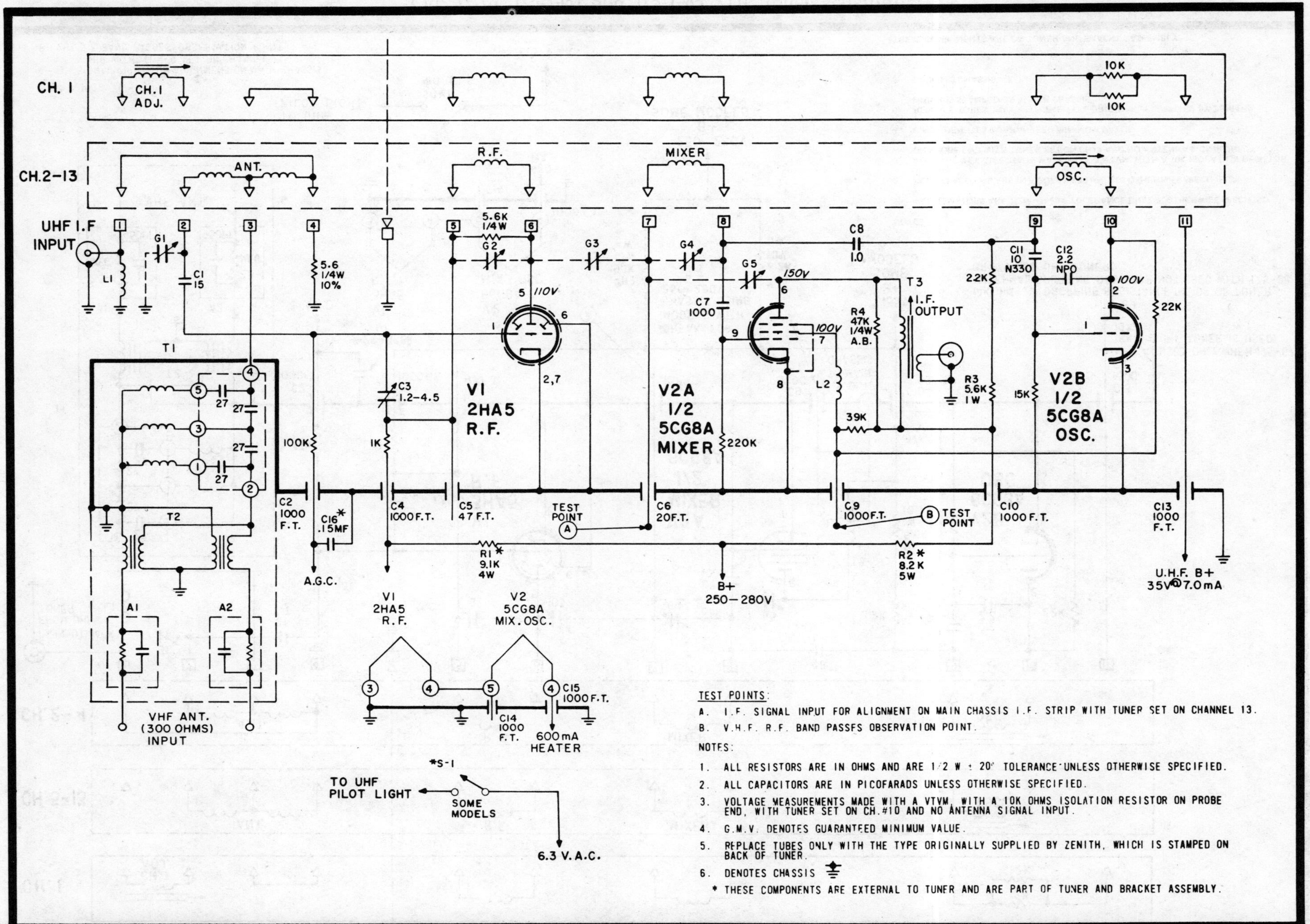

TEST POINTS:

A. I.F. SIGNAL INPUT FOR ALIGNMENT ON MAIN CHASSIS I.F. STRIP WITH TUNER SET ON CHANNEL 13.

B. V.H.F. R.F. BAND PASSES OBSERVATION POINT.

NOTES:

1. ALL RESISTORS ARE IN OHMS AND ARE 1/2 W ± 20% TOLERANCE UNLESS OTHERWISE SPECIFIED.
2. ALL CAPACITORS ARE IN PICOFARADS UNLESS OTHERWISE SPECIFIED.
3. VOLTAGE MEASUREMENTS MADE WITH A VTVM WITH A 10K OHMS ISOLATION RESISTOR ON PROBE END, WITH TUNER SET ON CH. #10 AND NO ANTENNA SIGNAL INPUT.
4. G.M.V. DENOTES GUARANTEED MINIMUM VALUE.
5. REPLACE TUBES ONLY WITH THE TYPE ORIGINALLY SUPPLIED BY ZENITH, WHICH IS STAMPED ON BACK OF TUNER.
6. DENOTES CHASSIS ⏚

* THESE COMPONENTS ARE EXTERNAL TO TUNER AND ARE PART OF TUNER AND BRACKET ASSEMBLY.

Fig. 12-25. 175-1402 VHF tuner schematic.

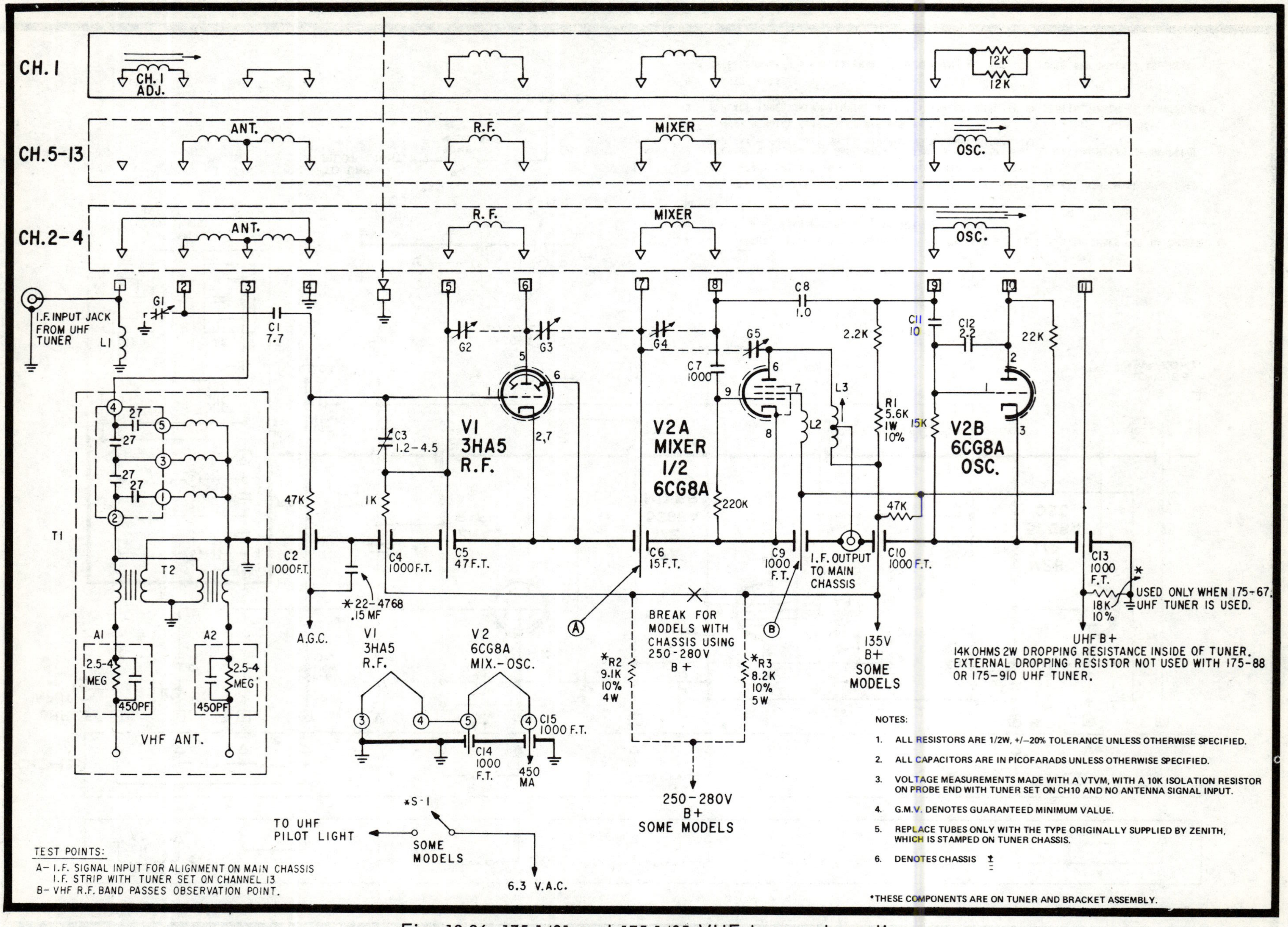

Fig. 12-26. 175-1401 and 175-1405 VHF tuner schematic.

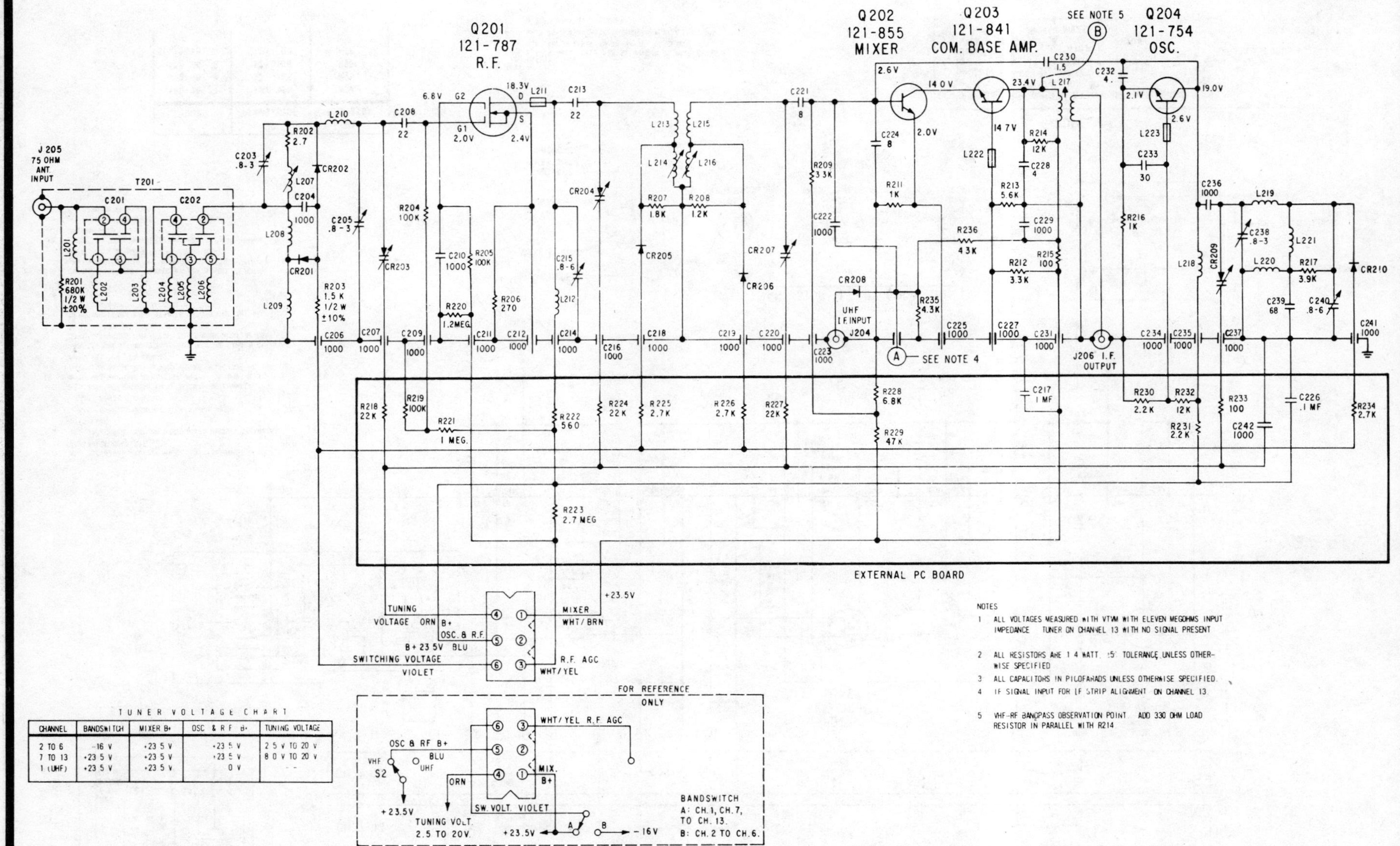

NOTES

1 ALL VOLTAGES MEASURED WITH VTVM WITH ELEVEN MEGOHMS INPUT IMPEDANCE TUNER ON CHANNEL 13 WITH NO SIGNAL PRESENT

2 ALL RESISTORS ARE 1 4 WATT ±5 TOLERANCE UNLESS OTHERWISE SPECIFIED

3 ALL CAPACITORS IN PICOFARADS UNLESS OTHERWISE SPECIFIED.

4 IF SIGNAL INPUT FOR IF STRIP ALIGNMENT ON CHANNEL 13.

5 VHF-RF BANDPASS OBSERVATION POINT ADD 330 OHM LOAD RESISTOR IN PARALLEL WITH R214

TUNER VOLTAGE CHART

CHANNEL	BANDSWITCH	MIXER B+	OSC & R F B+	TUNING VOLTAGE
2 TO 6	-16 V	+23 5 V	+23 5 V	2 5 V TO 20 V
7 TO 13	+23 5 V	+23 5 V	+23 5 V	8 0 V TO 20 V
1 (UHF)	+23 5 V	+23 5 V	0 V	- -

Fig. 12-27. 175-1502 varactor tuner schematic.

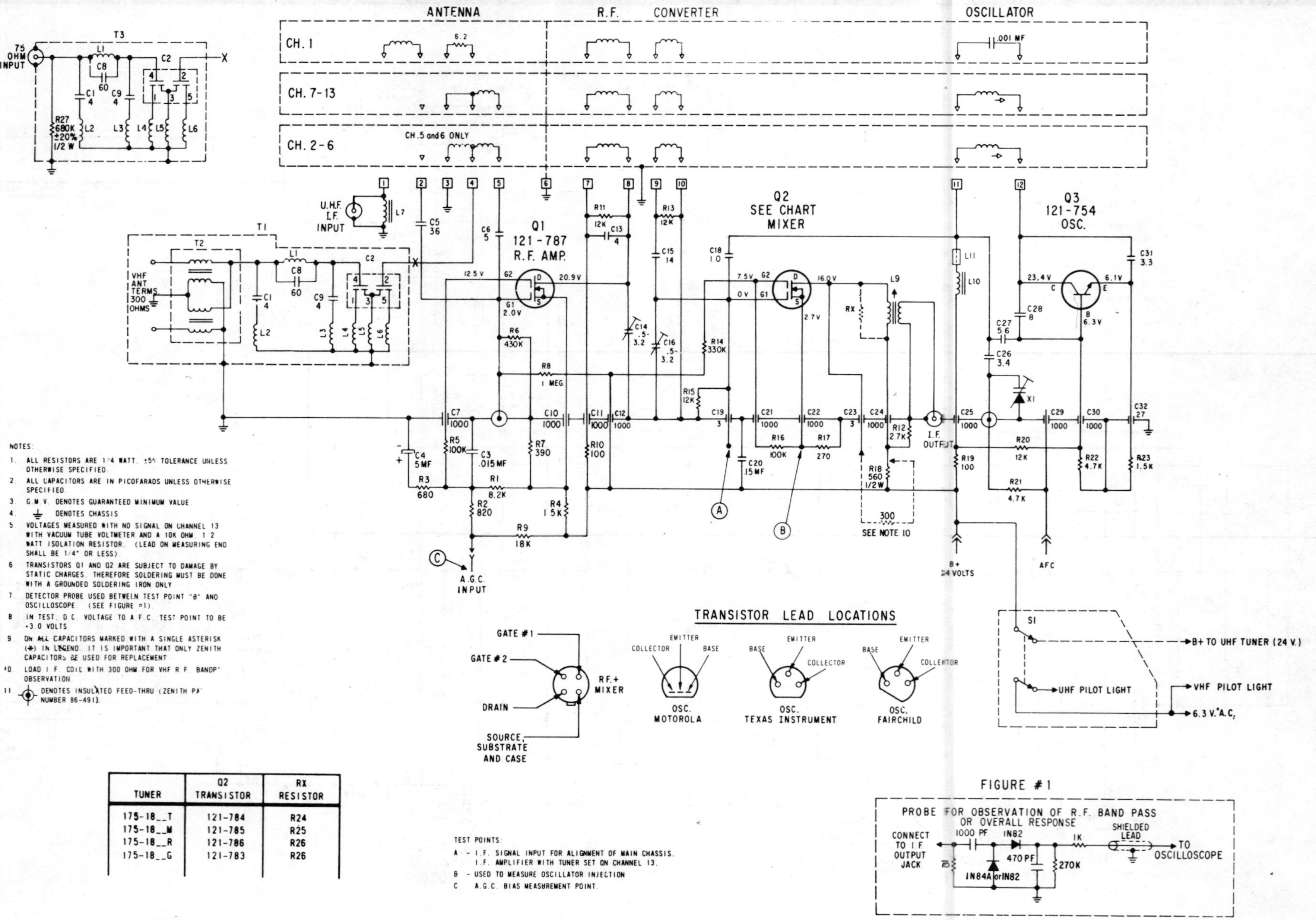

NOTES:

1. ALL RESISTORS ARE 1/4 WATT, ±5% TOLERANCE UNLESS OTHERWISE SPECIFIED.
2. ALL CAPACITORS ARE IN PICOFARADS UNLESS OTHERWISE SPECIFIED.
3. G.M.V. DENOTES GUARANTEED MINIMUM VALUE.
4. ⏚ DENOTES CHASSIS
5. VOLTAGES MEASURED WITH NO SIGNAL ON CHANNEL 13 WITH VACUUM TUBE VOLTMETER AND A 10K OHM, 1/2 WATT ISOLATION RESISTOR. (LEAD ON MEASURING END SHALL BE 1/4" OR LESS).
6. TRANSISTORS Q1 AND Q2 ARE SUBJECT TO DAMAGE BY STATIC CHARGES. THEREFORE SOLDERING MUST BE DONE WITH A GROUNDED SOLDERING IRON ONLY.
7. DETECTOR PROBE USED BETWEEN TEST POINT "B" AND OSCILLOSCOPE. (SEE FIGURE #1).
8. IN TEST, D.C. VOLTAGE TO A.F.C. TEST POINT TO BE +3.0 VOLTS.
9. ON ALL CAPACITORS MARKED WITH A SINGLE ASTERISK (*) IN LEGEND, IT IS IMPORTANT THAT ONLY ZENITH CAPACITORS BE USED FOR REPLACEMENT.
10. LOAD I.F. COIL WITH 300 OHM FOR VHF R.F. BANDP' OBSERVATION.
11. ⊕ DENOTES INSULATED FEED-THRU (ZENITH PA' NUMBER 86-491).

TUNER	Q2 TRANSISTOR	RX RESISTOR
175-18__T	121-784	R24
175-18__M	121-785	R25
175-18__R	121-786	R26
175-18__G	121-783	R26

TEST POINTS:

A - I.F. SIGNAL INPUT FOR ALIGNMENT OF MAIN CHASSIS. I.F. AMPLIFIER WITH TUNER SET ON CHANNEL 13.
B - USED TO MEASURE OSCILLATOR INJECTION
C A.G.C. BIAS MEASUREMENT POINT.

Fig. 12-28. 175-1800 series (AFC) VHF tuner schematic.

UHF CONTINUOUS TUNER	TRANSISTOR	C_P	R_{BG}	R_B	R_C	R_{SERIES}	B+ INPUT NOMINAL (SEE NOTE #6)
175-84LA (A.F.C.)	TR1	C6	R6	R7	R3	R8	265 V @ 7 ma
175-85LA	TR1	C11	R6	R7	R3	R8	265 V @ 7 ma
175-88L	TR1	C12	R6	R5	R3	NONE	35 V @ 7 ma
175-88U	TR2	C13	R6	R7	R3	NONE	35 V @ 7 ma

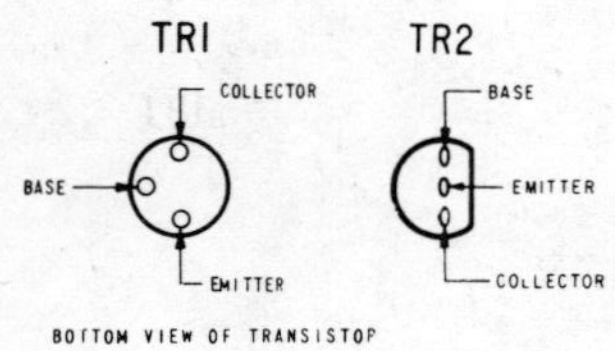

NOTES:

1. G.M.V. DENOTES GUARANTEED MIN. VALUE.
2. DENOTES CHASSIS. ——
3. FOR SERVICE TRANSISTOR REPLACEMENT, USE ONLY THE TYPE THAT WAS ORIGINALLY SUPPLIED IN TUNER.
4. ALL CAPACITORS IN PICOFARADS UNLESS OTHERWISE SPECIFIED.
5. ALL RESISTORS ARE 1/2 WATT, ±10% TOLERANCE UNLESS OTHERWISE SPECIFIED.
6. THE SPECIFIED UHF TUNER B+ INPUT MUST COINCIDE WITH THE B+ SUPPLY AVAILABLE FROM THE VHF TUNER AND TV CHASSIS

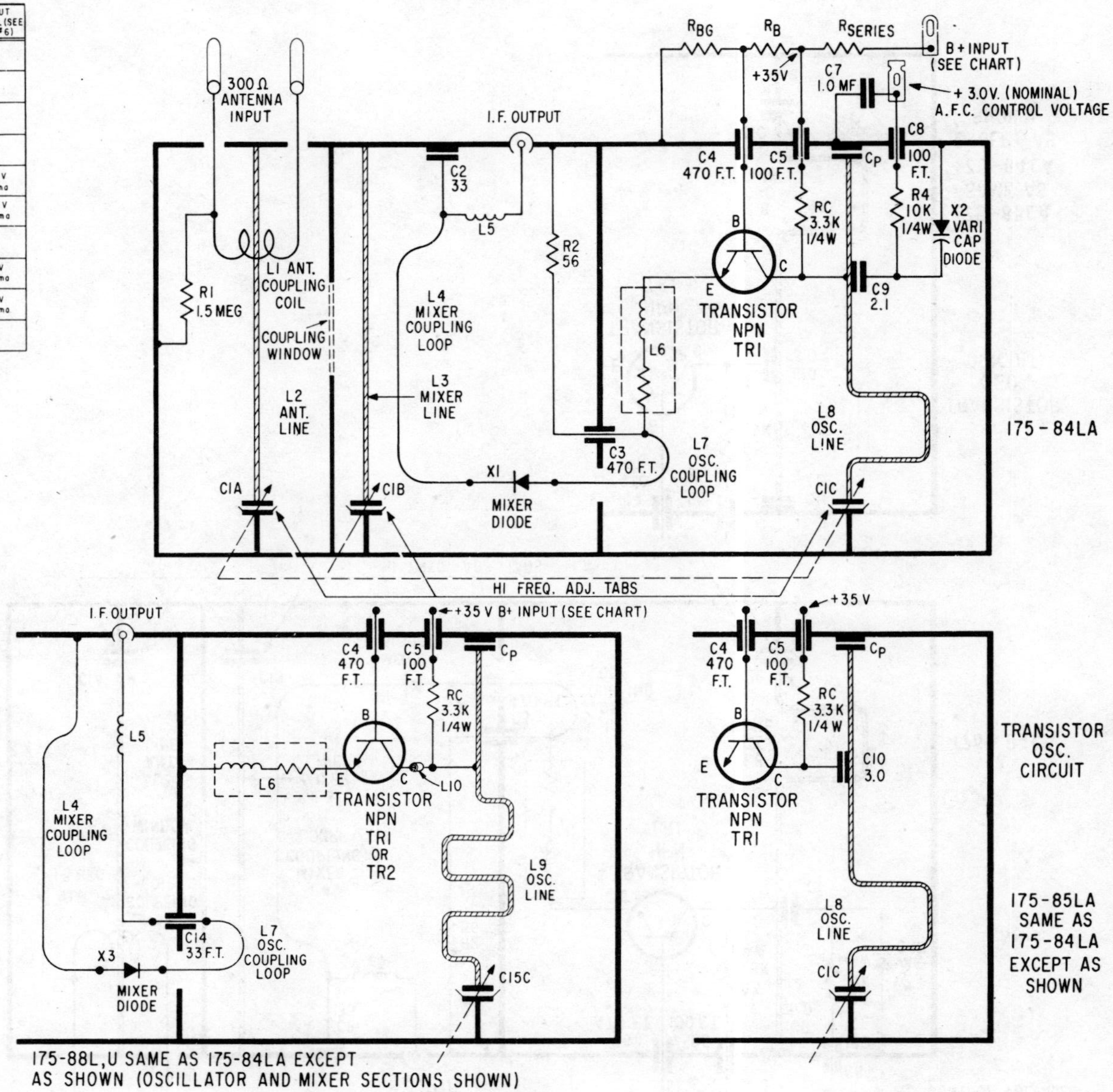

Fig. 12-29. 175-82 through 175-89 UHF tuner schematic.

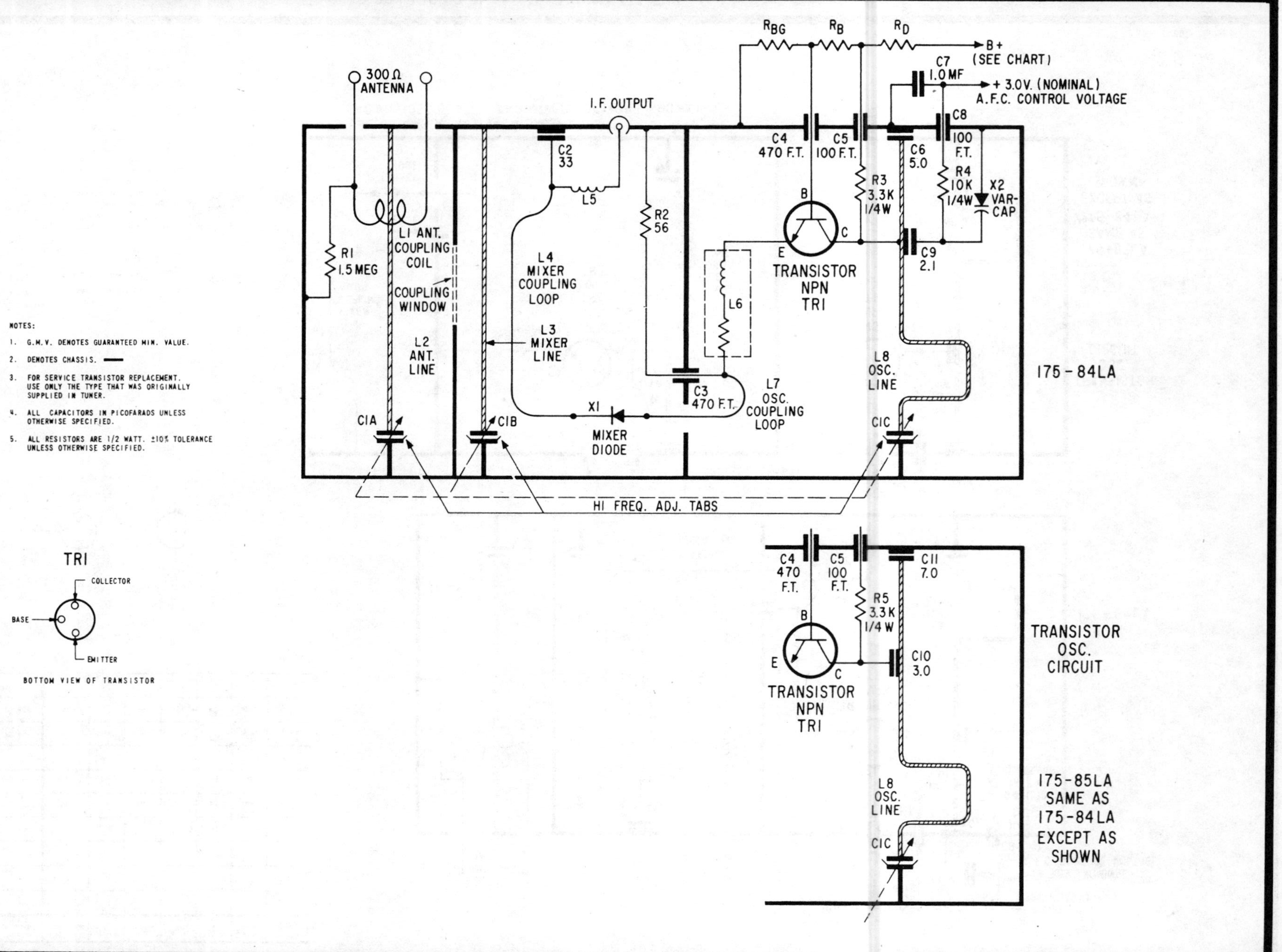

Fig. 12-30. 175-84 and 175-85 UHF tuner schematic.

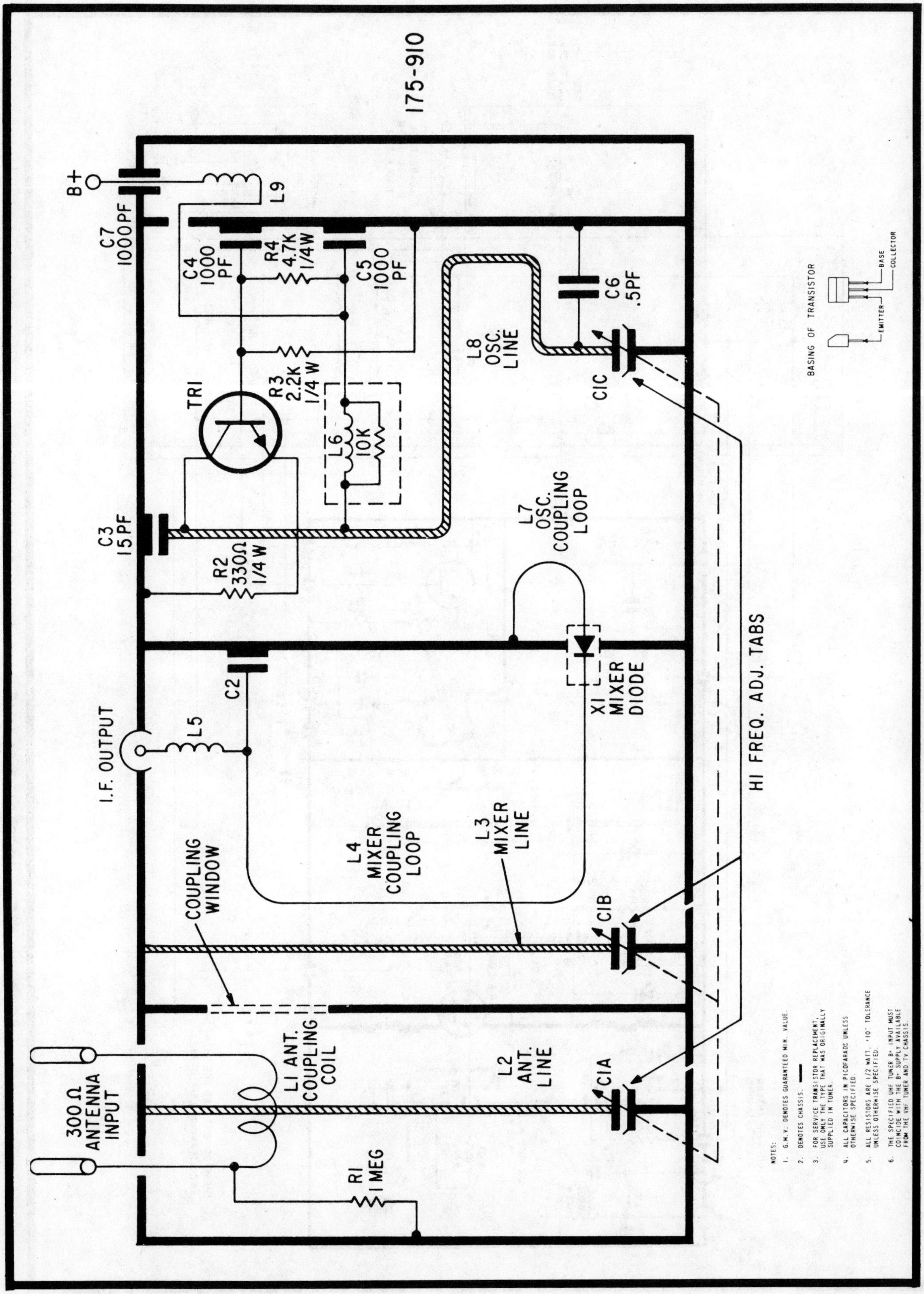

Fig. 12-31. 175-910 UHF tuner schematic.

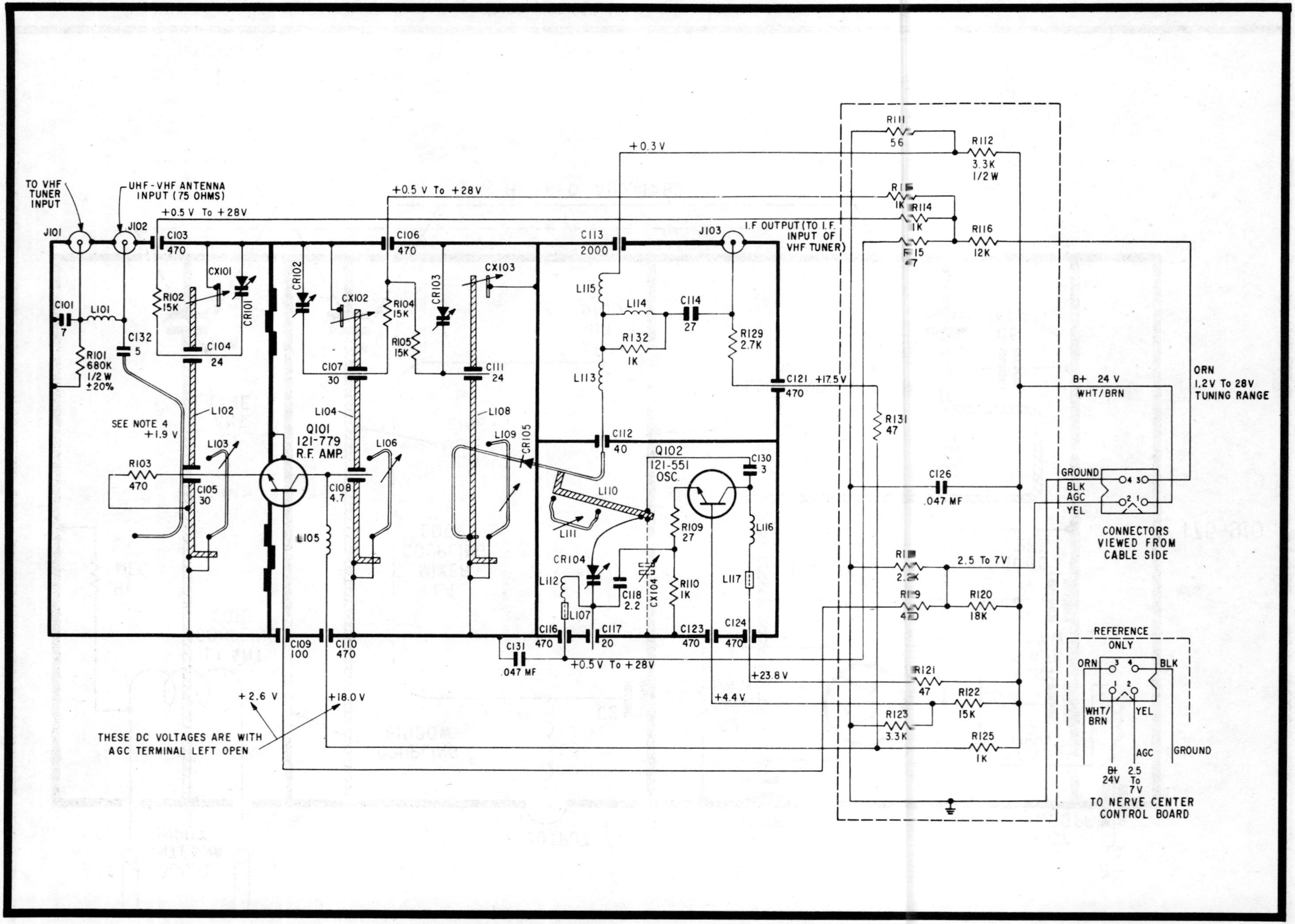

Fig. 12-32. 175-1606 UHF varactor tuner schematic.

APPENDIX

MANUFACTURERS' SERVICE NOTES

PERTAINING TO SUSPECTED TUNER PROBLEMS

Admiral

94A286-10 & 94A330-3—Tuner Replacement

When you replace the above tuners in 1971 color TV sets, check the antenna terminal board on the cabinet back. If it has a VHF Balun attached, you must disconnect and ground the Balun on the replacement tuner. Then connect the 75Ω coax from the terminal board to the replacement tuner in the same manner as the original tuner was connected. The schematic shows this arrangement.

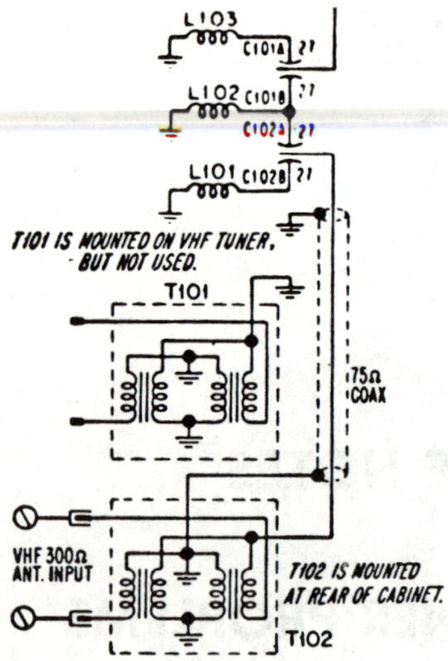

If you do not make this correction on the replacement tuner when an external balun is used, you will cause partial loss of signal and ghosting.

An instruction sheet will be included with these tuners in the future.

Color-TV Chassis K20—Service Hint

After early production models, the following changes were made to eliminate the possibility of the VHF tuner and the channel indicator disc getting one-half turn out of step.

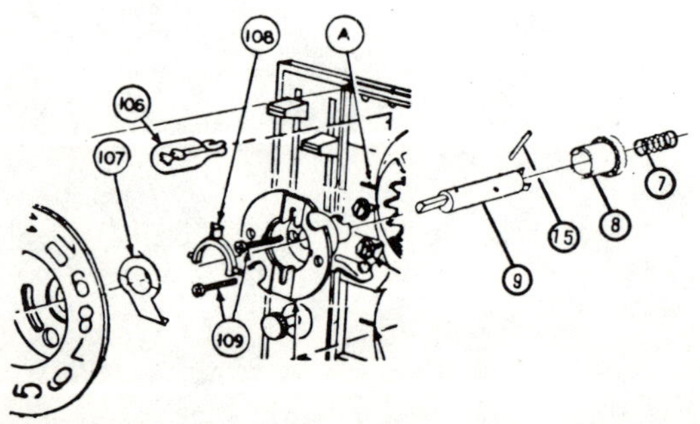

- Spring No. 107 was changed from 19A384-1 to a heavier spring No. 19A384-2 (or two 19A384-1). See illustration.
- Compression spring No. 7 in the VHF fine tuning shaft was changed from 19A248-2 to a heavier spring, 19A248-1. This spring is already in sets using the 94C381-5 VHF tuner.

Color-TV Chassis M20—Service Hint

A problem of UHF channels detuning when BRIGHTNESS, COLOR or CONTRAST control settings are changed—found only in models with varactor UHF tuners—can probably be caused by a defective dc regulator diode.

This diode, IC100, Part No. 56A7-1, regulates the 35v supplied to the preset UHF tuning controls. If the regulator is not functioning, a very small change in the B+ supply voltage will change the tuning of the varactor tuner.

To correct this condition, replace the diode, which is mounted on a terminal strip on the rear of the UHF tuner.

Color-TV Chassis K20—Service Hint

When a problem of greenflesh tones is found in the color picture and the TINT control has no effect, a probable cause is a poor connection at the TINT control leads in the Molex connector.

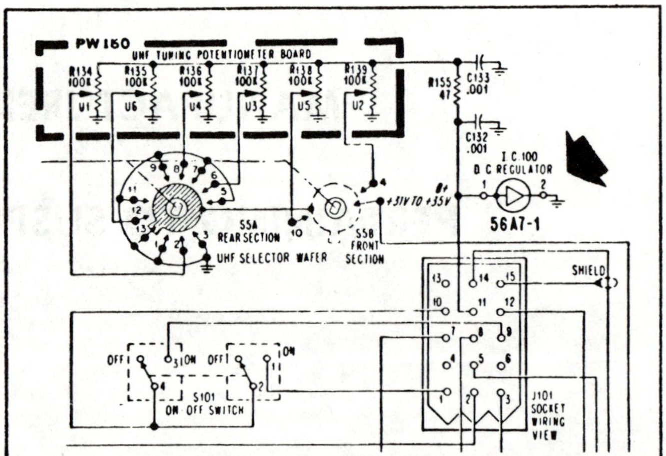

PARTIAL SCHEMATIC OF UHF TUNER SHOWING IC100

To correct this condition, locate the leads from the TINT control at the Molex connector. Remove the pins in both the chassis plug and the plug on the control cluster. Solder the crimped connections and reinsert pins, making sure that they seat firmly.

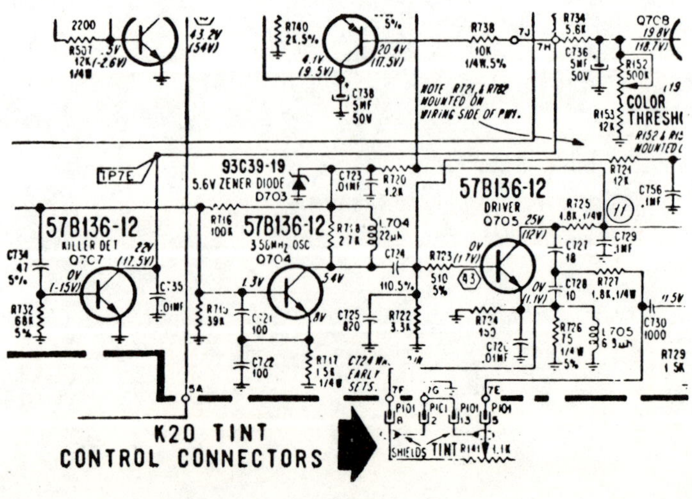

General Electric

Color TV Chassis KC Model M278CWD — Loose Antenna

There have been a few reports of loose VHF antennas on this model.

If you encounter such a complaint, it may usually be corrected by pushing the barbed retaining clip up tight to the underside of the cabinet. This is best done by using a length of metal tubing which slides over the antenna cartridge and presses evenly against the retaining clip.

If the antenna is still loose, use another clip (catalog No. ET3X618) inserted on the top of the original one and pressed up tight against it.

Returning Tuners for Repairs

We wish to stress the importance of proper handling of VHF tuners (and UHF Tuners) which you return for repairs. Improper or careless handling and packing of tuners means additional work for the repair service plus the possible introduction of extra defects, including intermittents. Therefore, in the interest of improving the quality and prompt return of tuners from the repair service, the following points should always be observed: (1) Clip all wires at feed-through capacitors. (Unsoldering frequently damages the feed-through capacitor.) (2) Do not clip IF line cables. Unsolder carefully to preserve the cable length and prevent damage to tuner feed-through. (Unsolder at tuner end.)

Color TV Chassis KE—Obscure Problems

Problem: 1. Grayish hum bar floating vertically at low brightness. Cure: Replace C201 and/or C202 (either may be open). 2. No video, vertical retrace lines, no audio (short surge of normal audio immediately after set is turned off). Cure: Replace C202 (shorted) 3. Horizontal bending or pulling, black floating hum bar. Cure: Replace C152 ("B" section open).

Color-TV Chassis JA—Low Brightness and Overload with Strong Signals

Low-brightness and overload with a strong signal symptom is often caused by the scan rectified 22v source not regulating properly. A possible correction to the problem is to modify the 22v regulator circuit in the following manner:

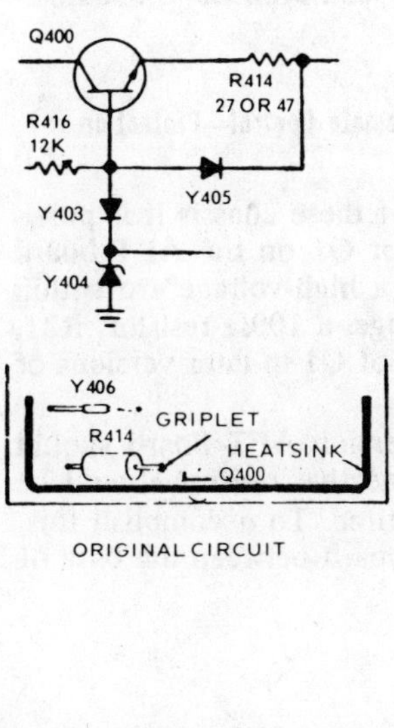

ORIGINAL CIRCUIT

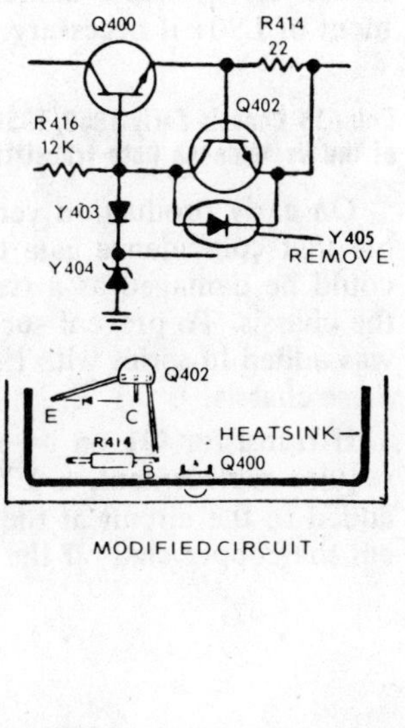

MODIFIED CIRCUIT

Remove diode Y405, resistor R414 and the griplet shown in the original simplified schematic. Replace, as shown in the modified circuit, resistor R414 with a 2.2Ω ½w resistor.

Color-TV Chassis MA—MOSFET Failure in Tuner No. EP86X19

The symptom of a shorted MOSFET RF Amplifier transistor in the tuner may be difficult to evaluate and lead to unnecessary adjustments and improper repair unless the following procedure is followed:

With the "Auto" button OFF (light out) and the receiver properly fine tuned, an interference or slightly high noise level may be observed on one or more channels. A weak to moderate signal level may show little or no interference. Retuning won't clear the problem. The stronger the signal, the worse the condition. The interference may look similar to mistuning into sound, as might occur with AFC misadjusted. Severe cases appear as streaks, like cross modulation.

Test the tuner to determine if the MOSFET is shorted. This can be done by unsoldering the AGC lead from the tuner, discharging the AGC tuner terminal to ground and

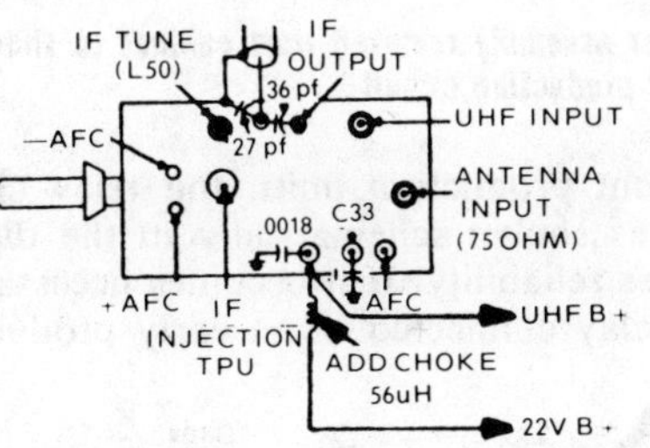

RED LEAD IN CABLE TO I-F MODULE CONNECTOR

EP86X19 VHF TUNER—NEW CONNECTIONS

then measuring the voltage at the open AGC terminal. If the voltage at this terminal is positive by more than a perceptible movement of the meter (the voltage should be zero), the MOSFET is shorted or partially shorted and the tuner should be replaced.

Another test is to adjust RF AGC. If the adjustment has little effect and AGC voltage at the tuner with a strong signal fails to drop below 9v positive, the MOSFET is defective and the tuner should be replaced.

To minimize a possible repeat of the problem, a small encased type choke coil (56μh EP36X33) should be soldered to the 22v B+ terminal on the tuner and in series with the red 22v B+ bus wire. This wire is the one in the cable group going to the IF module connector as shown in the illustration above.

New replacement tuners will have the choke coil and instructions included. Receivers manufactured with chassis code 5D4P and later have the choke.

Color TV Chassis T950 with 704059 Remote Control Receiver—Volume ON/OFF Stepping Relay Circuit Modification

Early production of the T950 chassis equipped with the 704059 Remote Receiver has the Volume ON/OFF relay, K401, connected in the rectifier bridge arrangement. The relay, Part No. 160418-6, and the bridge connected rectifiers, Part No. 530082-4, are physically located on the tuner mounting bracket as shown in the illustration (Fig. 1).

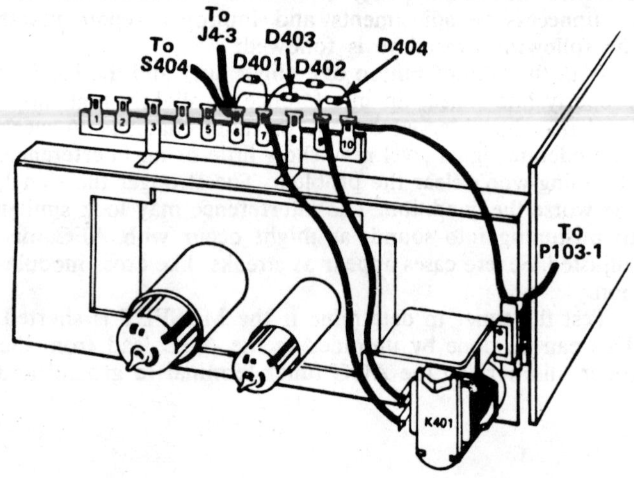

Fig. 1—Tuner assembly removed from cabinet to show component location of early production circuit.

In current production units, the relay circuit has been modified, as shown schematically in the diagram, to provide greater reliability. If it becomes necessary to replace a stepping relay connected in an early production circuit, it

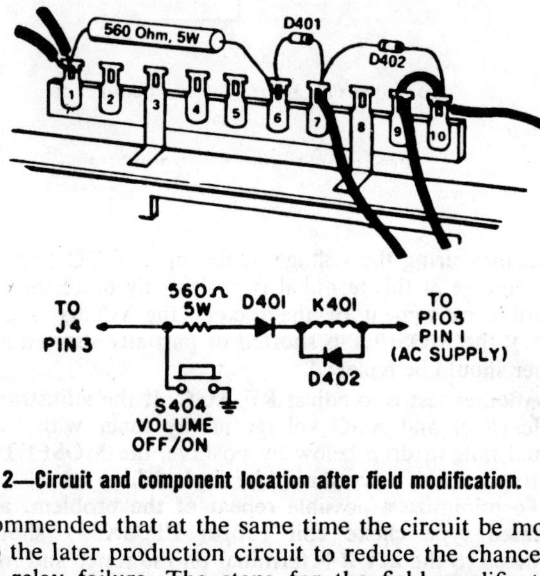

Fig. 2—Circuit and component location after field modification.

is recommended that at the same time the circuit be modified to the later production circuit to reduce the chance of future relay failure. The steps for the field modification are as follows: Fig. 1 illustrates the original parts location and Fig. 2 illustrates the parts location for the field modification. The terminals of the terminal strip have been numbered sequentially, 1 through 10, with terminal 1 nearest the front of the tuner assembly.

- Remove the discard diodes D403 and D404 (Fig. 1) which are installed between terminals 6 and 9, and between 9 and 10.
- Remove the two wires connected to terminal 6 (Fig. 1) and reconnect them to terminal 1 (Fig. 2).
- Connect an insulated jumper wire between terminals 9 and 10 (Fig. 2).
- Install a 560Ω, 5w resistor between terminals 1 and 6 (Fig. 2).

Electrically, the modified circuit should be as shown in the partial schematic in Fig. 2.

Color TV Chassis T936—AFT Field Adjustment

When field adjusting the AFT circuit, select a station broadcasting a color program, preferably VHF, adjusting the fine tuning for optimum picture and sound with the

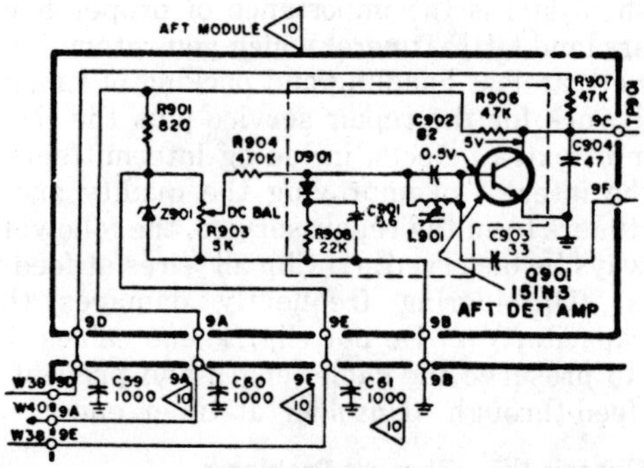

AFT switch in the OFF position. Connect a VTVM across points 9E and 9D of the AFT module and move the AFT switch to ON. Adjust the dc balance control, R903 for 0v.

Remove the VTVM and while watching the picture, move the AFT switch to OFF. If no change in the quality of the picture or chroma level is noted, no further adjustment is required. If a change is noted, switch the AFT switch to ON and adjust the slug in coil L901 for optimum picture and sound. Then switch the AFT OFF and then ON—while observing the picture. Retouch the adjustment of coil L901 until little or no difference in picture quality is noted between the OFF and ON positions of the AFT switch. Check all available stations and compromise the adjustment of L901 if necessary.

Color-TV Chassis Early T950/T951 with Remote Control—Protection of the Coincidence Gate Transistor

On early production versions of these chassis it is possible that coincidence gate transistor Q1 on the AFT board could be damaged as a result of a high-voltage arc within the chassis. To prevent such damage, a 100Ω resistor, R21, was added in series with the base of Q1 in later versions of these chassis.

If transistor Q1 on an early version AFT board should require replacement, a 100Ω protective resistor should be added to the circuit at the same time. To accomplish this, cut the copper clad off the AFT board between the base of

AUTOMATIC FINE TUNING (AFT) BOARD (EARLY PRODUCTION)

Q1 and the junction of C14, as shown in the illustration, install a 100Ω, ½w resistor, R21, in series between the base of Q1 and point B on the AFT board.

Remote Control Receivers 704058/704064/704065—Faulty Capacitor Causing Distorted Audio in TV Set

Distorted audio from the TV receiver can be caused by a faulty 0.47μf capacitor (C35 in models 704058 and 704064, C34 in model 704065) located in the collector circuit of the sound mute control transistor (Q19 in models 704058 and 704064, Q20 in model 704065).

Color-TV Models IT5052 and IT5054—UHF Tuner Transistor Failure

In cases of transistor failure in the UHF tuner on these models, check to see if there is a ground strap connected from the upper right side of the tuner mounting assembly to the picture tube shield. This ground strap was employed in only a small quantity of these models and you are requested to remove the strap on those models you service. It is suspected that a momentary arcing in the picture tube could result in sufficient conduction through this strap to damage the UHF transistor. The tuner assembly is grounded by other means and removal of this strap will not affect set operation.

Color-TV Chassis T979—White Balance Adjustment

The procedure listed in the company's T979 chassis Service Manual (7339) neglects to mention that it may be necessary in strong signal areas to switch to an unused channel or disconnect the antenna when making white balance adjustments. It is possible under strong signal conditions to have sufficient conduction in the IF stages to cause the luminance amplifier collector voltage to decrease, resulting in an incorrect white balance set-up. Removing the signal eliminates this possibility.

Color-TV Chassis T979/989—Board and Module Modifications

Engineering evaluation of certain audio and vertical problems occurring in these chassis have been traced to a resistive contact between the module and its socket or the panel and chassis connector. In the past removing and cleaning the connector in some cases resulted in only a temporary cure for the problem. A recommendation made earlier to solder a jumper across the faulty connector solved the problem but defeated the plugability feature.

To solve both problems, alterations have been made to the modules and panels involved. A pigtail lead has been attached to the module or panel and a push-on connector fastened to the free end of the lead. This permits bridging the connector at fault while retaining its interchangeability feature.

Motorola

Motorola's Solid-State Color Chassis Fine Tuning Indicator and Fine Tuning Lock Circuit

In some solid-state color receiver models, an FTL (fine tuning lock) AFC circuit is employed to assure correct fine tuning. FTL compensates for normal tuner drift and aging of components. Also, an indicator light operates in conjunction with the FTL to signal the customer when fine tuning is necessary.

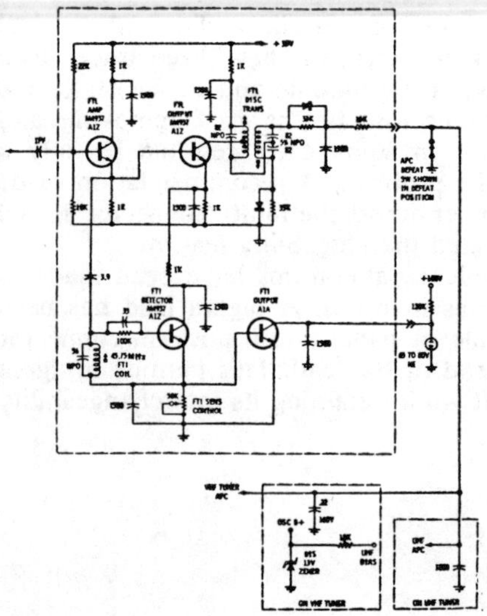

Four NPN transistors, two diodes and a neon lamp make up the active components in the fine tuning indicator and lock circuits. This complete network is located on a single replaceable panel located on top of the video IF panel.

A selected portion of the 45.7MHz video IF carrier is coupled from the 3rd video IF collector through a 1pf capacitor to an FTL amplifier stage. Operating as a class "A" emitter-follower, the FTL amplifier minimizes loading on the video IF, sending the IF signal to the FTL output and fine tuning indicator (FTI) detector.

Located across the input of the FTI detector emitter-follower is a high "Q" 45.75MHz parallel tuned resonant tank (FTI coil). The tank selects the video IF carrier and presents the carrier to the FTI detector for detection and current amplification. The 45.75MHz video IF carrier is only present when fine tuning is correct. Here the 45.75MHz carrier is converted to a dc voltage and directly coupled to the FTI output.

Connected as a common-emitter, the FTI output is in shunt with the FTI neon indicator lamp. Conduction of the transistor extinguishes the neon lamp.

When 45.75MHz is present, indicating correct fine tuning, the FTI detector and output both conduct to extinguish the neon lamp.

If 45.75MHz is not present, indicating incorrect fine tuning, the FTI detector and output will become nonconductive, allowing the neon lamp to light. This signals the customer to re-adjust the fine tuning control.

Directly coupled from the FTL amplifier stage, the video IF signal is presented to a class "A" operated common-emitter FTL output. A discriminator transformer tuned to 45.75MHz center frequency recovers the amplified IF signal.

The discriminator secondary feeds two diodes. Rectification of the IF signal by the diodes produces opposite voltages across balanced diode load resistors. Across both diodes a dc correction voltage is coupled through a "pi" filter to a varactor (voltage-variable capacitor) across the tuner oscillator.

Tuner drift is counteracted by a varying dc correction voltage applied to the varactor from the FTL circuit

and R11 in the base circuit and resistor R14 in the emitter. This stage acts as a limiter because of the high value of the 10K collector load resistor, R13 and the fact that the signal having passed through the previous amplifying stages now has sufficient amplitude to drive transistor Q3 from cutoff to saturation on alternate half cycles. R15, the 10K pot, operates as a range control by changing Q3's collector voltage and in turn its output voltage. With the arm at ground, the control is set for maximum output and range.

Color-TV Chassis TS-929/931/934/938—INSTA-MATIC Switch will not Function Properly

If the INSTA-MATIC switch locks down but will not release or switch will not lock down and there is erratic, mechanical action, make the following corrections:

The switch body may not be properly positioned relative to the mounting bracket. Align the switch body perpendicular to the bracket and tighten the four mounting tabs as shown in the illustration. Turn the coil spring to place one or two turns over latch spring, whichever obtains the best action. The latch spring may be distorted. If so, replace it with latch spring Part No. 42P63085.

If the switch has intermittent electrical action, make the following corrections: The replacement of switch contacts may be avoided by spraying inside of the switch with good grade of tuner or control cleaner (Part No. 11P65176A04).

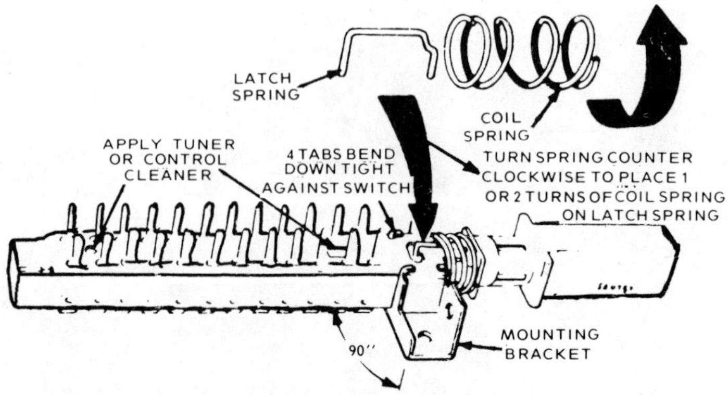

Philco-Ford

Color TV Chassis 14M91/15M91/16M91 —High Voltage Adjustment

Measured AC Line Voltage	High Voltage Adjustment Tolerances (-1000v +0v)
105	21kv
110	22kv
115	23kv
120	24kv
125	25kv

The chart provides information for adjusting high voltage under various ac line conditions. To use the chart, first determine the ac line voltage with an accurate ac voltmeter at the outlet to which the set will be connected. After measuring the voltage at the outlet, refer to the chart and set the HIGH VOLTAGE ADJ. control (VR8 14M91). (VR21 15M91 and 16M91) for the high voltage corresponding to the ac line voltage. Under no circumstances should the high voltage exceed the rated value indicated the the chart. Note that the tolerance of the high voltage adjustment is − 1kv and + 0v. If there is any question as to the voltage setting, adjust the high voltage slightly below the rated setting.

Use an accurate, reliable, high voltage meter and high voltage probe.

If the ac line voltage measured is somewhere between two of the voltages given in the chart, use the nearest lower ac voltage. As an example, if the line voltage measures 112vac — that is, part way between 110 and 115-vac as shown in the chart — use the 110vac figure and adjust the high voltage for 22kv. The high voltage adjustment is always made with the brightness control set at minimum (zero beam current).

Do not exceed 30kv on high voltage probe.

Color TV Chassis 20QT88/20QT90—Excessive Tuner Torque and Ratchet Type Noise When Changing Channels

Because of the physical arrangement of the tuner drive assembly used in the 20QT88 and 20QT90 chassis, it became necessary to add additional shipping protection to the drive train assembly to prevent the gears from disengaging during shipments to the field.

The addition of neoprene pads between the channel drum bracket and the VHF and UHF idler shafts, plus beaded wire dress ties between the VHF and UHF drive gears and idler shafts.

In the event of field complaints of tight channel changing or ratchet type noise, the neoprene pads and beaded ties should be removed.

Chassis 20QT88 and 20QT90 is used in the following Models:

C7260TWA	C7280TWA	C9350TWA
C7260TWA-1	C7280TWA-1	C9351TMA
C7261TMA	C7292TPC	C9352TPC
C7261TMA-1	C7292TPC-1	C9360TWA

Color-TV Chassis 3CS90/3CS91/3CY90/3CY91—Service Adjustments UHF Tuner 70 Position (Detent)

Chassis employing the 70 position detent UHF tuner (76-14326-1) may require readjustment of the FINE TUNING control when changing channels. This is characteristic of this tuner and should not be classified as a defective tuner.

Color TV Chassis CTC40 – the MOS Field-Effect Transistor

The KRK142 tuner used in the CTC40 color chassis features an RF amplifier stage using a special type dual-gate MOS-FET transistor. To those not acquainted with this terminology, the MOS-FET is a Metal Oxide Semiconductor-Field Effect Transistor. Because a solid-state device of this type has not been previously used, it would be wise to acquaint the reader with some interesting facts about this new transistor.

The MOS-FET combines the advantages of solid-state devices (small size, low power consumption and mechanical ruggedness) with a very high input resistance and linear over-all transfer characteristics that closely resemble those of a pentode vacuum tube. In many respects an MOS-FET controls electron flow by the action of an electrostatic field. This implies that the signal input need not supply current, as is necessary in the case of driving the base of a transistor.

The input element of the MOS-FET consists of a metal control electrode (gate) that acts as one plate of a capacitor – the other plate consists of the semiconductor material. A charge applied to the gate electrode produces an equal but opposite charge in the semiconductor layer (channel) located directly beneath the gate. This charge in the channel area controls conduction between two additional electrodes known as the source and drain, which

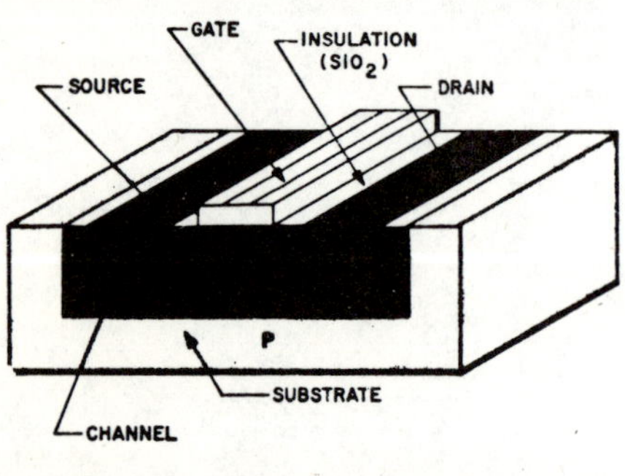

are applied to the opposite ends of the channel. As shown in the illustration, the gate electrode of a MOS-FET is comprised of a metallic contact separated from the semiconductor channel by an extremely thin layer of silicon oxide which acts as an insulator (capacitor dielectric). In the depletion type device illustrated, a negative charge applied to the insulated gate electrode will deplete the source-to-drain channel of electrons. When a positive charge is applied to the gate electrode, conductivity will increase, resulting in increased electron flow. Because the source-drain current is controlled entirely by varying the electrostatic charge on the gate control electrode, no current is required to bias the input of the device. Therefore, the MOS-FET has an extremely high input resistance.

N-CHANNEL DEPLETION TYPE

The manufacturing processes used to produce MOS transistors are similar to those used for high-speed silicon switching transistors. Production starts with a lightly doped wafer of P type silicon material. After a polishing operation, the wafer is oxidized in a furnace and photo-lithographic techniques are used to etch away the oxide coating, exposing bare silicon in the source and drain regions of each of the multitude of transistors that are being formed on the wafer. The source and drain regions are next formed by diffusing an N type material such as phosphorus into the wafer.

This diffusion operation forms a bridge (channel) between the source and drain electrodes.

The wafer is again oxidized to cover the bare silicon regions and a second photo-etching step removes the oxide overlaying the contact regions. In the next step, metal is evaporated over the entire wafer and another photo-etching step removes all metal not needed to produce ohmic contacts to the source, drain and gate electrodes. Finally, the many MOS-FET's on the wafer are mechanically separated and mounted on individual headers. Connection wires are then bounded to the metalized regions and each unit is hermetically sealed in its case. Courtesy of RCA Sales Cor.

Non-remote Power Tuning CTC30 Chassis — Channel Skipping

Some models with nonremote power tuning may occasionally skip channels. The following corrective action should clear this problem.

Condition "A." Where there is a tendency to skip a VHF channel and an Electro Counter motor is used,

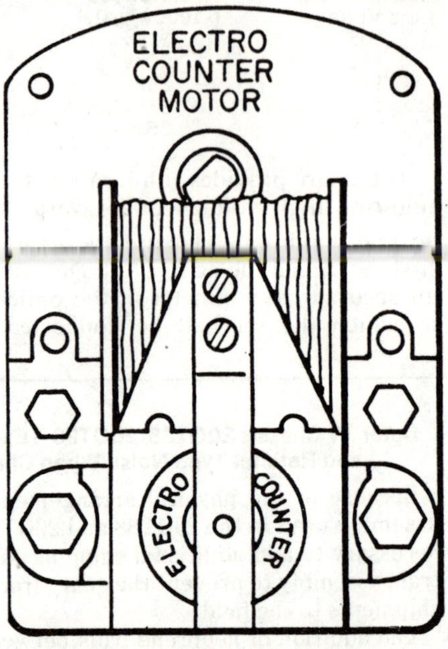

replace "leaf switch" S1002, stock number 115497. **Condition "B."** Where CR1202 diode has failed and an Electro Counter motor is used: (1) Add 10 μf capacitor C1201 — if it is missing, stock number 118832. (2) Replace S1002 switch, stock number 115497. (3) Replace CR1202 diode, stock number 116052. **Condition "C".**

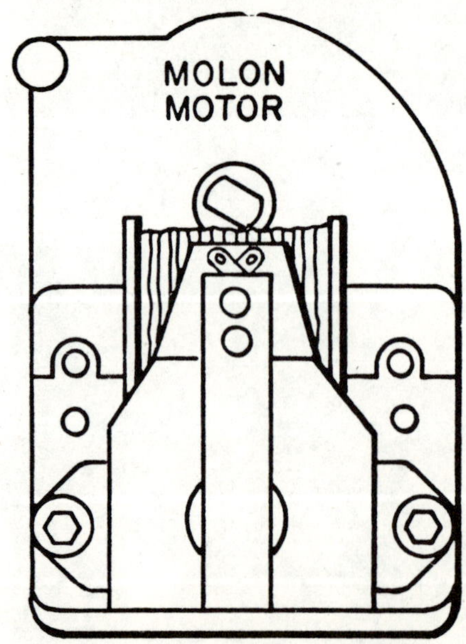

Where there is a tendency to skip VHF channel or jam and a Molon motor is used, replace motor with Electro Counter motor stock number 119680.

Color TV Chassis Employing U Tuner (KRK170)—UHF Channel Indicator

In areas of multiple UHF channel reception (where the TV station channels used are relatively close together), some difficulty may be encountered identifying specific channels on instruments utilizing the six-detent U tuner (KRK170).

Optimum channel identification can be achieved by either of the following methods. Both require removal of the tuner mounting assembly.

First, if the error is slight and the instrument is non-remote:

- Tune the mechanism so that the lowest channel in the area is indicated on the drum.
- Loosen three ¼-in. U tuner mounting screws and slide the tuner (holes are slotted) until that channel is received, then tighten the screws.

If the error is greater, or the instrument is a remote type:

- Tune the KRK170 for reception of a known channel, preferably one between channels 45 and 55.
- While holding the indicator drum, spring the nylon retainer out until the teeth of the drum gear are disengaged from the teeth of the rack gear arm.
- Turn the U drum until the indicator shows the channel being received.
- Mesh the teeth of the drum gear and the rack gear, then return the nylon retainer to its original position.

UHF Traps

In areas where one or more of the lower UHF channels are received at an exceptionally high level resulting in AGC overload or interference, a UHF trap may be helpful in attenuating the signal. A trap cut for a particular channel can be positioned on the 300Ω lead-in near the antennal terminals. This trap should be moved and retuned until the optimum position along the lead-in is found. Secure the trap by taping along its entire length with vinyl tape.

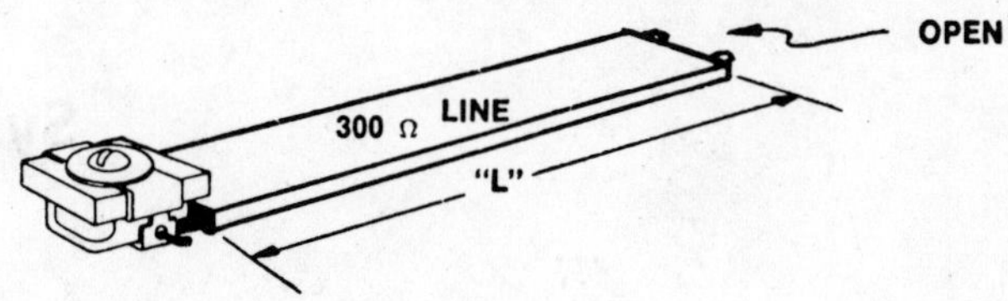

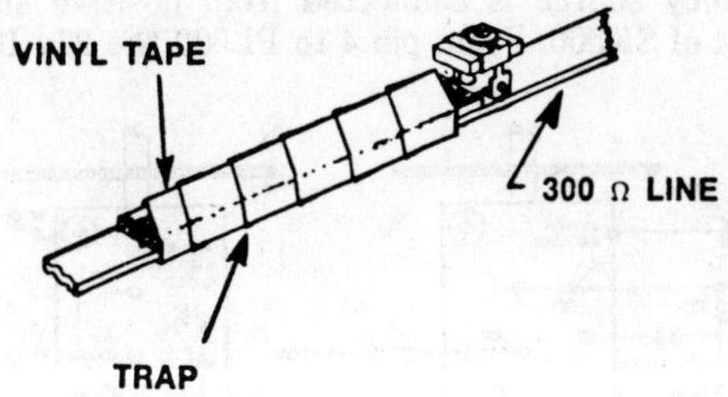

UHF traps can be easily made by cutting the Stock No. 78818 FM Trap to a specific length as noted below:

UHF Channel	"L" Inches	UHF Channel	"L" Inches
14-15	4⅜ in.	25-26	3⅝ in.
16-17	4¼ in.	27-28	3½ in.
18	4⅛ in.	29-30	3⅜ in.
19-20	4 in.	31	3¼ in.
21-22	3⅞ in.	32-33	3⅛ in.
23-24	3¾ in.		

Sylvania

Color TV Chassis D12/D13—Local Oscillator Stabilization on 54-23857-3 Tuner

To stabilize local shift with line voltage, a 27v B+ oscillator frequency source is connected from positive side of C522 to pin 4 of SK500. From pin 4 to PL500 the 27v B+ is

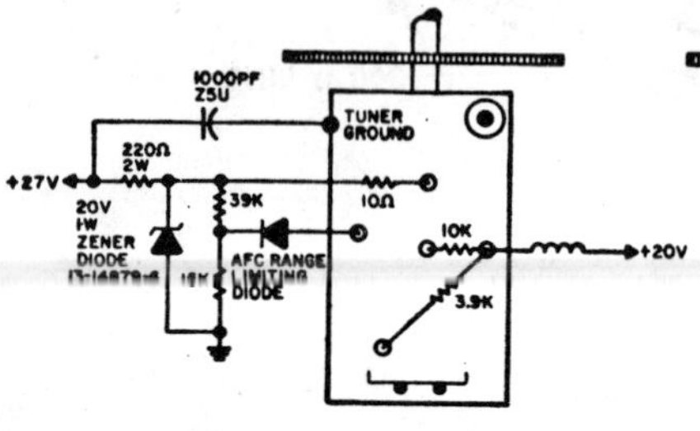

NEW CIRCUIT

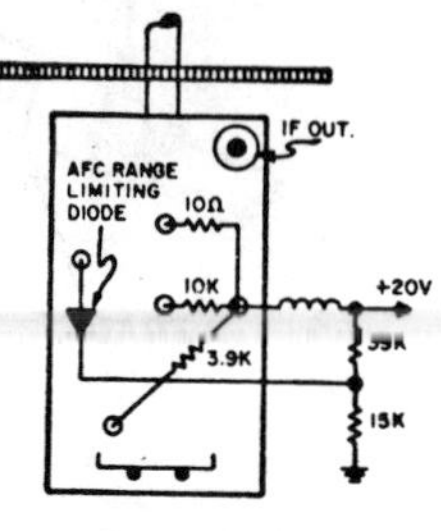

ORIGINAL CIRCUIT

wired to the tuner through the circuit shown.

To correct for local oscillation radiation problems, a 1000pf-Z5U capacitor has been added from the + 27v side of a 220 Ω 2w resistor to tuner ground.

Color-TV Chassis D14, D15 and D16, Early Chassis Without Video Buffer Stage—Intermittent Loss of Horizontal Sync When Changing Channels

When all checks of the horizontal and sync circuits show normal operation, and advancing the AGC to the point just before overload corrects the horizontal problem but results in vertical jitter, several modifications can be made to boost the sync level and eliminate these problems.

Change the emitter resistor to the first IF transistor, Q200, from 15K to 8.2K. In many cases this resistor has already been changed in production. Then modify the AGC amplifier circuitry as shown in the diagram below.

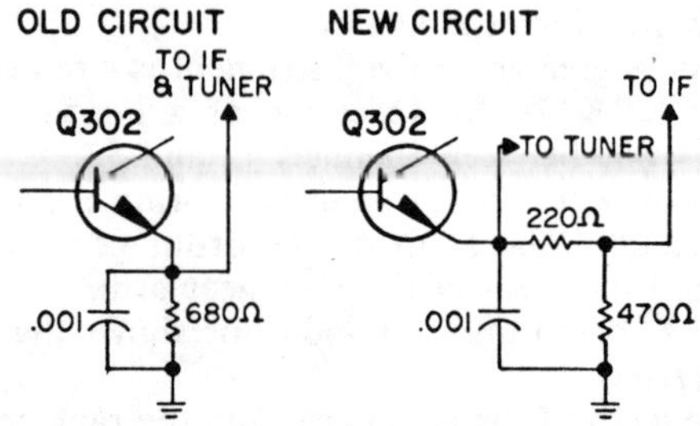

Also change resistor R334 in base circuit of transistor Q308 from 470K to 390K. Any lower value than this will affect horizontal phasing and could cause excessive horizontal jitter on weak stations. Change capacitor C304, the AGC filter, to approximately twice its original value. Too much capacity will cause airplane flutter and increase effects of co-channel interference.

Zenith

The AFC (automatic fine tuning control) permits the receiver to lock on the proper color tuning when the fine tuning perma-set control is reasonably close to frequency.

Circuitry is basically an FM discriminator network of two diodes, center-tuned to 45.75MHz, at the output of a transistor. Since 45.75MHz is on the slope of the IF response curve, a correction has to be introduced in the form of L1, which is tuned above 45.75, reacting with the IF curve to create a flat area between 45 and 46MHz.

The discriminator transformer is tuned so the primary (top) is on 45.75MHz. The secondary is then adjusted so that, on an a... signal, with the perma-set tuning adjusted for the best picture, there is no shift in tuning frequency as the AFC switch on the front panel is switched from "off" to "on."

When making frequency adjustments of L1 and primary of T1, the AFC output white lead is disconnected to improve sensitivity. It is then reconnected while making the final "Zero Shift" adjustment of the T1 secondary.

This unit is a plug-in type, with a mounting screw to secure firm ground contact. It may be removed from the set for servicing without upsetting receiver operation, other than deleting the AFC function.

The discriminator response curve for the AFC unit is rather typical,

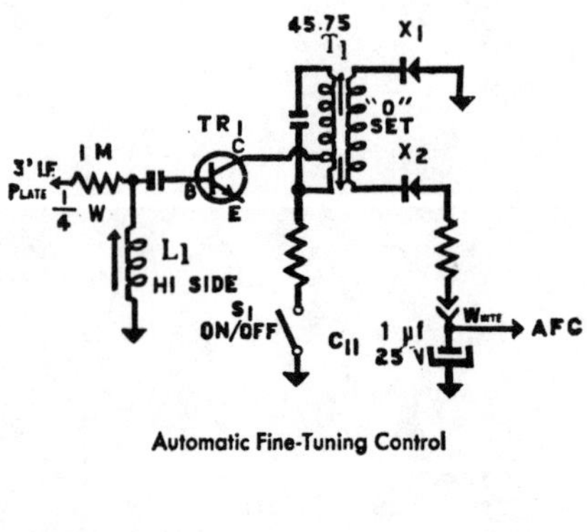

Automatic Fine-Tuning Control

with one exception. Note that the low frequency side of the response curve dips back below zero to the negative side in the vicinity of 42.75-MHz.

This was designed for a very good reason—there are two carriers involved in the transmitted television signal: the picture carrier at 41.75-MHz and the sound carrier at 41.25-MHz. Now with a normal discriminator response curve we would observe that as 45.75 shifted upward in frequency. it would move into the negative portion of the response curve. However, the 41.25 carrier would then be riding up the far left portion of the positive skirt so the two voltages would tend to counteract and cancel.

With the negative dip designed into the response curve at 42.75MHz, however, we find that the carriers work together: when the frequency shifts upward and the 45.75MHz picture carrier shifts into the negative voltage area, the 41.25MHz sound carrier also dips toward the 42.75-MHz minimum and develops a negative voltage, to aid in the AFC error voltage.

On the other hand, as frequency shifts downward, and 45.75MHz moves into the positive voltage area, the 41.25MHz signal drifts out to the left into an area of no correction voltage so it has no effect and can be disregarded.